The Chemistry of Hydrocarbon Fuels

Related books

Introduction to Carbon Science
Edited by Harry Marsh

Combustion of Liquid Fuel Sprays
Alan Williams

Coke – Quality and Production (2nd edition)
Roger Loison, Pierre Foch, André Boyer

Industrial and Marine Fuels Reference Book
George H. Clark

Reviews in Coal Science
The problems of sulphur
IEA Coal Research Ltd

Trace Elements in Coal
Dalway J. Swaine

Gas-Liquid-Solid-Fluidization Engineering
Liang-Shih Fan

Related journals

Fuel

Fuel and Energy Abstracts

Gas Separation and Purification

The Chemistry of Hydrocarbon Fuels

Harold H. Schobert BS, PhD
Associate Professor and Chairman,
Fuel Science Program,
The Pennsylvania State University, USA

Butterworths
London   Boston   Singapore   Sydney   Toronto   Wellington

PART OF REED INTERNATIONAL P.L.C.

First published, 1990

© Butterworth & Co (Publishers) Ltd, 1990

British Library Cataloguing in Publication Data

Schobert, Harold H.
  The chemistry of hydrocarbon fuels.
  1. Fuels. Hydrocarbons
  I. Title
  662.6

ISBN 0-408-03825-X

Library of Congress Cataloging-in-Publication Data

Schobert, Harold H., 1943–
    The chemistry of hydrocarbon fuels / Harold H. Schobert.
      p   cm.
    Includes bibliographical references.
    ISBN 0-408-03825-X
    1. Fuel. 2. Hydrocarbons. I. Title.
  TP318.S37   1990

  662'.6—dc20   90-1311

Printed and bound by Hartnolls Ltd, Bodmin, Cornwall

Contents

TP318
S37
1990
CHEM

Preface

Fuel science is the discipline that studies the origins, chemical constitution and physical properties, and reactions of fuels, and materials derived from fuels. The principal sources of energy in our society are the hydrocarbon fuels: natural gas, coal, and petroleum. Fuels are too often considered only from a technological or utilitarian perspective. Any use of fuels necessarily involves chemical reactions. To make the best use of fuels, in terms both of the efficiency of their conversion to energy or to other materials, and of their use in ways compatible with maintaining environmental quality, means that we must understand the chemistry of fuel utilization processes. However, the chemical reactions that take place during fuel use are governed by the types of bonds and molecular structures present in the fuel. A good understanding of fuel use must therefore be based on the relationships of molecular structures, and the relationship of structures to physical properties, in the fuel. To obtain a full understanding of fuels, it is important to understand that the structures and properties of fuels are some random accident of nature, but rather are the direct result of chemical processes applied to the accumulated organic matter that was eventually converted to fuel. Thus the full scope of fuel science must start with the composition of the organic substances from which the fuels derive, consider the processes of fuel formation in nature and then the relationships between composition and properties, and finally examine processes for fuel utilization.

As a practicing fuel scientist and a fuel science educator, it has seemed to me that there is a great lack of books that bridge the gap between standard textbooks in the sciences or engineering on the one hand and specialized monographs for professionals in the field on the other hand. Of course no single book could of itself fill that gap, but it is my intention in writing this volume to provide at least one stepping stone along the way. I am not aware of other books that provide an introduction to the three major hydrocarbon fuels in a single volume. In bringing coal, gas, and oil together into a single introductory text, I have tried to point out both the differences and similarities among these fuels.

This book has been written for several kinds of potential readers: those who are new to the field of fuel science and seek an introduction to fuel chemistry; practising scientists or engineers who would find a knowledge of some aspects of fuel chemistry to be helpful in their work; and fuel scientists specializing in one fuel but wanting an overview of the chemistry of another - the petroleum chemist, say, who wants to learn a bit about coal. In any case I have assumed that the reader has had an introductory course in organic chemistry, and is familiar with organic nomenclature and the common reactions of organic functional groups. I have also assumed that the reader has some knowledge of the descriptive inorganic chemistry of some of the major elements, and some knowledge of the basic principles of physical chemistry. All chapters have a list of suggestions for further reading, and in most cases I have included references both to textbooks and to specialist's monographs. Much of this book is based on organic chemistry, but organic chemistry of a very different kind from that introduced in textbooks. Most introductory organic chemistry texts nowadays give scant mention - if any - to natural gas, petroleum, or coal. The three major topics of this book - organic geochemistry, structure-property relationships in complex organic mixtures, and industrial organic chemistry - are topics not ordinarily encountered in organic chemistry texts. I hope that some chemists will find this book a useful expansion of the fundamentals of organic chemistry into new areas.

This chapters in this book fall into three units. Chapters 1 through 3 treat the formation of fuels from accumulated organic matter, and essentially provide an introduction to the organic geochemistry of coal, oil, and gas. Chapters 4, 5, and 6 discuss the composition of these fuels and some of their important physical properties. The remaining chapters - 7 through 14 - cover some of the chemistry of separation, refining, and use of the fuels. Obviously it is not possible in this relatively short book to provide a comprehensive treatment of any one of these areas, let alone all three. In selecting topics for inclusion I have tried to select ones that build on a knowledge of organic chemistry, that are instructive or illustrative examples of fuel chemistry, and that will convey enough familiarity with the field of fuel chemistry so that the reader than then follow his or her particular interests into the specialized literature.

Acknowledgments

This book has been developed from notes for a course in "Chemistry of Fuels" that I teach at The Pennsylvania State University. When I began teaching the course, I was assisted enormously by inheriting the lecture notes from two of my predecessors at Penn State: the late Peter Given, and Frank Derbyshire, now at the University of Kentucky. I am very indebted to Drs. Given and Derbyshire for their kindness and help. Many of my colleagues in fuel research at Penn State have at one time or another contributed ideas, suggestions, or comments which have been useful in preparing this book. I have also been helped by several informative discussions with Harry Marsh, of the University of Newcastle upon Tyne. As the ideas and planning for this book began to take form, I received encouragement and useful comments from Robert Jenkins, now at the University of Cincinnati.

The manuscript could not possibly have arrived at its final form, let alone have arrived at the publisher, without the extraordinary assistance of Karen Copenhaver. Karen produced all of the structural formulae in the text; took care of the page set-up, to insure that figures or tables were not cut off mid-page; printed the manuscript on the publisher's forms; and assisted in preparing the index, surely the most mind-numbing task in the universe. I am sure that Karen would join me in thanking our local wizard of the Macintosh, Maxine Deeslie, who kindly helped both of us in installing and using software. Jackie Kunes helped with several of the less pleasant tasks involved in the actual production of the final version submitted to the publisher, managed to put up with us during the many moments of crises, and learned many new and interesting words in the process.

Of course any errors or short-comings are entirely my own responsibility.

Note for the reader

The units used in this book generally follow SI practice. The only significant exception is the calorific value of coals when used for classification by rank, where the English unit Btu/lb is used. **All temperature data in this book are given in degrees Celsius.** Although there seems to be an increasing usage of IUPAC nomenclature for organic compounds in textbooks, the fuel and chemical industries still very heavily use common or informal names for many compounds, *e.g.*, ethylene for ethene. For many compounds I have introduced both names the first time the compound is mentioned in the text, but then have used either name in subsequent discussion. I realize that this practice may introduce some slight confusion, but I find it a useful device to help students, and, I hope, the reader, in developing the ability to "translate" facilely between the two systems.

Each chapter ends with a list of textbooks and professional monographs that I have found particularly relevant to the material in that chapter. This list should point the way for the reader wishing to delve further into a particular topic. The data used in the tables and graphs are widely available in a variety of standard references, such as the *Handbook of Chemistry and Physics* (CRC Press), *Perry's Chemical Engineers' Handbook* (McGraw-Hill), and *Lange's Handbook of Chemistry* (also McGraw-Hill). Useful data on coal are tabulated in the *Coal Conversion Systems Technical Data Book* (U.S. Government Printing Office).

The figures in this book are ones I have used for several years in teaching Chemistry of Fuels at Penn State and a graduate course in Fuel Technology at the University of North Dakota. They were produced on a Macintosh II computer using Macdraw, Macpaint, and Cricketgraph software.

Chapter 1

The origin of fuels in nature

This chapter introduces the concept that the principal hydrocarbon fuels are are materials which originated in nature from once-living organisms. The accumulation of organic matter in the natural environment and the types of compounds which are the precursors to fuels are discussed.

Introduction

A fuel is a substance which is burned to produce energy. In many practical situations in which fuels are used, it is helpful first to carry out some type of chemical transformation on a fuel before it is actually burned, to improve the yield of the fuel from its source, or to improve the performance of the fuel during combustion, or to mitigate potential environmental problems. Examples of these types of transformations include processes to increase the yield of gasoline from crude oil, to improve the performance of gasoline in automobile engines, to convert coal to gaseous or liquid fuels, and to removal sulfur from fuels prior to combustion. Some fuels - most notably natural gas and petroleum - are not only a source of energy, but are also very important feedstocks for the organic chemical industry, where they are converted into a host of useful materials. A fuel might therefore actually be used in at least three different ways: direct reaction with oxygen in a combustion process to liberate useful energy; chemical transformation to cleaner or more convenient fuel forms; or conversion to other, non-fuel materials such as polymers.

All of the potential uses of fuels, regardless of how different they might appear at first sight, are related by the fact that they must involve in some fashion the making and breaking of chemical bonds and the transformation of molecular structures. Therefore the uses to which we put fuels, and the behavior of fuels during utilization or conversion, will depend on their chemical composition and molecular structure.

The main focus of this book is on the three principal naturally occurring hydrocarbon fuels: natural gas, petroleum, and coal. Natural gas is of course a gas; as it is delivered for use it generally contains greater than 90% of a single chemical compound - methane. Petroleum is a homogeneous liquid solution containing several hundred to several thousand individual chemical compounds. Coal is a heterogeneous solid having an ill-defined and quite variable macromolecular structure. Clearly these substances are very diverse, and obviously substantial differences in composition and structure must exist among them. In order to appreciate why three hydrocarbon fuels of such diverse composition and structure exist, we must begin by developing an understanding of how the fuels were formed in nature.

The processes involved in the origin and formation of the fuels establish their composition and structure. The composition and structure in turn establish the kinds of chemical reactions involved in the conversion and use of the fuels. This relationship among origin, composition and structure, and use represents the foundations of the discipline of fuel science. In essence, fuel science may be defined as the study of the origin, composition and properties, and fundamental chemical reactions of fuels. This book is divided into three major units which deal in turn with these three cornerstones of fuel science.

The origins of fuels

Natural gas, petroleum and coal are materials which appear to be radically different in physical properties and molecular composition. They are all linked, however, by the fact that in each case the predominant element (on a weight basis) is carbon. Natural gas contains about 75% carbon; petroleum usually ranges from 82 to 87% carbon; and coals might contain 65 to 95% carbon. Therefore we can gain an understanding of the formation of these fuels in nature by examining the biological and geological behavior of carbon and its compounds.

Not only are these fuels linked by containing a preponderance of carbon, but it is also possible to establish the fact, especially for petroleum and coal, that they derive from once-living organisms. Several lines of evidence support this conclusion. First, it is possible to extract from coal or petroleum compounds which have molecular structures identical to, or obviously derived from, compounds produced by living organisms. These compounds are sometimes called biological markers or biomarkers, because they indicate an origin from living matter. An especially important group of biomarkers extracted from petroleum providing strong evidence for a biological origin are the porphyrins. Two of the most important chemical compounds in the world are porphyrins - chlorophyll and heme. Chlorophyll is the green pigment of plants which catalyzes the photosynthesis process; heme is a constituent of hemoglobin, which is responsible for the transport of oxygen in the bloodstream. Second, petroleum contains compounds which are optically active. Optically active compounds found in nature derive only from living organisms. Third, coal contains readily identifiable plant remains which have been converted to coal (i.e., coalified). These plant remains range in size from spores and pollen through leaves and branches to, in extreme cases, entire coalified tree trunks. For these reasons, it is particularly important to focus on the transformations of carbon in biological systems. To do this, we consider the global carbon cycle (Fig. 1.1).

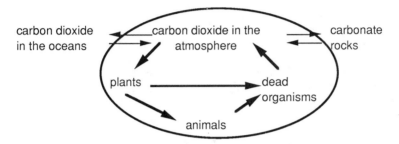

Fig. 1.1 The global carbon cycle is illustrated by atmospheric carbon dioxide being taken into plants, passing through animals, accumulating as dead organisms which decay to release carbon dioxide back to the atmosphere.

For studying the formation of fuels, the slow equilibria between atmospheric carbon dioxide and CO_2 dissolved in the oceans, and between atmospheric carbon dioxide and carbon held in carbonate rocks, are not of importance. The first reaction of concern for the biological carbon cycle is photosynthesis, which extracts carbon dioxide from the air. The photosynthesis reaction can be represented as

$$6CO_2 + 6H_2O \rightarrow C_6H_{12}O_6 + 6O_2$$

In this equation the compound $C_6H_{12}O_6$ is a six-carbon atom sugar, a class of compounds known as hexoses, of which glucose is an example. Photosynthesis serves as the starting point for introducing carbon into biological systems; the hexoses formed during photosynthesis are used by plants as the energy source for the biosynthesis of the many other compounds necessary for their life processes.

A remarkable feature of the photosynthesis reaction as written above is that the free energy change, ΔG, at 298 K is +2720 kJ. The equilibrium constant for the reaction is about 10^{-47} ! The reaction will not take place spontaneously, and the equilibrium is very far to the left-hand side. Photosynthesis can only proceed by the absorption of large quantities of

3

energy from the sun. The subsequent biochemical transformations of the hexose into other plant components such as wood, fats, and proteins do not involve large changes in free energy. Thus if the accumulated organic matter escapes degradation and is preserved as fuel, the preserved material still contains large amounts of energy which can be released for our use at some later time when the fuel is burned. In this regard the hydrocarbon fuels can be thought of as being a stored supply of solar energy.

The carbon, once incorporated into plants by photosynthesis, then passes through the biological food chain. The plant itself may live out its life cycle and die, or it may be eaten by an animal. The plant-eating animal may itself live out a normal life, or it may be consumed by another animal higher on the food chain. Regardless of which of these possible fates takes place, the eventual and inevitable result is that the organisms die and begin to decay.

The decay process can be thought of as the reverse of photosynthesis. Decay involves the reaction of organic compounds with oxygen to form carbon dioxide and water. In principle a decay reaction could be written for any organic compound, but it is particularly illustrative to consider the decay of a hexose:

$$C_6H_{12}O_6 + 6O_2 \rightarrow 6CO_2 + 6H_2O$$

For a hexose, the decay reaction is the exact reverse of the photosynthesis reaction. (In fact, to have a truly cyclic process, the decay of a hexose *must* be the reverse of its formation.) With the complete decay of the dead organisms, the carbon accumulated in the organisms during their life processes is returned to the atmosphere as carbon dioxide, completing the carbon cycle.

As shown in Fig. 1.1, the global carbon cycle accounts, in principle, for all of the carbon in the world. Why then do we have large accumulations of carbon-based fuels? The answer to this question derives from observations often made by those who have experienced an organic chemistry laboratory course: Most organic reactions do not go to completion, i.e., to 100% yield, and many organic reactions do not proceed cleanly to a single product. In nature over 95% of the organic matter accumulated in dead organisms returns to the atmosphere as carbon dioxide via the decay process, but a small portion, less than 5%, is preserved as accumulated organic sediments. In other words, the reason we have large, commercially useful deposits of gas, petroleum, and coal is that the decay reaction is only about 95% complete. We must now examine how the decay process is modified to produce accumulations of fuels.

On a molecular level, the mechanisms of the decay reactions of biological materials are quite complex. However, it is helpful to consider a simplistic view of the decay process as a "black box" (that is, a schematic concept in which the performance is known, but the means of operation are not) reactor (Fig. 1.2):

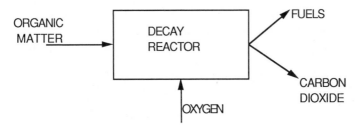

Fig. 1.2 A conceptual reactor system for the conversion of accumulated organic matter into fuels and carbon dioxide.

In this reactor, we have a system that produces two products - carbon dioxide and a fuel. For purposes of obtaining useful energy or chemical materials, we would like to maximize the formation of one of the products, the fuel, at the expense of the other (the carbon dioxide). To maximize formation of a desired product, two strategies are available. First, we could maximize the input of reactants, in this case, the organic matter. Second, we could intervene in some way to alter the reaction conditions to favor formation of the desired product at the expense of the product not wanted.

Both strategies are followed in nature. A small quantity of organic matter of which 100% is preserved would not be adequate to provide a large, commercially useful, accumulation of natural deposits of fuel. On the other hand, an enormous quantity of organic matter which was not preserved at all (or in other words experiences 100% decay) would not provide a large accumulation of fuel either. To maximize formation of organic matter we need abundant light (because *photo*synthesis is driven by light), moisture (recall the presence of water as a reactant in the photosynthesis equation), and warmth (because chemical reactions, including biochemical processes, are speeded up by increasing temperature). To alter the reaction conditions to reduce carbon dioxide formation, we need to minimize the interaction of accumulated organic matter with oxygen, the key reactant in the decay equation.

The requirements of maximizing photosynthesis while minimizing contact with oxygen have important implications for deducing the types of ancient environments in which the organic precursors to fuels must have accumulated. The need for abundant light implies latitudes which provide long growing seasons. The need for abundant warmth suggests a subtropical or tropical climate. The need for abundant moisture, either as liquid or vapor, implies a humid, watery environment such as rivers, lakes, swamps, or shallow seas. Oxygen can participate in the decay reaction in several ways: by direct reaction of atmospheric oxygen, by direct reaction of oxygen dissolved in water, or indirectly via aerobic bacteria which cause decay. Protection against atmospheric oxygen is provided by covering the accumulated organic matter. Either water or a layer of inorganic sediment, such as mud, might serve as a cover. The effects of dissolved oxygen are minimized if water in the environment is not highly oxygenated (*i.e.*, does not contain a high concentration of dissolved oxygen) and is not likely to replenish dissolved oxygen as it is consumed in the decay process. These conditions are met by slow moving, poorly stirred bodies of water such as swamps, lagoons, or some river deltas. These same strategies will help keep oxygen away from aerobic bacteria, thus restricting the growth of large bacteria populations. In addition, the decay-causing bacteria produce intermediate products of decay, such as phenols, which are toxic to themselves. In rapidly moving water the phenols are diluted and washed away as they form. In still or sluggishly moving water the phenol concentration can increase to a point which limits bacteria growth. In summary, accumulation of precursors to fuel formation occurred in environments that are subtropical or tropical, humid, and have abundant shallow, slow-moving water. Such conditions are met by tropical swamps, tidal pools, lagoons, or river deltas.

Plankton and algae live in aquatic environments and flourish in the upper layers of the water, where there is ample light, warmth, and dissolved oxygen. Dissolved oxygen makes the water *aerobic*. The accumulation of organic matter in aerobic environments is not an appropriate route to fuel formation, because the organic matter can be destroyed by the dissolved oxygen. Accumulation of organic matter in aquatic environments requires a zone or layer in the water which contains no dissolved oxygen (*anaerobic*) and requires that the organic matter settle through the water fast enough to reach the anaerobic zone before it is completely oxidized by the dissolved oxygen in the aerobic zone. As plankton and algae die, they settle downward through the water in a process sometimes called "organic rain."

Several kinds of environments can provide the aerobic/anaerobic layering needed to support life and then preserve the dead organic matter. Examples are freshwater lakes where a bottom layer of relatively cold, anaerobic water lies under a layer of relatively warmer, aerobic water (thermal stratification) shown in Figure 1.3, and marine environments where up-welling of water spreads an anaerobic layer over the continental shelf, Figure 1.4.

The formation of fuels at the expense of decay to carbon dioxide does not mean that the carbon cycle has somehow been permanently broken. Eventually the fuels will be burned in a combustion process. For example, isomers of heptane, which might be components of gasoline, react as

$$C_7H_{16} + 11O_2 \rightarrow 7CO_2 + 8H_2O$$

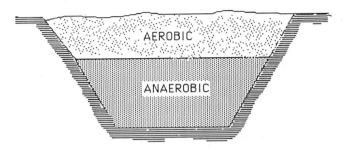

Fig. 1.3 A sketch of a freshwater lake, showing an aerobic water layer (*i.e.*, with a relatively high concentration of dissolved oxygen) lying over an anaerobic layer.

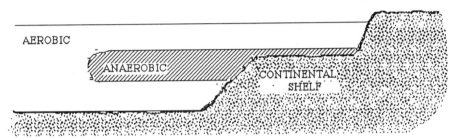

Fig. 1.4 A sketch of an anaerobic layer of water extending outward from a continental shelf. The anaerobic layer helps preserve organic matter on the shelf; this is one reason why some off-shore oil wells are located on continental shelves.

Thus fuel formation represents only a detour in the carbon cycle, and eventually the carbon is indeed returned to the atmosphere as carbon dioxide. In fact, current concern is that the release of CO_2 to the atmosphere via combustion of fuels is occurring much faster than photosynthesis and the slow equilibria with carbonate rocks or the oceans can respond. The increasing amount of carbon dioxide in the atmosphere may be responsible for a general increase in temperatures around the world, the so-called greenhouse effect.

Structures of fuel precursors

This section introduces some of the molecular structures found in the organic matter which eventually transforms to fuels. At the end of the discussion it will be evident that these materials can be placed into two general categories.

As we will see in later chapters, these two categories of fuel precursors will transform into different types of fuels.

We have seen from the global carbon cycle that the organic matter which is the starting material for fuel formation is composed of the dead remains of plants and animals. These organisms include phytoplankton and zooplankton, algae, and the higher plants (e.g., shrubs and trees) - in essence, all types of living organisms except the higher animals. The biochemistry of any of these organisms is quite complex, but the categories of molecular components of the organisms are relatively few: proteins, carbohydrates, lipids, glycosides, resins, and, in the higher plants, lignin. Some of the important features of the composition and structures of these components are discussed in the subsections which follow.

The reason for examining the structures of these components in detail derives from our ultimate concerns of understanding the chemistry of the utilization of fuels, and whether a particular fuel would need refining or "upgrading" before it can be used. Both of these concerns are dictated by the composition and molecular structure of the fuel. In turn, the composition and structure of the fuel are established by the processes in which the fuels was formed, specifically, the molecular structures of the materials entering the decay process, and the chemical transformations occurring to these materials as the organic matter is converted to fuel. It is not critical for understanding the chapters which follow that specific structures presented in this section be committed to memory. What is important is to recognize the principal structural characteristics of these precursors, and the structural differences among them.

Proteins

Proteins are ubiquitous in living matter; indeed the name derives from the Greek *proteios*, primary, indicating the importance of proteins in biochemical processes. Proteins serve many functions, including the transport of nutrients, participation in structural components of cells, and action as biochemical catalysts, the enzymes.

The basic building blocks of proteins are amino acids. One of the simplest of amino acids is alanine.

$$CH_3-CH-COOH$$
$$|$$
$$NH_2$$

The key structural feature of an amino acid is the presence of an amine and a carboxyl functional group in the same molecule. A general reaction of amines is their reaction with acids to produce amides, as shown by the reaction of methylamine with acetic acid to form N-methylacetamide:

$$CH_3NH_2 + CH_3COOH \rightarrow CH_3NHOOCCH_3 + H_2O$$

The important structural feature of the product is the amide linkage.

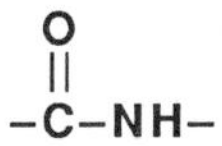

$$\begin{matrix} O \\ \| \\ -C-NH- \end{matrix}$$

Because amino acids contain both an amine and an acid group in the same molecule, they can react with themselves, as in the reaction of two molecules of alanine:

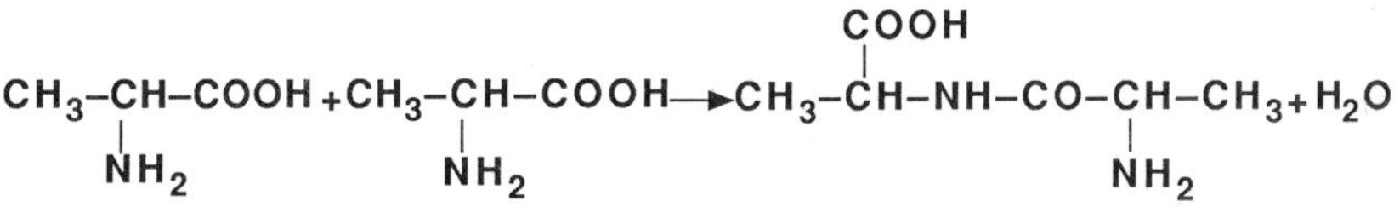

$$CH_3-\underset{\underset{NH_2}{|}}{CH}-COOH + CH_3-\underset{\underset{NH_2}{|}}{CH}-COOH \rightarrow CH_3-\underset{\underset{NH_2}{|}}{\overset{\overset{COOH}{|}}{CH}}-NH-CO-CH-CH_3 + H_2O$$

The product of this reaction is itself an amino acid, having a free amine group at one end of the molecule and an acid group at the other. Therefore, this reaction product is capable of reacting with yet another molecule of an amino acid. In the special case of the reaction of amino acids with themselves, the resulting amide functional group is called a peptide. Continued reaction leads to di-, tri- .... and ultimately polypeptides. When the molecular weight of a polypeptide exceeds 10,000 amu, the material is by definition called a protein. There are 26 naturally occurring amino acids, any one of which can in principle react with itself or any of the 25 others. Thus the possible number of different proteins is enormous. (For example, if an average molecular weight of 166 is assumed for a "typical" amino acid, there would be 60 amino acid units in the protein. The number of possible ways in which 26 different amino acids can be linked in a 60-acid protein is 8×10^{84}.) The key features of protein composition and structure are that peptide formation is reversible and that, on a weight basis, proteins contain significant amounts of oxygen and nitrogen.

Carbohydrates

Carbohydrates are broadly defined as compounds of carbon, hydrogen, and oxygen in which the hydrogen and oxygen occur in 2:1 atomic ratio. An example is the hexose produced in the photosynthesis reaction, $C_6H_{12}O_6$. The concept of these compounds being "hydrates" of carbon (i.e., carbo-*hydrates*) is illustrated by an older and obsolete way of writing the molecular formulas, for example, $C_6(H_2O)_6$. Carbohydrates are very widespread in nature; they include sugars, starches, pectins, and cellulose, which is the principal structural material of plants and may account for up to 80% of the dry weight of plants.

A structure of a specific hexose, glucose, is

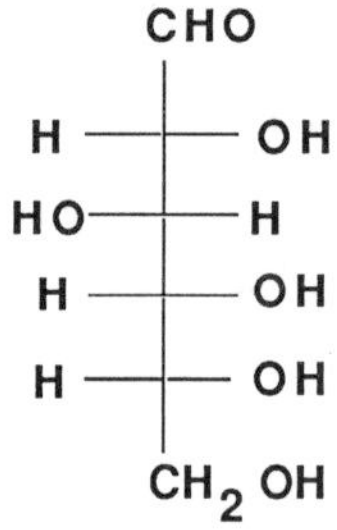

$$
\begin{array}{c}
\text{CHO} \\
\text{H} \!\!-\!\!|\!\!-\!\! \text{OH} \\
\text{HO} \!\!-\!\!|\!\!-\!\! \text{H} \\
\text{H} \!\!-\!\!|\!\!-\!\! \text{OH} \\
\text{H} \!\!-\!\!|\!\!-\!\! \text{OH} \\
\text{CH}_2\text{OH}
\end{array}
$$

In this "cross" notation (the Fischer projection) it is assumed that a carbon atom is present at each of the intersections of the lines. Simple sugars are also called monosaccharides. Monosaccharides can be characterized by specifying the number of carbon atoms in the chain, by indicating whether the carbonyl oxygen is an aldehyde or ketone functional group, or by both methods. Thus in addition to glucose being a hexose, it can also be said to be an aldose (implying the existence of the aldehyde group). The full class name would be aldohexose. Most naturally occurring sugars are aldopentoses and aldohexoses.

As shown in the cross structural notation, monosaccharides contain an aldehyde or ketone (as in fructose below) functional group

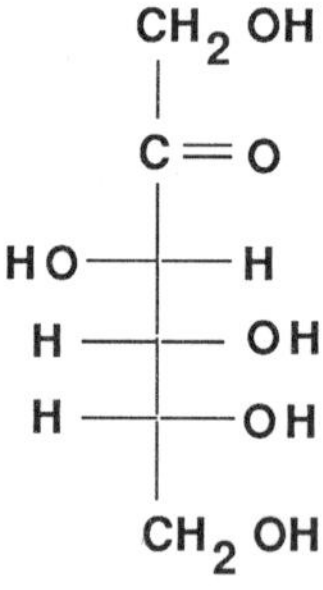

$$
\begin{array}{c}
\text{CH}_2\text{OH} \\
| \\
\text{C} \!=\! \text{O} \\
\text{HO} \!\!-\!\!|\!\!-\!\! \text{H} \\
\text{H} \!\!-\!\!|\!\!-\!\! \text{OH} \\
\text{H} \!\!-\!\!|\!\!-\!\! \text{OH} \\
\text{CH}_2\text{OH}
\end{array}
$$

as well as abundant alcohol functional groups. Aldehydes react with alcohols to form *hemiacetals*, as shown by the reaction of acetaldehyde with methanol:

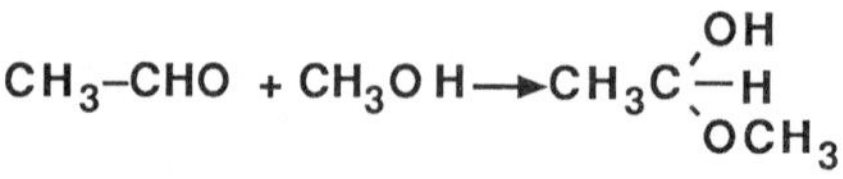

$$
CH_3\text{--}CHO + CH_3OH \longrightarrow CH_3C \overset{OH}{\underset{OCH_3}{-}} H
$$

10

Hemiketals can form by analogous reactions. It is reasonable to expect that a molecule containing both an alcohol and an aldehyde or ketone functional group could react with itself, and indeed most simple saccharides exist as cyclic acetals or ketals

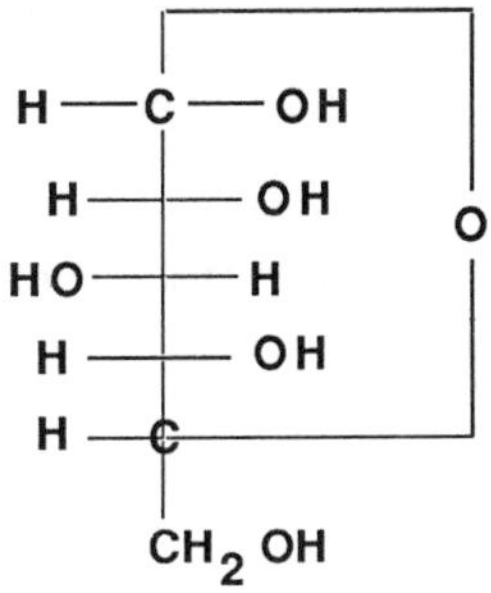

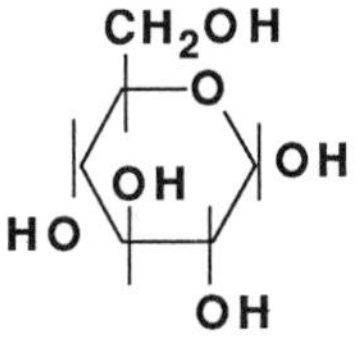

A hemiacetal (or hemiketal) can react with a second molecule of an alcohol to form an acetal (or ketal).

The cyclic hemiacetal structure could in principle exist as either a five- or a six-membered ring. Most simple sugars exist in the six-membered ring form. The carbon atom at which the hemiacetal or hemiketal forms is called the anomeric carbon. For each hemiacetal or hemiketal structure two *anomers* exist, differentiated by the positions of the -OH at carbon atom 1 and the -CH_2OH on carbon atom five. For aldohexoses, if these two groups are trans- to each other, with respect to the plane of the ring, the anomer is said to be the α structure. If the two groups are cis-, the structure is then the β anomer.

A nomenclature system based on corresponding simple heterocyclic compounds is used to indicate the cyclic structure of a sugar. Thus a six-membered ring sugar is called a pyranose, by comparison with the heterocyclic compound pyran. Consequently the structure shown above could be called a β-D-glucopyranose. Similarly, sugars existing in a five-membered ring structure are called furanoses, by analogy with the heterocyclic compound furan.

A monosaccharide can react with another monosaccharide when the hemiacetal form of one monosaccharide unit reacts with an alcohol functional group on the other monosaccharide. Compounds containing up to eight monosaccharide units are known as *oligosaccharides*. Molecules with more than eight monosaccharides are the *polysaccharides*. Most naturally occurring polysaccharides contain 100-3000 monosaccharide units. In the saccharides the acetal linkage between two monosaccharide units is given the general name of glycoside linkage; when glucose units are involved, this linkage is called the glucoside linkage. Sucrose, which is common table sugar, is a disaccharide. Cellobiose is a disaccharide produced during the hydrolysis of cellulose. A naturally occurring trisaccharide is raffinose, found in sugar beets.

Naturally occurring polysaccharides may contain 100 - 3000 monosaccharide units. The most important polysaccharide is cellulose. Cellulose is the most abundant organic compound on earth; up to 80% of the dry weight of plants might be cellulose. The monosaccharide components of cellulose are glucose units. Cellulose characterized in the laboratory contains about 3000 glucose units hand has a molecular weight of about 500,000 amu. (The isolation procedure used to separate cellulose from a plant for study in the laboratory may partially degrade the structure; cellulose as it actually occurs in plants may have 10,000-15,000 glucose units and a molecular weight up to 2,400,000.) A portion of the cellulose structure is shown below

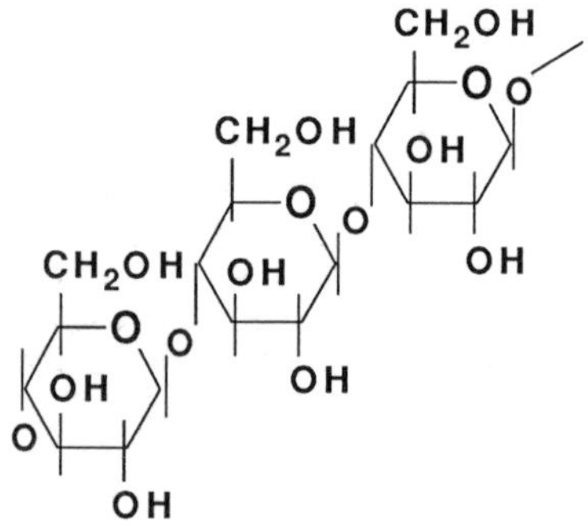

Cellulose is a polymer of D-glucose. Each glucose monomer is linked to the next through a β-glucoside linkage to the -OH group on carbon atom 4 of the next ring.

In some plants other polysaccharides are associated with cellulose. An example is xylan, which occurs in amounts up to 40% in some brans, and the related compound poly(glucuronic acid).

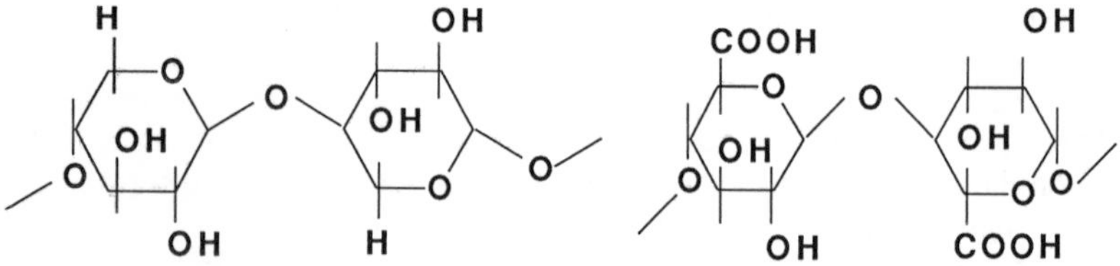

Starches are very similar in structure to cellulose except for the configuration of the glucoside linkage.

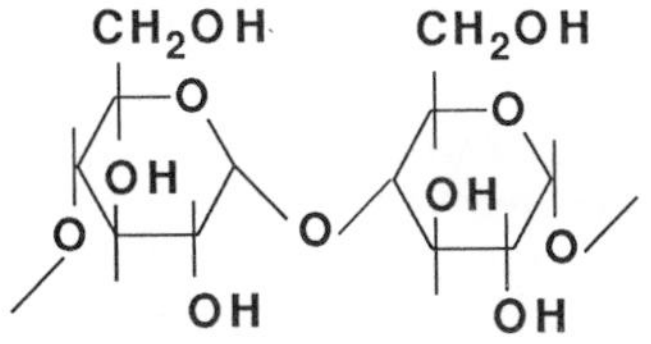

As in cellulose, the glucoside bond occurs through the -OH on carbon atom 4, but is in the α configuration rather than β, and several thousand glucose units are linked to form the starch molecules. Starch is second to cellulose in abundance among the naturally occurring polysaccharides. Starch functions as a reserve carbohydrate in plants; that is, the production of starch represents a way of storing chemical energy which can be released as needed by hydrolysis of the starch. As in cellulose, the only monosaccharide unit in starch is glucose. Naturally occurring starch contains two fractions, amylopectin and amylose. The amylopectin has a branched structure in which -OH groups in the monosaccharide units of one linear chain of monosaccharides form glucoside linkages to a second chain of monosaccharides. The branches occur through the -OH groups on carbon atom 6. Amylose has a linear structure.

Key features of the carbohydrates are the existence of cyclic structures, a very high oxygen content on a weight basis, and the fact that oxygen is bonded both to carbon and to hydrogen (in contrast to peptides, where oxygen is bonded only to carbon). Like peptides, the formation of glucosides is also reversible.

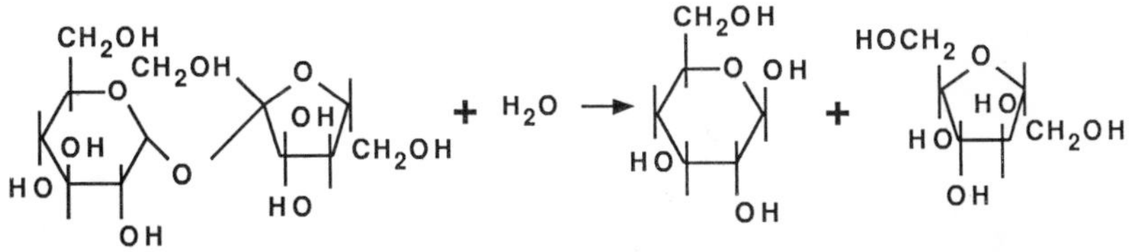

If the carbonyl functional group in a sugar is reduced to the corresponding alcohol group, the reduced compound is no longer a sugar but rather a polyalcohol. This class of compounds is the *alditols*. A naturally occurring alditol is D-mannitol, which is found in algae living in marine environments. This compound also occurs in onions, olives, and mushrooms.

13

Glycosides and tannins

Glycosides
The structure of glycosides is somewhat similar to that of disaccharides, the important distinction being that one of the two units linked via the glucoside bond is an alcohol. Aldoses react readily with alcohols in the presence of acid catalysts to form cyclic acetals called glycosides. Many naturally occurring glycosides are characterized by the alcohol portion of the structure deriving from a hydroxyaromatic compound. An example of a glycoside is arbutin, the structure of which is shown below.

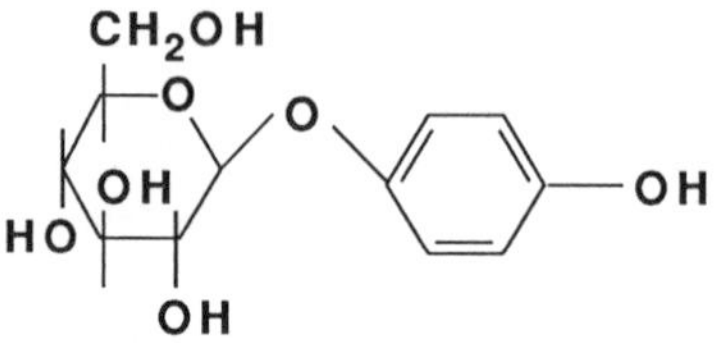

Arbutin occurs in the leaves of the cranberry and bearberry plants and is used as a diuretic. Other glycosides are found as pigments in plants. The term glycoside is the general term for the family of compounds; when the sugar component is known, the compounds can be named as derivatives of the sugar (*e.g.*, glycosides derived from mannose would be mannosides). The sugar portion of the molecule is called the glycon and the alcohol portion the aglycon.

Anthrocyanins are glycosides which serve as plant pigments. The aglycons of the anthrocynanins are the *anthrocyanidins*. Three anthrocyanidins are important in nature: cyanidin, delphinidin, and pelargonidin. The structure of pelargonidin is

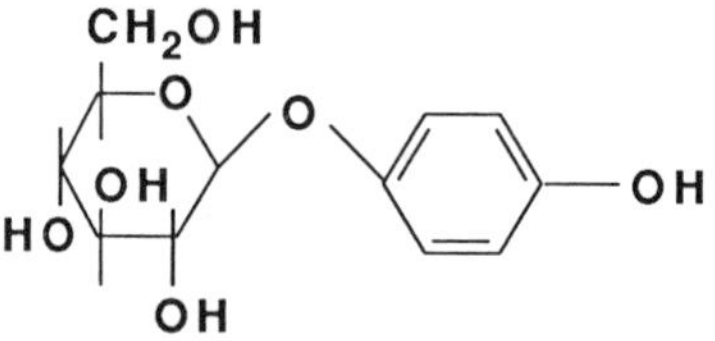

The glycoside salicin is found in nature in the bark of willow trees. Studies of willow extract led to the discovery that compounds which are derivatives of salicylic acid possess pain-reducing properties. The medicinal behavior of salicylates led in turn to the research which eventually discovered the compound familiarly known as aspirin.

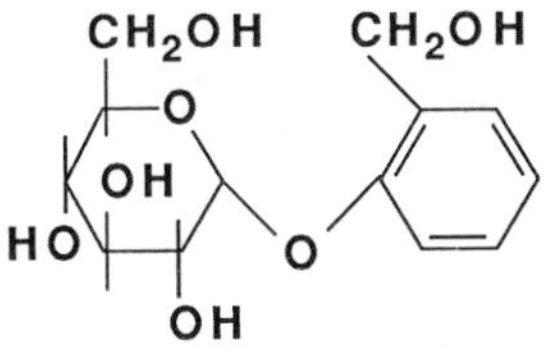

The essential features of the glycoside structure are the high content of oxygen on a weight basis, and the predominance of cyclic structures, including aromatic materials. The formation of glycosides is also reversible.

Tannins
Tannins are structurally similar to the glycosides, but in most tannins the linkage to the sugar component is an ester. The acid component of the ester is often gallic acid or m-digallic acid.

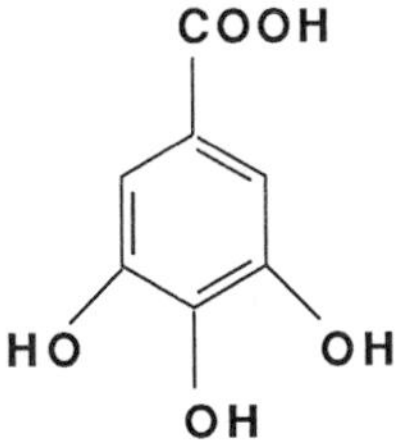

gallic acid 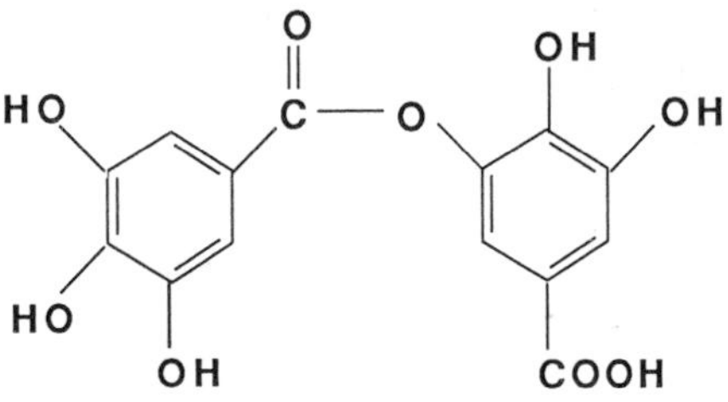   m-digallic acid

Tannins occur in plant leaves and derive their name from their use in tanning animal hides. An example is 1-galloyl-β-D-glucoside, a compound present in the leaves of rhubarb.

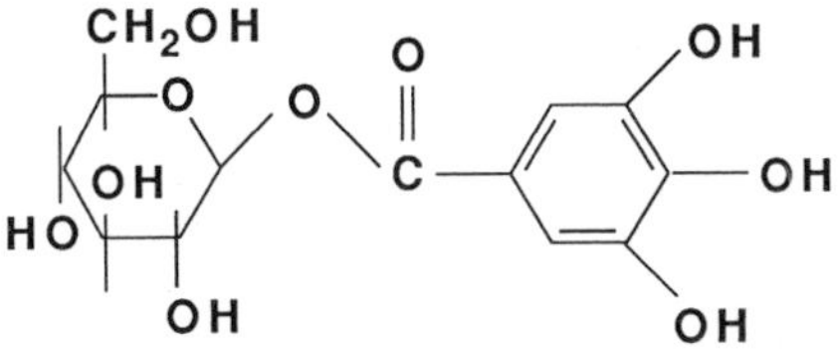

This compound is an ester of glucose and gallic acid. More complex tannins are esters in which all of the hydroxy groups on the sugar are esterified, usually with gallic acid or m-digallic acid.

A second type of tannins is comprised of phenolic structures joined via ether linkages. An example is the structure shown below, which can be isolated from green tea.

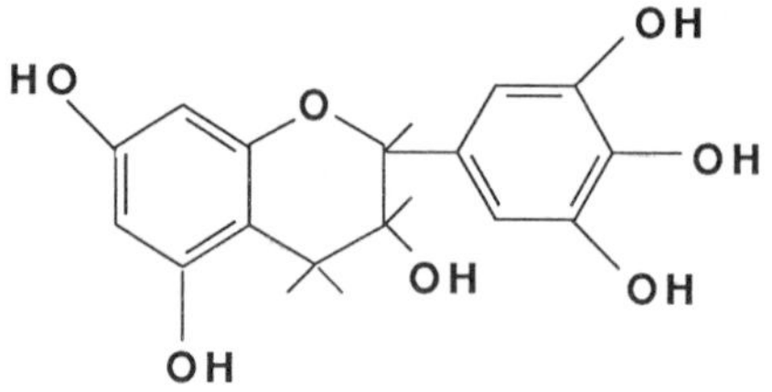

Lipids

In broadest terms, lipids are naturally occurring substances which are soluble in hydrocarbon solvents but insoluble in water. Biochemists use a more restricted definition, defining lipids as naturally occurring compounds which yield fatty acids (i.e., carboxylic acids with long carbon chains) upon hydrolysis. Families of compounds which may be classified as lipids include fats, oils, waxes, terpenes, and a variety of other hydrocarbons including alkanes. The first three families fall into the biochemical classification of lipids; terpenes and other hydrocarbons are sometimes called *nonsaponifiable lipids* since they do not hydrolyze to fatty acids.

Waxes
Waxes are esters of long chain alcohols with fatty acids. Plants produce waxes for use as sealants, to retard water evaporation, for example. (The use of waxes as sealants is of course a way that we, as well as plants, utilize waxes.) An example of a naturally occurring plant wax is carnauba wax, which has the general formula $CH_3(CH_2)_xCOO(CH_2)_yCH_3$, where x may have values of 22 and 26 and y, 30 and 32. Carnauba wax occurs as the coating on leaves of Brazilian palm trees.

Hydrocarbons found in waxes are generally of 27 to 37 carbon atoms, and those containing an odd number of carbon atoms predominate. For example, the waxy cuticle on apples contains heptacosane, $C_{27}H_{56}$, and nonacosane, $C_{29}H_{60}$. These compounds most likely were formed by the loss of a carboxyl group from a fatty acid, since virtually all naturally occurring fatty acids contain an even number of carbon atoms.

Fats and Oils
Fats and oils are also esters of long chain carboxylic acids, but in this case the alcohol portion of the molecule is glycerol. The distinction between fats and

oils is based on physical properties, fats being solids at room temperature and oils being liquids. This distinction is not important for understanding the chemistry of these materials in fuel formation processes. An example of a naturally occurring oil is palm oil, which has the structure

$$CH_2OOCC_{15}H_{31}$$
$$|$$
$$CHOOCC_{17}H_{33}$$
$$|$$
$$CH_2OOCC_{17}H_{33}$$

The two 18-carbon acids in palm oil are unsaturated.

The most abundant saturated acids in fats are compounds with 8, 10, 12, 14, 16, or 18 carbon atoms. They are often known by their trivial names: caprylic, capric, lauric, myristic, palmitic, and stearic acids, respectively. The most abundant unsaturated acids are all derivatives of the C_{18} acid, with double bonds at the 9-; 9,12-; or 9,12,15- positions. These compounds have the trivial names oleic, linoleic, and linolenic acids, respectively. The greater the number of unsaturated acid structures in the molecule, the more likely the compound is to be a liquid at room temperature (and hence classified as an oil).

Terpenes

The terpenes are widely distributed in plants. The terpenes are all characterized by repeating units of the same carbon skeleton as 2-methyl-1,3-butadiene, *isoprene*. Thus the terpenes may be considered to be polymers of isoprene. This structural relationship leads to the alternative name, isoprenoids, for the terpenes.

Isoprene can polymerize into a structures which may contain carbon chains or rings, or sometimes both. However, the characteristic isoprene structure with the terminal methyl groups is preserved throughout. The classification of the naturally occurring terpenes is based on the number of isoprene monomers incorporated into the molecule (Table 1.1):

Table 1.1 Classification of terpenes

Number of carbon atoms	Classification
10	Monoterpene
15	Sesquiterpene
20	Diterpene
25	Sesterterpene
30	Triterpene

17

Compounds containing only a single isoprene unit (*i.e.,* C_5) are very rare in nature. The monoterpenes are found in many plants, in both linear and cyclic structures. Linear monoterpenes include citronellal, the compound which gives lemon oil its distinctive odor, and geraniol, which occurs in geraniums.

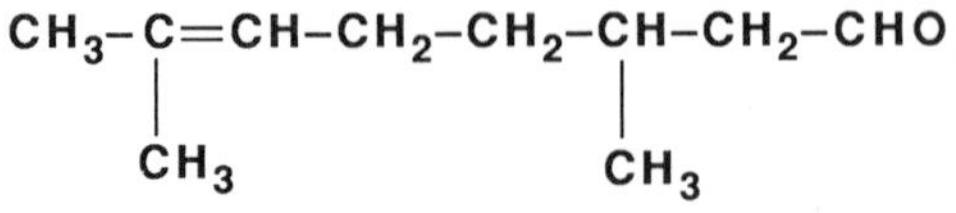

CH3–C=CH–CH2–CH2–CH–CH2–CHO
CH3
CH3

CH3–C=CH–CH2–CH2–C=CH–CH2OH
CH3
CH3

Cyclic monoterpenes include camphor, used in the past in the manufacture of celluloid; menthol, which has a very characteristic peppermint odor; and β-pinene, which is found in turpentine. The structure of menthol is

CH3
OH
CH
CH3
CH3

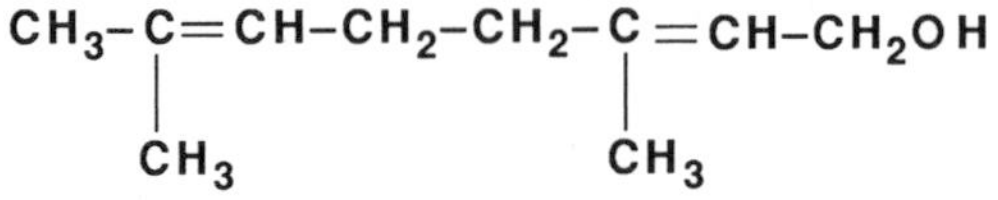

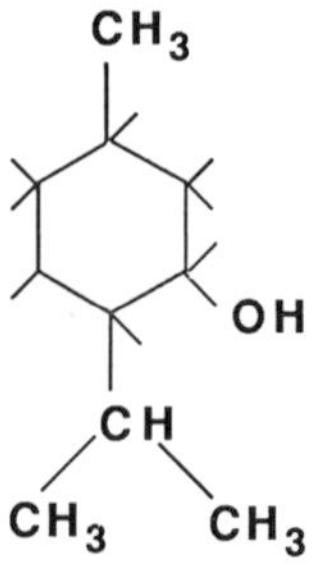

The sesquiterpenes also have linear or cyclic structures. Rose oil contains the sesquiterpene alcohol farnesol, while oil of myrrh contains bisabolene.

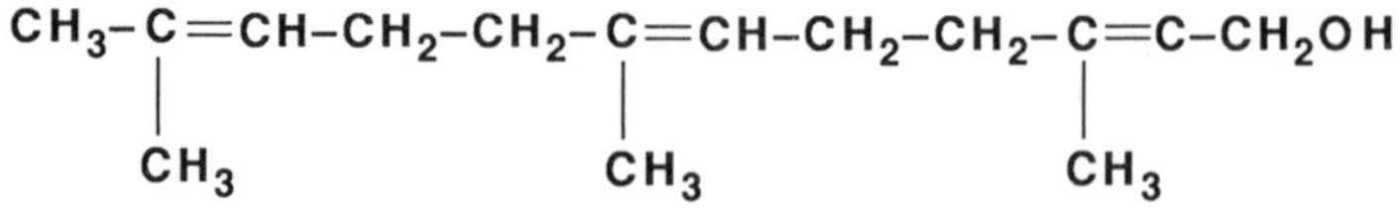

CH3–C=CH–CH2–CH2–C=CH–CH2–CH2–C=C–CH2OH
CH3
CH3
CH3

CH3–C=CH–CH2–CH2–C=
CH3
CH3
CH3

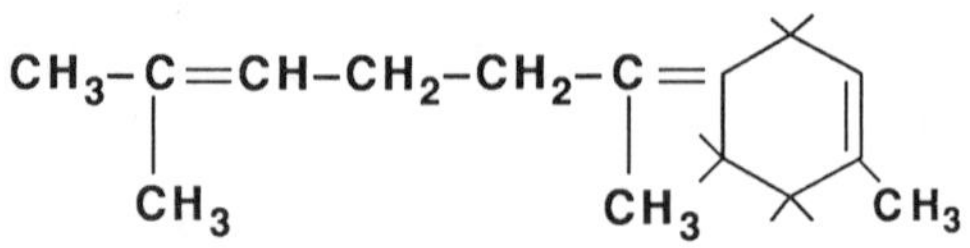

Diterpenes include retinal, a compound important in vision, and abietic acid, a component of plant resins.

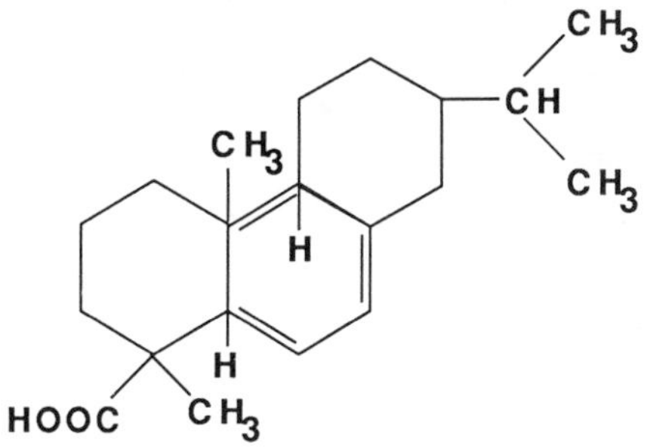

The sesterterpenes are rarely encountered in nature. An example of a naturally occurring triterprene is β-carotene,

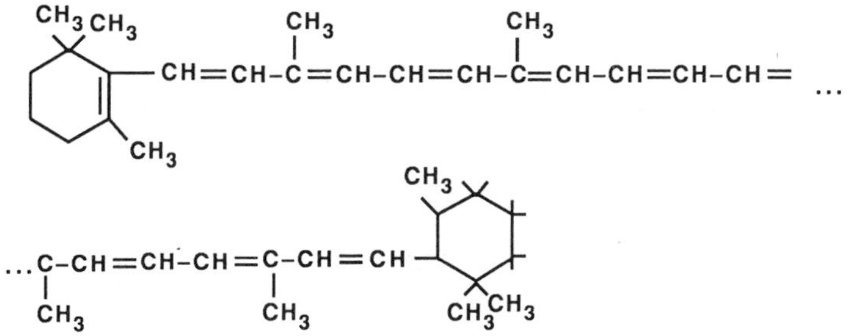

which gives carrots their characteristic color, and is the biochemical precursor of vitamin A. β-Carotene is also used commercially in some sunscreen preparations. Carotene is thought to be a source of some of the smaller aromatic molecules found in fuels. A structurally related triterpene is lycopene.

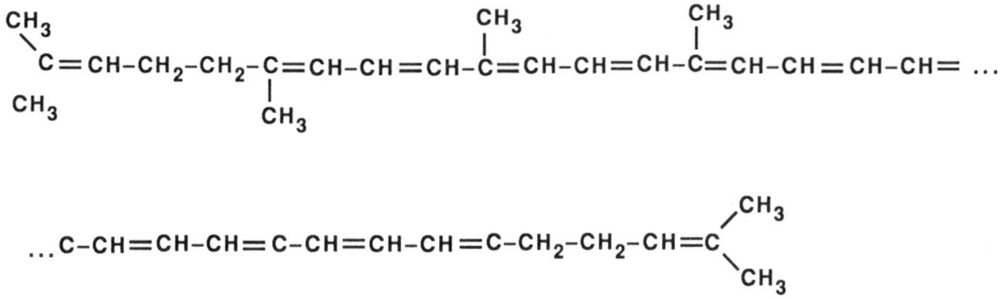

Polyterpenes also occur in some plants, the most important example being natural rubber. Dextropimaric acid is isolated from resins exuded by pine trees to heal wounds:

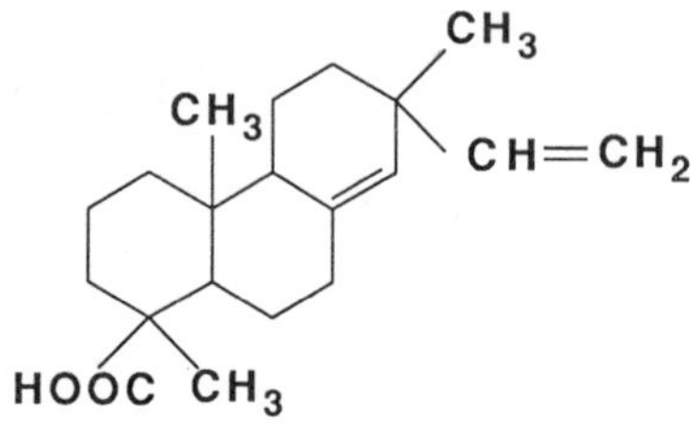

The characteristic structural feature of the terpenes is the terminal methyl group acting as a "flag" on every fifth carbon atom in the ring or chain.

Steroids
The steroids are natural products containing four fused rings, three cyclohexane rings and one cyclopentane ring. The steroids are somewhat analogous in structure to the diterpenes. The basis ring system has two methyl substituents and an additional side chain, denoted by R- in the general structure below.

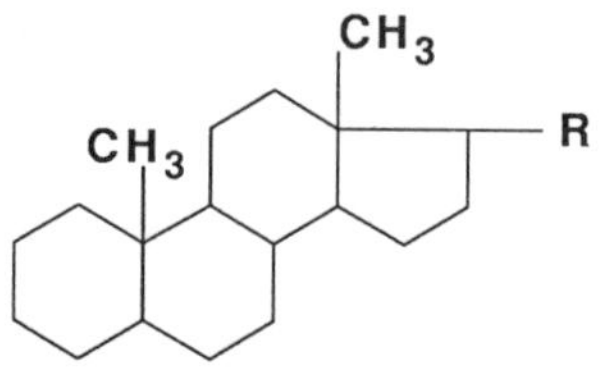

The structural relationship of terpenes and steroids is illustrated by the conversion of the triterpene squalene to the steroid lanosterol.

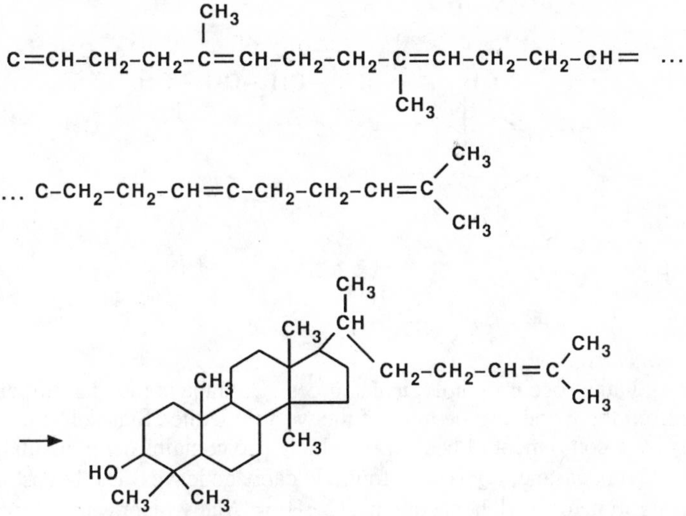

$$CH_3$$
$$C{=}CH{-}CH_2{-}CH_2{-}\underset{\underset{CH_3}{|}}{\overset{\overset{CH_3}{|}}{C}}{=}CH{-}CH_2{-}CH_2{-}C{=}CH{-}CH_2{-}CH_2{-}CH{=}\ \cdots$$

$$\cdots\ C{-}CH_2{-}CH_2{-}CH{=}C{-}CH_2{-}CH_2{-}CH{=}C\overset{\nearrow CH_3}{\underset{\searrow CH_3}{}}$$

Lanosterol is a precursor in the biosynthesis of cholesterol. The steroid digitoxigenin is a component of digitalis, which is used as a heart stimulant.

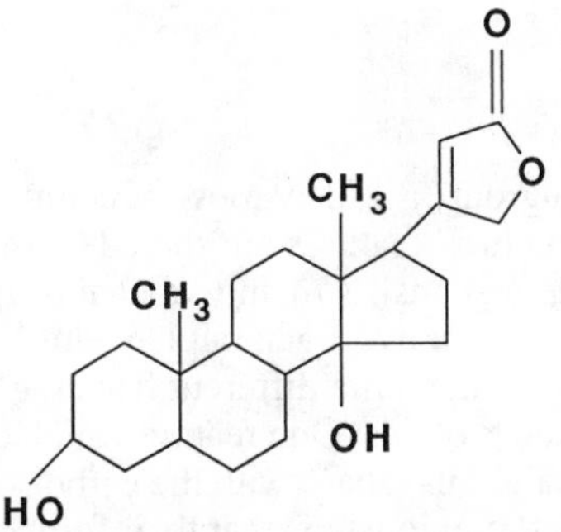

Although today digitalis is used as a treatment for cardiac problems, one of the major health problems in our society, in simpler times it was used as one of the active ingredients on poisoned arrows. Cholesterol, a compound of interest because of its medical effects in humans, is the most abundant non-saponifiable lipid in animal tissue.

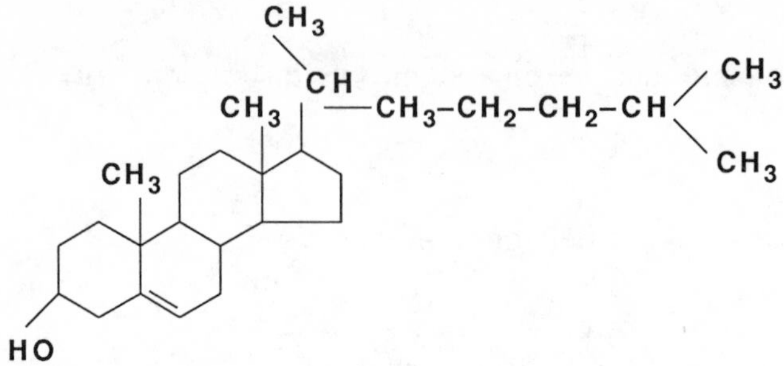

CH_3
CH
CH_3 —CH_3–CH_2–CH_2–CH CH_3
CH_3 CH_3
CH_3
HO

Hydrocarbons

Some n-alkanes occur in biological sources. Heptane is found in turpentine made from the wood of pine trees of the western United States. Heptacosane, $C_{27}H_{56}$ is a component of bees' wax, which also contains hentriacontane, $C_{31}H_{64}$. Nonacosane, $C_{29}H_{60}$, is found in cabbage leaves. The largest alkane occurring in plants is dohexacontane, $C_{62}H_{126}$. Many plant waxes contain alkanes in the range C_{25} to C_{35}. In plant waxes alkanes with odd numbers of carbon atoms predominate by about a factor of ten over those with even numbers of carbons. The amount of aromatic hydrocarbons in living organisms is negligible.

Summary

Although the lipids contain, as a group, a wide variety of compounds seemingly diverse, some key structural features are shared by most of the lipids. As a rule, the carbon skeleton consists mainly of chains of carbon atoms. Lipids are comparatively rich in hydrogen but have low contents of oxygen. In these respects, the lipids are quite different from the carbohydrates, glycosides, and the proteins. One feature which the biochemical lipids (waxes, fats, and oils) share with the carbohydrates, glycosides, and proteins discussed previously is that their formation is reversible.

Resins

Resins are compounds synthesized by plants for special purposes, such as for sealing wounds. An example is dextropimaric acid, a component of pine resin.

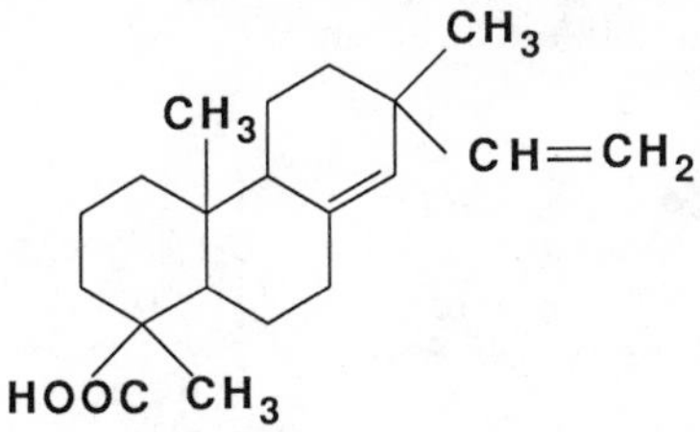

Pine resins may be familiar as the sticky substance sometimes encountered in handling the annual Christmas tree. Resins are generally characterized as having multiple systems of carbon rings and low oxygen contents.

The resin called storax is the principal component of incense. About 50% of storax is composed of the simple compound cinnamic acid.

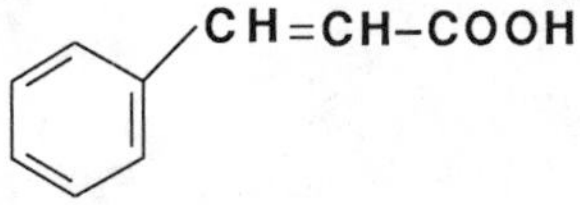

Lignin

Lignin occurs only in the "higher" (woody) plants. Lignin is the constituent of cell walls that provides mechanical rigidity to the plant. The exact structure of lignin is not known, and the lignin structure may vary from one plant species to another. Despite these uncertainties, lignin is considered to be a biopolymer of three compounds, coniferyl alcohol, sinapyl alcohol, and p-coumaryl alcohol.

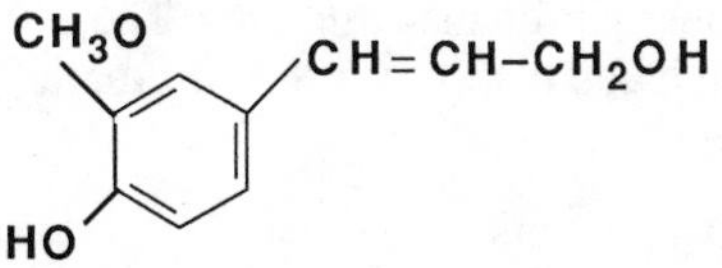

coniferyl alcohol

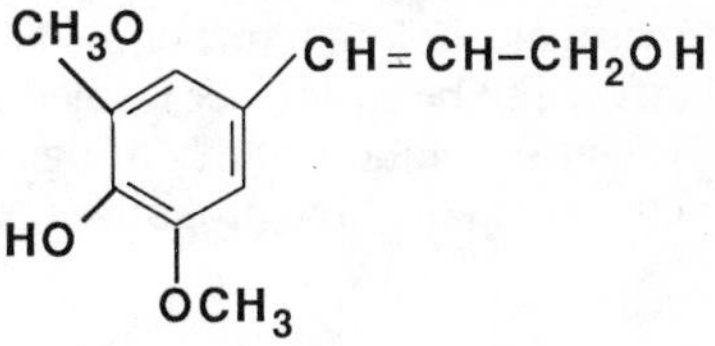

sinapyl alcohol

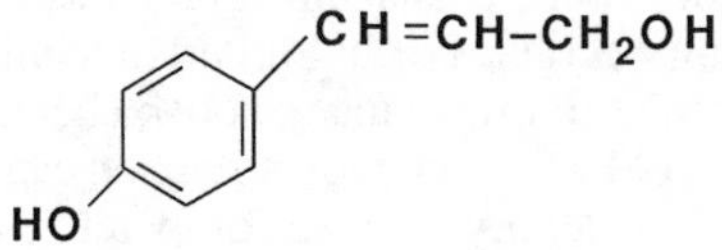

p-coumaryl alcohol

An example of a proposed lignin structure is

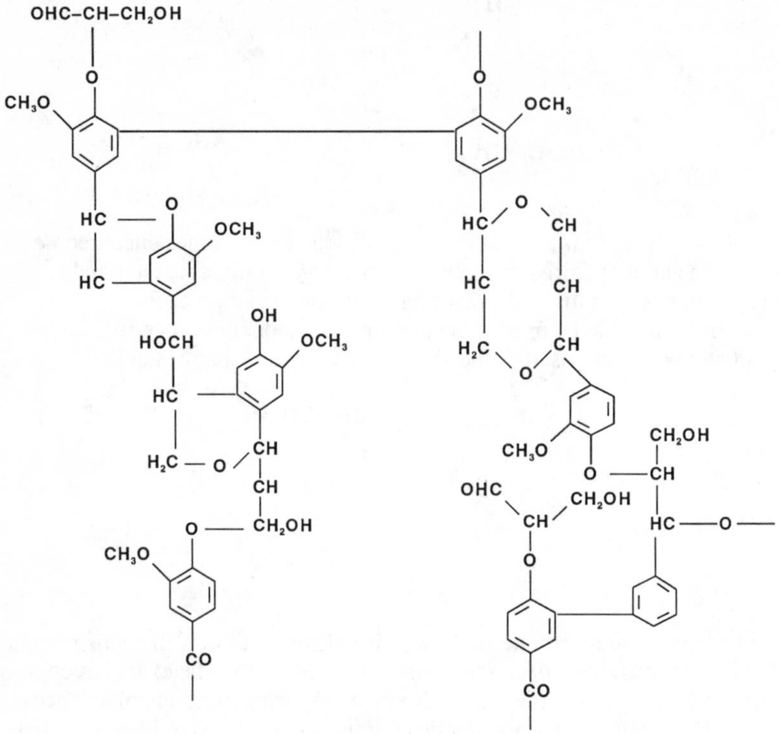

OHC–CH–CH₂OH

CH₃O   O   OCH₃

HC — O   OCH₃   HC   O   CH
HC   HC   CH

                OH
HOCH   OCH₃   H₂C   CH
HC   O

                            CH₃O   CH₂OH
H₂C — O   CH   O — CH
           CH   OHC   CH₂OH
      O — CH₂OH   CH   HC — O —
CH₃O   O

           CO   CO

The so-called guaiacyl lignins are primarily polymers of coniferyl alcohol, with some incorporation of sinapyl alcohol. All lignins contain 1-5% of p-coumaryl alcohol. In terms of the arrangement of carbon atoms, lignin can be considered to be a polymer formed from phenylpropane structures. The important structural features of lignin are the abundance of cyclic aromatic structures and the high proportion of oxygen.

Summary

Again, it is not important that the fine details of the molecular structures presented in the preceding sections be committed to memory. For the purposes of the discussion to follow in the next two chapters, the most important point is that all of the compounds discussed can be put into two broad categories. The first category is those compounds having relatively high concentrations of hydrogen and low concentrations of oxygen, and the carbon atoms mainly being in chain structures. The materials which fall into this category include the terpenes, hydrocarbons, fats, oils, and waxes. The second category is the compounds characterized by high oxygen contents and carbon ring structures, which include the carbohydrates, glycosides, and lignin. The proteins and resins are not as easy to incorporate in this simple scheme, but, on the basis of composition, the resins would be included with the category one high-hydrogen, low-oxygen compounds, and the proteins with the category two high-oxygen compounds.

The occurrence of the compounds discussed in this chapter in living organisms is summarized in Table 1.2.

Table 1.2 Distribution of fuel precursors among living organisms

	Zooplankton	Phytoplankton	Algae	Higher plants
Proteins	Major	⇒ ⇒ Decreasing importance ⇒ ⇒		
Carbohydrates		⇒ ⇒ Increasing importance ⇒ ⇒		
Lipids	⇒ ⇒ ⇒ ⇒ ⇒ Decreasing importance ⇒ ⇒ ⇒ ⇒			
Glycosides	⇐ ⇐ ⇐ ⇐ ⇐ ⇐ ⇐ ⇐ ⇒ ⇒ ⇒ ⇒ ⇒			
Resins			⇒ ⇒ Increasing ⇒	
Lignin			Only in higher plants	

As a rule, the compounds characterized by carbon chains, rich in hydrogen and relatively low in oxygen, predominate in the plankton and algae. These organisms thrive in aquatic environments. The second class of compounds, with carbon rings, rich in oxygen and relatively low in hydrogen, predominate in the higher plants. The higher plants are terrestrial organisms.

It should seem reasonable that if two compounds, one with a carbon chain structure, low oxygen content, and relatively high hydrogen content, and the other with cyclic carbon structures, high oxygen and low hydrogen contents, are subjected to the same sequence of reactions, both the behavior of the two compounds in the reactions and the kinds of products formed should be different. In essence, there are two different kinds of starting materials which are the input to the "black box" reactor for conversion of organic matter to fuels. One type is the low-oxygen chain structures, predominant in organisms thriving in aquatic environments, and the other type is the high-oxygen cyclic structures of the higher plants. It is reasonable to expect that the products of the formation of fuels will therefore be different, depending upon the nature of the original organic matter. In fact, we will see that one type of starting material will lead primarily to the formation of petroleum and natural gas, and the other, mainly to coal.

The conversion of organic matter into fuels proceeds in two major phases. These two phases are the subjects of the next two chapters. By the conclusion of Chapter 3 we will have seen how the different starting materials have been converted to petroleum, gas, and coal. Subsequent chapters will then show how the compositions of these fuels differ from each other, and how the composition and properties of the fuels derive from the organic materials from which they had formed.

Further reading

Barker, C. (1979). *Organic Geochemistry in Petroleum Exploration.* Tulsa: American Association of Petroleum Geologists; Chapter 3.

Berkowitz, N. (1979). *An Introduction to Coal Technology.* New York: Academic Press Chapter 1.

Fieser, L. F. and Fieser, M. (1961). *Advanced Organic Chemistry.* New York: Reinhold Chapters 2, 29.

Haun, J. D. (1976). *Origin of Petroleum. I.* Tulsa: American Association of Petroleum Geologists p 162ff.

Loudon, G. M. (1988). *Organic Chemistry,* 2d edn. Menlo Park, CA: Benjamin/ Cummings; Chapter 17.

McMurry, John. (1988). *Organic Chemistry*, 2d edn. Pacific Grove, CA: Brooks/Cole; Chapter 24.

Streitwieser, A. Jr. and Heathcock, C. H. (1985). *Introduction to Organic Chemistry*, 3rd edn. New York: Macmillan; Chapters 28, 34.

Van Krevelen, D. W. (1961) *Coal: Typology - Chemistry - Physics - Constitution* Amsterdam: Elsevier; Chapter V.

Chapter 2

Diagenesis

This chapter begins the discussion of the processes which are responsible for the conversion of organic matter into fuels. The processes discussed in this chapter are collectively known as diagenesis. The reactions occur in the shallow subsurface where the temperature and pressure are near ambient. Many of the processes discussed in this chapter are facilitated by bacteria; consequently diagenesis is sometimes called the biochemical stage of fuel formation. The main products of diagenesis are methane, carbon dioxide, water, and the immediate precursor to fuels, kerogen.

The end of accumulation of organic matter

At some time the accumulation of organic matter (that is, the remains of dead organisms) will stop, usually as a result of some radical change in the environment. Considering for example a tropical swamp as a place where large amounts of organic matter could accumulate, a change to a dryer climate could occur to cause the swamp to dry out, a change to much colder temperatures could greatly restrict the growth of organisms, or perhaps land subsidence could allow the ocean to flood the swamp, with the influx of saline water killing the freshwater organisms.

Regardless of the specific environment and the reasons for the ending of build-up of organic matter, one of two possible fates is likely to follow. In some cases, the accumulated organic matter may be exposed to the air. Alternatively, the accumulated organic material may be covered by a layer of stagnant water or inorganic sediment (*e.g.*, silt, sand, or mud). In the former case, exposure of organic matter to oxygen will result in its being slowly but surely destroyed. (We can realize this, for example, on a walk through a forest, where it is not necessary to be clambering over the remains of trees which died decades or centuries in the past.) The destruction of the organic matter eliminates the prospect of its being converted to fuel, so from the perspective of fuel chemistry this situation is of no further interest. In the latter case, where the accumulated organic matter is at least partially protected against the ravages of oxygen, conversion to fuels can begin. We must consider what happens to the organic matter as it experiences burial and the first stages of conversion to fuels.

In making this consideration, two points must be kept in mind. First, most organic compounds preserved in sediments and converted to fuels are structures held together by relatively strong covalent bonds. Most organic compounds are not soluble in water, and most do not ionize. Consequently, reactions of organic compounds at moderate temperatures prevailing in the environment tend to be very slow in comparison with reactions of inorganic ions in solution or with reactions carried out in the laboratory at elevated temperatures. Second, the starting materials for the reactions are usually complex mixtures rather than pure compounds, and in some instances the structures of the starting materials are not known exactly. Most of the components of the mixtures have numerous C-C, C-H, or C-O bonds which are, roughly, of comparable bond energy. Hence a given substance may be able to react in several different ways which are thermodynamically similar. Those reactions which do occur are slow and may not reach completion. The practical effect of these two points is that it may not be possible to follow the detailed reaction chemistry, step by step, for a particular compound as it is transformed to become a component of a fuel. Rather, the focus must be on understanding how various classes of compounds - those introduced in Chapter 1 - respond to the reaction variables of temperature, pressure, composition, and time during natural geological processes.

Rotting

The first reaction zone occurs at depths of 0 - 1 m below the surface. The interface with the atmosphere is still close enough that oxygen can diffuse through the covering sediments to react with the organic matter. An example is the reaction introduced previously, the oxidation of a simple hexose

$$C_6H_{12}O_6 + 6O_2 \rightarrow 6CO_2 + 6H_2O$$

The rotting zone essentially represents a race between the rate at which the organic matter is being oxidized and the rate at which it is being buried more deeply, and thus becoming better protected. Fortunately for the eventual formation of fuels, many organic reactions at near-ambient temperatures are slow, for the reasons explained above. Even though oxidation of the organic matter to carbon dioxide and water occurs in the rotting zone, the rate of burial of the organic matter can outstrip the rate of oxidation, so that the organic matter survives to pass into the next zone, where the oxygen content is lower.

Moldering

This reaction zone is generally 1 - 2 m below the surface. In this region the oxygen concentration is low enough to allow reactions other than oxidation to occur. An important reaction is the hydrolysis of biopolymers such as cellulose and proteins. The hydrolysis reaction is thermodynamically favored, but kinetically is very slow in nature. For example, the hydrolysis of polypeptides, discussed below, proceeds at rates on the order of 10^{-6} to 10^{-7} sec^{-1}, which in effect means that one peptide linkage is cleaved per month.

Proteins

Proteins are particularly likely to undergo some type of reaction or transformation. Protein structure can be altered by heat, exposure to acid or base, or exposure to ultraviolet light, x-rays, or even high pressure. In the laboratory numerous reagents can affect the protein structure in some fashion. The alteration of protein structure is known as *denaturation*. A denatured protein is always of lower solubility than the unaltered structure.

Proteins undergo hydrolysis to smaller polypeptides, and, on extensive hydrolysis, decompose all the way to the constituent amino acids. This reaction can be illustrated for a simple peptide as

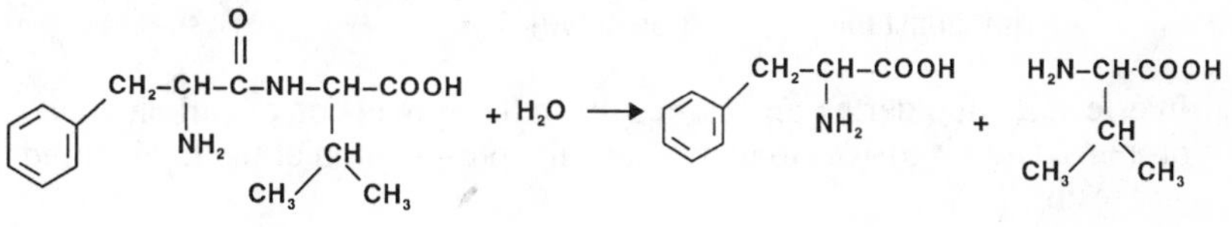

Carbohydrates and glycosides

Analogously, carbohydrates also undergo hydrolysis to smaller polysaccharides and eventually to simple sugars:

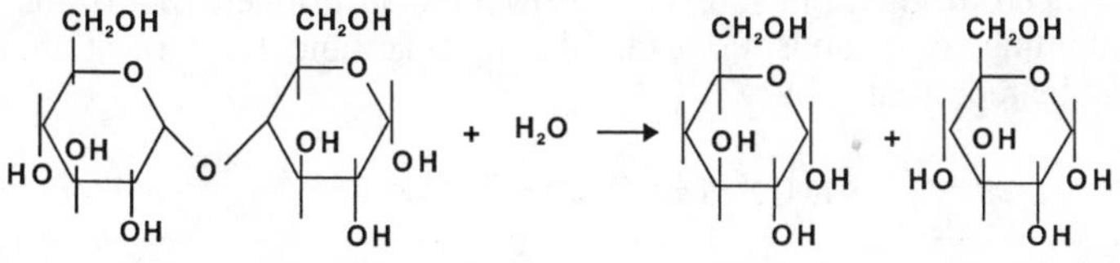

Glycosides are hydrolyzed similarly:

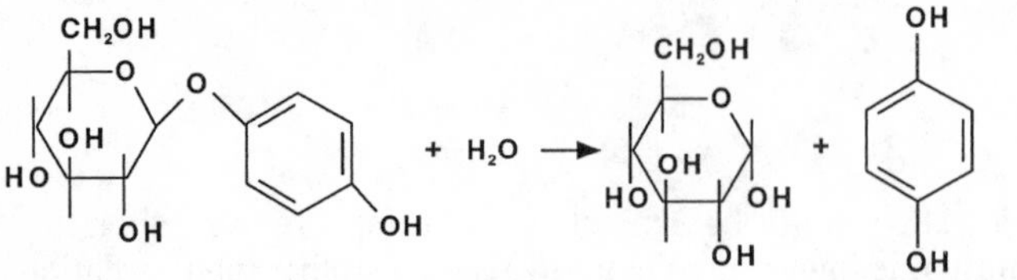

Formation of humic acids

The products of the decomposition of proteins, carbohydrates glycosides recombine, by reactions which are not well understood, to form compounds called humic acids.

    Humic acids are black, macromolecular solids of undefined structure soluble in aqueous sodium hydroxide but reprecipitated when the solution is acidified. Humic acids are a principal component of a soft, wet, amorphous, dark brown to black organic material called *humus*. It should be noted that this definition of humic acids is not an indication of their composition and structure, but rather of their behavior. Such definitions are sometimes called *operational definitions*. As we will see in later chapters, operational definitions occur frequently in fuel chemistry.

    As a result of the processes ocurring in the moldering zone, a relative ranking of the stability of the components of organic matter begins to emerge:

proteins < carbohydrates, glycosides <<< lignin, lipids, resins
(quite unstable)   (unstable)   (very stable)

In effect, the moldering process results in the conversion of carbohydrates, proteins, and glycosides to humus, and the preservation of the lignin, lipids, and resins.

Putrefaction

In the region 2 - 10 m below the surface, the oxygen concentration is so low that the conditions are effectively anaerobic. In this region, bacteria obtain the oxygen they need for their metabolism not from the atmosphere, but from oxygen containing compounds. For example, the anaerobic decomposition of a hexose can be represented by

$$C_6H_{12}O_6 \rightarrow 3CO_2 + 3CH_4$$

This reaction should be compared with the aerobic destruction of hexoses shown previously. The key distinction is the emergence of a new product, methane.

Proteins

The decomposition of proteins occurs slowly from the chemist's perspective, but rather rapidly on a geological time scale. Proteins are not found in rocks older than the Pleistocene, that is, a maximum of about two million years old.

The first step in the decomposition of proteins is their denaturation and hydrolysis into smaller structures, including the constituent amino acids. Individual amino acids vary in their resistance to further degradative reactions. The degradation of the amino acid components of proteins proceeds via oxidative deamination. During putrefaction, proteins lose nitrogen via deamination. The process is illustrated using alanine as an example:

$$CH_3-\underset{\underset{NH_2}{|}}{CH}-COOH \longrightarrow CH_3-\underset{\underset{NH}{||}}{C}-COOH \longrightarrow NH_3 + CH_3-\underset{\underset{O}{||}}{C}-COOH$$

A product of this reaction is ammonia, the vehicle by which nitrogen is removed. This reaction is responsible for the odor of ammonia noticeable where protein-rich organic wastes are putrefying, as for example around stables or outdoor toilets.

Deamination proceeds via the intermediate imino acid, which is fairly unstable and is seldom isolated as a reaction product. The subsequent decomposition of the imino acid results in the formation of an α-keto acid. The α-keto acids themselves are somewhat unstable and may decompose further to an aldehyde and carbon dioxide:

$$CH_3-\underset{\underset{O}{||}}{C}-COOH \longrightarrow CH_3-CHO + CO_2$$

Carbohydrates

Carbohydrates are transformed to methane by the reaction illustrated previously for anaerobic decomposition of a hexose. The methane formed this way is called *biogenic methane* or biochemical methane to distinguish it from methane formed by the thermal reactions to be discussed in Chapter 3. Most biogenic methane arises from anaerobic bacterial decomposition of cellulose.

If sufficient biogenic methane accumulates in the environment, it can form a useful fuel resource. Deliberate generation of biogenic methane for fuel use is now being practiced, particularly in areas where abundant organic

31

matter - such as agricultural wastes - is available. For example, farmers can collect the wastes from their animals and convert these wastes to methane via anaerobic digestion. The methane can be collected for use as a fuel in farm vehicles, for domestic heating and cooking on the farm, or for burning to power an electric generator. Biogenic methane from agricultural wastes offers farmers a way to become energy self-sufficient.

In nature biogenic methane sometimes bubbles to the surface of the earth. The observation of such bubbles of gas in swamps or bogs is the source of one of the trivial names for methane - marsh gas. If the methane migrates to an aerobic region, it can be destroyed by oxidation:

$$CH_4 + 2O_2 \rightarrow CO_2 + 2H_2O$$

Occasionally the methane bubbling to the surface of the water in a swamp or bog will catch fire, a phenomenon sometimes known as will o'the wisp. The observation of flashes of light from the methane combustion has been the origin of numerous lurid folk tales about ghosts or other swamp creatures.

A second important reaction of carbohydrates is the reaction with sulfate:

$$3SO_4^{-2} + C_6H_{12}O_6 \rightarrow 6HCO_3^- + 3H_2S$$

The sulfate ion may be present in the environment if an incursion of sea water has occurred (sulfate is the second most prevalent anion in seawater) or may have formed from oxidation of sulfur in sulfur-containing amino acids during moldering.

Hydrolysis of polysaccharides in acidic conditions leads to the formation of heterocyclic oxygen compounds, the furans. An example is the formation of furfural, which is prepared industrially by acidic hydrolysis oat hulls, straw, and similar agricultural wastes. The polysaccharides in these materials are made of pentoses, which dehydrate to furfural in acidic media:

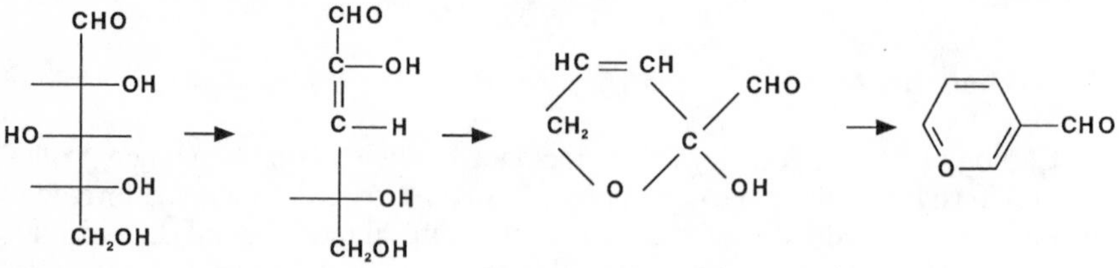

Lipids

During putrefaction some lipids begin to decompose. Fats and oils are hydrolyzed, as, for example, in the reaction of glyceryl palmitate to glycerol and palmitic acid:

$$CH_2-OOC-C_{15}H_{31}$$
$$CH-OOC-C_{15}H_{31} \quad + \quad 3H_2O \longrightarrow$$
$$CH_2-OOC-C_{15}H_{31}$$

$$CH_2OH$$
$$CHOH \quad + \quad 3C_{15}H_{31}COOH$$
$$CH_2OH$$

This hydrolysis produces long chain carboxylic acids. As we will see later, these compounds subsequently decompose via loss of the carboxyl group to produce the alkanes which are major components of petroleum.

The waxes, however, are generally resistant to hydrolysis at this stage. A structural factor responsible for this difference between waxes and fats or oils is the position of the ester group in the molecule. In fats and oils the ester is at one end of the molecule, relatively accessible to the attack of water molecules. In a wax, though, the ester group is effectively "buried" in the interior of a long chain of hydrophobic methylene groups. Entanglement of the long hydrocarbon chains also makes it difficult for water molecules to reach and attack the ester linkage.

Lignin

Lignin begins to undergo some decomposition during putrefaction. The products of the breakdown of lignin are phenols, aromatic acids, and quinones.

The behavior of these compounds during putrefaction allows expanding the preliminary ordering of stability developed earlier. The list of relative stabilities can now be written as

proteins < carbohydrates, glycosides < fats, oils < lignin < waxes, resins

>------------- increasing stability ----------------->

The formation of kerogen

At the end of putrefaction, the original organic matter has been converted into an "organic soup" of humic acids, products of the mild decomposition of lipids and lignin, and relatively unaltered waxes, resins, and hydrocarbons.

This mixture can polymerize to an insoluble, high molecular weight material called *kerogen*. A possible simplified model for the formation of kerogen is the reaction of phenol and formaldehyde:

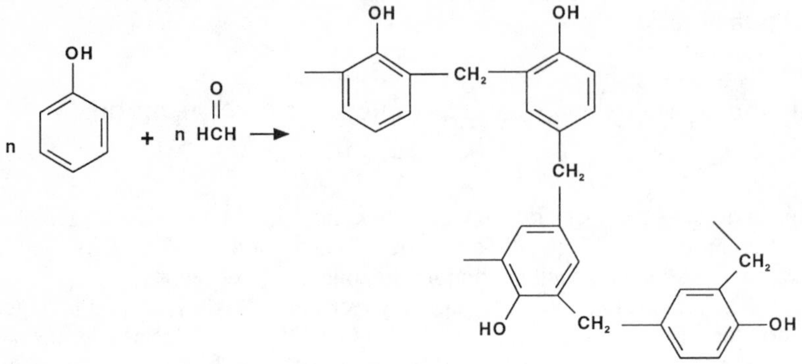

Although this analogy cannot be pursued too far, it should be recalled that aldehydes may be produced from the decomposition of proteins (via α-keto acids) and carbohydrates, and that phenolic compounds can be produced from the decomposition of lignin and glycosides.

It is actually more appropriate to refer to kerogen*s*, because different types of organic matter may have predominated during kerogen formation. Kerogen is divided into three classes, based on the nature of the original organic source material (Table 2.1).

Table 2.1 Classification of kerogens by source material

Type	Name	Source
I	Algal kerogen	Mainly algae
II	Liptinitic kerogen	Mainly plankton, some contribution from algae
III	Humic kerogen	Mainly higher plants

Types I and II kerogen are hydrogen-rich, with predominantly straight chain compounds. Type I kerogen, which derives mainly from algae, has an H/C atomic ratio of 1.2 - 1.7, and an H/C ratio of about 1.6 - 1.7. Lipids, particularly fats and oils and waxes, are the predominant contributors to this type of kerogen. Type II kerogen is also highly aliphatic, containing material deriving both from algae and from plankton. Type III kerogen is oxygen-rich, with mainly cyclic carbon structures. The H/C ratio is lower than in the other types of kerogen, being less than 0.85. Type III kerogen tends to be rich in aromatic compounds, and derives mainly from the lignin of the higher (woody) plants. Because the chemical composition of the original organisms - algae, plankton, and higher plants - is different, the composition of each of the types of kerogen will be different. In turn, since the compositions of the kerogens are different, so too will be the ways in which they react in subsequent processes and the kinds of products they form.

As organic matter accumulates and is converted to kerogen during diagenesis, it will become mixed with various proportions of inorganic sediments. The sedimentary deposit which results can be classified both in terms of the proportion of kerogen in the deposit (the balance being inorganic material) and the type of kerogen present. Three major classifications are recognized (Table 2.2).

Table 2.2 Classification of deposits by amount and type of kerogen

Amount of kerogen	Principal kerogen types	Type of deposit
~1%	I, II	Oil source rock
< 50%	I, II	Oil shale
> 50%	III	Coal

Type III kerogen seldom yields oil, but is the source of most of the world's coal. (Actually there is now some interest among geochemists for the potential role of coal as an oil source rock; that is, the possible generation of oil from sediments containing >50% of type III kerogen.) Most oil and natural gas are produced from types I and II kerogen. Some very special coals, of minimal commercial value, also derive from types I and II kerogen.

The van Krevelen diagram

The formation of kerogen brings diagenesis, the biochemical stage of fuel formation, to a close. The subsequent conversion of kerogen to coal, petroleum, and gas is driven by exposure to the temperatures and pressures in the geological environment. The formation of fuels from kerogen is therefore called the geochemical stage, or catagenesis. This process is the subject of the next chapter.

In Chapter 1 it was pointed out that petroleum contains several hundred to several thousand individual compounds, and that coal is a heterogeneous material of unknown structure. Fortunately it is not necessary to track the fate of each component of the original organic matter or of kerogen through diagenesis and catagenesis to understand the fuel formation process. Rather, changes in composition can be followed in simplified fashion by tracking changes in elemental, rather than molecular composition. Of the elements which occur in fuels, three are of primary importance for understanding the chemistry of fuel formation: carbon, hydrogen, and oxygen. In principle, changes in composition could be plotted on a three-dimensional diagram. However, such diagrams, regardless of their utility, can be difficult to draw and read. Changes in composition can be plotted as only two variables by using ratios of the three elements. Since carbon is the most important, a standard approach is to plot the H/C and O/C ratios. By convention, this is done using atomic ratios, rather than weight ratios, and the graph is drawn by showing H/C as the abscissa and O/C as the ordinate. Such a plot is known as a van Krevelen diagram, in honor of the great Dutch coal scientist D. W. van Krevelen. An example of a van Krevelen diagram with points for typical kerogens plotted, is shown in Figure 2.1. The van Krevelen diagram will be used extensively in the next chapter to track composition changes during fuel formation.

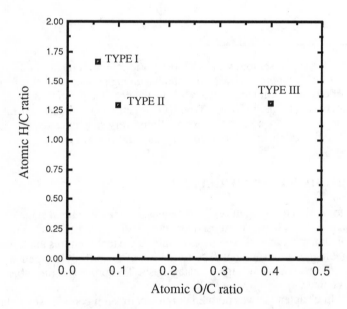

Fig. 2.1 van Krevelen diagram showing the H/C vs O/C relationships for examples of the three types of kerogen.

Finally, the relationships among the requirements for growth, the types of organisms and their chemical components, and the kerogens formed are summarized in the chart of Figure 2.2.

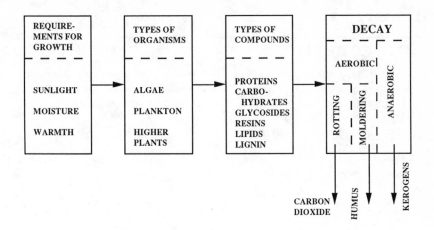

Further reading

Bustin, R. M., Cameron, A. R., Grieve, D. A., and Kalkreuth, W. D. (1983). *Coal Petrology: Its Principles, Methods, and Applications.* St. John's, Newfoundland: Geological Association of Canada.

Fox, R. F. (1988). *Energy and the Evolution of Life.* San Francisco: Freeman; Chapter 1.

Krauskopf, K. B. (1979). *Introduction to Geochemistry.* New York: McGraw-Hill; Chapter 11.

Selley, R. C. (1985). *Elements of Petroleum Geology.* San Francisco: McGraw-Hill; Chapter 5.

Streitwieser, A. and Heathcock, C. H. (1985). *Introduction to Organic Chemistry.* New York: Macmillan. Chapters 29, 31.

van Krevelen, D. W. (1961). *Coal : Typology - Chemistry - Physics - Constitution..* Amsterdam: Elsevier; Chapter V.

Chapter 3

Catagenesis

This chapter discusses the conversion of kerogen to coal, petroleum, and gas. The process is known as catagenesis, or the geochemical stage of fuel formation. Two key concepts, which will recur throughout the book are introduced in this chapter: hydrogen redistribution to a carbon-rich, hydrogen-poor product and a hydrogen-rich, carbon-poor product; and some of the important processes in free radical reactions.

Hydrogen redistribution

Catagenesis reactions are driven by temperature and pressure conditions prevailing inside the earth. At this point in the formation of fuels from organic matter the organic sediments are now buried so deeply that reactions are completely anaerobic. In the absence of oxygen, the thermodynamically stable end products of the reactions of organic compounds are graphite and methane. If we consider a reaction for a simple alkane such as octane,

$$2C_8H_{18} \rightarrow 9CH_4 + 7C$$

this reaction is strongly favored thermodynamically. The $\Delta G°$ for this reaction at 298 K is -58.5 kcal per mole of octane.

During catagenesis there is a tendency to redistribute the hydrogen available in the kerogen to form a hydrogen-rich fraction, which would ultimately become methane, and a hydrogen-poor fraction, which ultimately becomes graphite. This hydrogen redistribution process is illustrated in the schematic diagram of Figure 3.1.

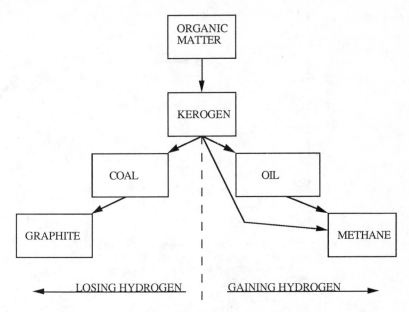

Fig. 3.1 The redistribution of hydrogen in natural organic materials, with the formation of hydrogen-rich products, oil and natural gas, and hydrogen-poor (carbon-rich) products, coal and graphite.

Obviously, the existence of coal and petroleum shows that not all the kerogen in the world has transformed to graphite and methane. The major limitation preventing the redistribution from going to completion is kinetics, that is, the very long reaction times required. In some cases, the formation of significant amounts of methane is also limited by stoichiometry, in that there may not be sufficient hydrogen available in the kerogen to provide much methane.

The hydrogen redistribution scheme shown in Figure 3.1 can be expressed as reaction pathways on a van Krevelen diagram (Figure 3.2) where, for a generalized kerogen composition, we see one pathway leading to methane (H/C = 4) and a second to graphite (at the origin).

We will also encounter hydrogen redistribution later in this book, in connection with production of synthetic liquid fuels from coal and the formation of cokes of catalyst surfaces during petroleum processing.

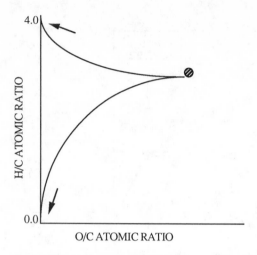

Fig. 3.2 The hydrogen redistribution process illustrated on a van Krevelen diagram. A kerogen (indicated by the shaded circle) transforms to hydrogen-rich methane and hydrogen-poor graphite.

Catagenesis of humic kerogen

Kerogen will be covered by succeeding layers of inorganic sediments and will gradually become buried to greater and greater depths. The chemical

reactions during catagenesis are very complex, but like any chemical reaction we might carry out in the laboratory, the reactions are affected by three major variables: temperature, pressure, and time. Of these, time and temperature are the most important for the catagenesis processes.

The cessation of peat formation and the beginning of the transformation of peat to coal can be represented by Figure 3.3, in which the peat is shown covered by a layer of inorganic sediment. Gradually the layer of sediment on top of the peat thickens. This can happen if, for example, the land slowly sinks, allowing further accumulation of sediment on top of the buried peat (Fig. 3.4). All aerobic processes cease. The buried peat is exposed to increasing temperatures, which come mainly from the natural thermal gradients of the earth. Ordinarily these gradients arise from heat generated by the decomposition of naturally occurring radioactive isotopes. The geothermal gradient varies from place to place on the earth, but typical values are 10° - 30° C/km. Then for example, assuming a surface temperature of 30° C, burial of a sediment to 2000 m would raise its temperature to about 100° C. In some unusual and very localized instances, very high thermal gradients

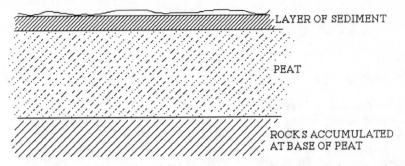

Fig. 3.3 Peat gradually becomes buried by layers of inorganic sediments.

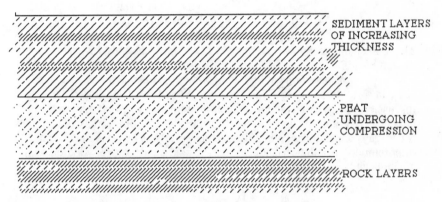

Fig. 3.4 As time passes, the buried layer of peat is covered by inorganic sediment layers growing in thickness, compressing the accumulated peat.

may be established by other geological processes, such as an intrusion of magma near the buried kerogen. At the same time, the weight of the overlying accumulating sediments begins to increase the pressure on the peat. In extreme cases, high pressures are generated by the folding of the rock sediments enclosing the kerogen.

The peat is now sealed inside the earth, where the very long time periods typical of geological processes can operate. Catagenesis of humic kerogen may require tens of millions of years. The long reaction times are an important feature of catagenesis, because reactions which would be immeasurably slow on a human time scale may still be able to occur. For example, suppose a reaction had a rate constant of 10^{-10} sec^{-1}. If this reaction were run in a laboratory, we might have to wait 300 years to take the first data point (depending of course on the type of measuring equipment used).

However, in 30,000,000 years there are 10^{15} sec - which provides a lot of time for even a very slow reaction to occur.

Reactions during early catagenesis

Dehydration and decarboxylation
Recall that during diagenesis fats and oils were hydrolyzed. As applied to lipids, the hydrolysis of esters is called saponification. (If the hydrolysis reaction were done in basic solution, the sodium or potassium salts of the fatty acids are produced. These salts of fatty acids are soaps, hence the name saponification.) During catagenesis, waxes may also be saponified. To illustrate this reaction for a possible component of carnauba wax:

$$CH_3(CH_2)_{22}COOCH_2(CH_2)_{30}CH_3 + H_2O \rightarrow$$
$$CH_3(CH_2)_{22}COOH + HOCH_2(CH_2)_{30}CH_3$$

The long chain alcohols are sometimes called fatty alcohols, by analogy with the fatty acids.

Alcohols dehydrate at high temperatures. This reaction is catalyzed by acids in solution (*e.g.*, H_3O^+) or by clays. We will see later during the discussion of catalytic reactions in petroleum processing that materials chemically similar to clays can be efficient catalysts for a variety of reactions. Since clays are abundant minerals in nature, catalysis by clays may be an important contribution to catagenesis. The acid-catalyzed dehydration of an alcohol can be represented as

$$\overset{H^+}{RCH_2CH_2OH \rightarrow RCH=CH_2 + H_2O}$$

When this reaction is run in the laboratory, typical temperatures are about 160° C. As a rough rule of thumb, the rate of a reaction doubles for each 10° increase in temperature, or is halved for each 10° drop. If we assume that kerogen is buried to a depth of 1000 m, so that the temperature would be roughly 60°, the reduction of 100° (that is, ten 10° increments) from laboratory reaction temperatures would reduce the reaction rate by $(1/2)^{10}$ or $1/1024$. Nevertheless, a reaction that requires, say, an afternoon in the laboratory would still be extremely fast on a geological time scale even if the rate were reduced by $1/1024$.

When carboxylic acids undergo decarboxylation at high temperatures. This reaction can be illustrated for palmitic acid as

$$CH_3(CH_2)_{13}CH_2COOH \rightarrow CH_3(CH_2)_{13}CH_3 + CO_2$$

42

Decarboxylation shortens the carbon chain by one atom. Most naturally occurring fatty acids contain an even number of carbon atoms (because the biosynthesis of these acids begins with acetic acid and lengthens the chain in increments of two carbon atoms at a time). Hydrocarbons produced from decarboxylation will therefore be predominantly those having odd numbers of carbon atoms. We will see later that the relative proportion of even and odd carbon atom chains provides a way to estimate the age of the sediment. Decarboxylation of humic acids in similar reactions gives rise to a substance called humin.

Thermal decarboxylation of acids is performed in the laboratory at 300-400°. However, petroleum is found in many sediments for which other geological information suggests that the temperature never exceeded 100-120°. This significantly lower temperature in nature again suggests that reaction times are very long compared to ordinary laboratory reactions. However, with thousands to millions of years for these reactions to take place, it is still possible for many reactions which are slow on a human scale to occur in nature.

Esters also undergo thermally induced decarboxylation. The immediate products of the thermal decomposition of an ester are an olefin and an acid:

$$R'CH_2CH_2OOCR \rightarrow R'CH=CH_2 + HOOCR$$

However, at high reaction temperatures, the acid will decarboxylate (assuming that it does not vaporize away from the region of high temperatures as rapidly as it forms). Thus

$$RCOOH \rightarrow RH + CO_2$$

With the decarboxylation of the acid produced from the decomposition of the ester, the final products of the process will be an alkane, an alkene, and carbon dioxide. The overall process can be illustrated using a carnauba wax component as an example:

$$CH_3(CH_2)_{22}COOCH_2CH_2(CH_2)_{29}CH_3 \rightarrow$$
$$CH_3(CH_2)_{21}CH_3 + CH_2=CH(CH_2)_{29}CH_3 + CO_2$$

(To illustrate the course of this reaction the formula for the carnauba wax component has been written in a slightly different fashion than used previously. Note, however, that it is still the same compound.)

Phenol condensations
Lignin is a biopolymer of three aromatic alcohols, sinapyl, p-coumaryl, and coniferyl alcohol. Partial degradation of lignin may result in the formation of

simpler compounds; for example, if lignin were partially decomposed to its constituents, those compounds could themselves be degraded further, as illustrated for the hypothetical conversion of sinapyl alcohol to pyrogallol:

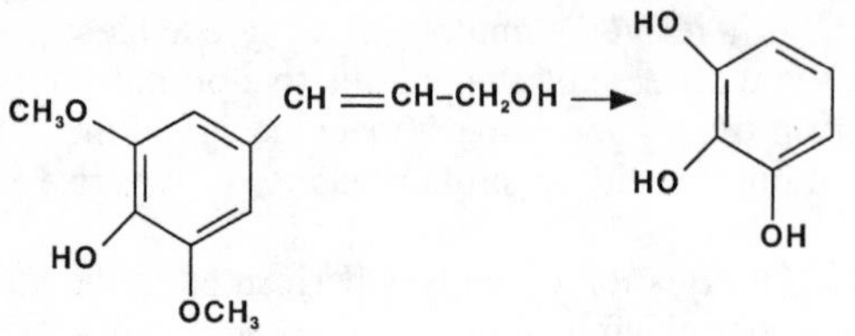

Compounds such as pyrogallol can condense with themselves to form larger structures, as, for example:

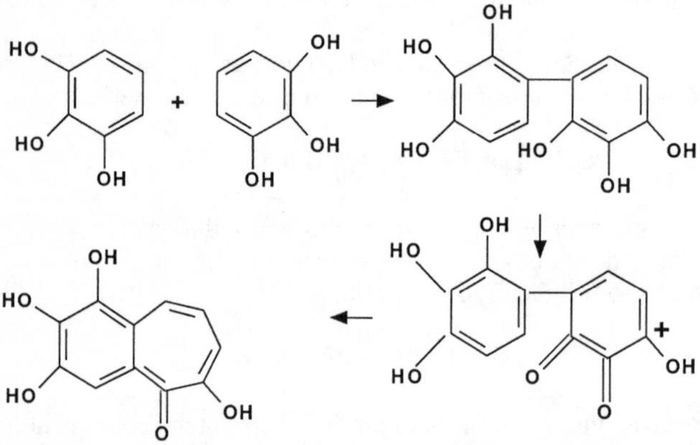

The overall process proceeds with the condensation of cyclic compounds into larger, multi-ring compounds accompanied by the loss of carbon dioxide.

Reaction pathways on the van Krevelen diagram
To review briefly the reactions just discussed, the following processes occur:

$$\text{Alcohols} \rightarrow \text{Alkenes} + H_2O$$
$$\text{Acids} \rightarrow \text{Alkanes} + CO_2$$
$$\text{Esters} \rightarrow \text{Alkenes} + \text{Alkanes} + CO_2$$
$$\text{Phenols} \rightarrow \text{Multi-ring compounds} + CO_2$$

The common feature of these reactions is that in each case substantial amounts of oxygen are lost, mainly as carbon dioxide.

Tracking the composition of a kerogen during catagenesis suggests that there should be a decrease in the O/C ratio, probably a large one, with a slight decrease in the H/C ratio (because of hydrogen removed as water). On a van Krevelen diagram we would expect a reaction pathway proceeding from the kerogen composition toward the abscissa (decreasing O/C) with a small slope, as shown in Figure 3.5.

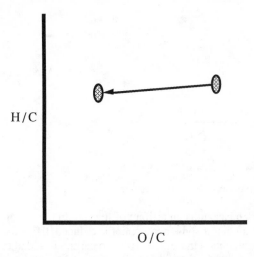

Fig. 3.5 A reaction pathway for Type III kerogen during the early stages of catagenesis, illustrating a small drop in H/C ratio accompanying a much larger drop in O/C ratio.

However, the hydrogen redistribution process suggests that the eventual product of the reaction will be a solid having nearly the composition of graphite. Since for graphite O/C = H/C = 0, at some point the reaction pathway shown in Figure 3.5 must undergo a change of slope to head in the direction of the origin (Figure 3.6).

Type III kerogen (peat) is a loosely consolidated material associated in the earth with large amount of water. In some cases, a sample of peat may contain 90% moisture; that is, the organic portion of the mass dug from the ground may represent only about 10% of the total weight of material. The composition of peat is not much different from lignin and cellulose, being about 55% C, 6% H, and 35% O on a weight basis. This similarity should not be surprising, since lignin and cellulose are the most important components of the higher plants, and radical changes in elemental composition do not occur during diagenesis.

As the weight of the overlying sediment increases, the peat becomes compressed. Catagenesis results in a loss of oxygen and an *apparent* increase

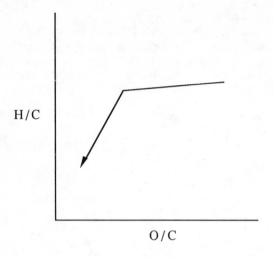

Fig. 3.6 As catagenesis of Type III kerogen progresses,
the reaction pathway undergoes a change in slope as significant
drops in H/C begin to occur with hydrogen redistribution.

in carbon. (The increase in carbon is an artifact of the requirement that
percentages sum to 100, so that if some component is reduced, the percentage
of the others increases to maintain the sum at 100. No carbon is added to the
system in a chemical sense.) The compression of peat to about half its original
thickness and reduction of the oxygen content results in a material having a
composition of about 65% C, 6% H, and 25% O. This material is *brown coal*.
The compaction during the formation of brown coal also reduces the moisture
content somewhat. As mined, brown coals may contain about 60% moisture.

A further compaction (resulting in a four-fold reduction in thickness
relative to peat) and further reduction of oxygen to give a composition of
roughly 72% C, 6% H and 20% O results in the formation of *lignite*. Many
brown coals are approximately 30 million years old; lignites are about 60 to
70 million years old. This process is also accompanied by a reduction in
moisture content, so that when mined lignites may contain 30 - 40% moisture.
The sequence of changes from peat to brown coal to lignite can be shown on a
van Krevelen diagram (Figure 3.7).

As catagenesis progresses beyond lignite, the next significant changes in
composition are a loss of carboxyl and methoxyl groups from the structure.
We have seen that lignin is somewhat resistant to chemical degradation during
diagenesis. Eventually, however, the lignin undergoes a partial
decomposition, in reactions which can be represented by

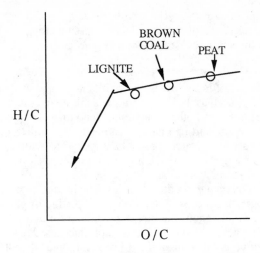

Fig. 3.7 The van Krevelen diagram showing the sequence of transformations from peat through brown coal to lignite.

Similarly, resins may lose carboxyl groups by thermal decarboxylation, as in the reaction of abietic acid:

The effect of the loss of methoxyl and carboxyl groups on the elemental composition is a decrease in the O/C ratio with a relatively small drop in the H/C ratio. At an elemental composition of about 75% carbon, 5% hydrogen, and 15% oxygen, the product of catagenesis is *subbituminous coal*. The formation of subbituminous coal still represents a progression along a nearly horizontal trajectory on the van Krevelen diagram (Figure 3.7).

The origin of heterogeneity in coals
Type III kerogen is a mixture of humic acids, the stable lipids, and other relatively stable compounds such as the resins. Humic acids are comparatively oxygen-rich, whereas the lipids and other compounds tend to be hydrogen-rich. Since chemically different starting materials subjected to similar reaction conditions will produce different products, it is reasonable to expect that the oxygen-rich and the hydrogen-rich components will transform during catagenesis into different materials.

Humic acids decarboxylate to produce humins. The humins interact with water to produce a gel, this process being called gelification. A gel is a stiff, semi-rigid material which retains the liquid in which it formed. The most familiar example of a gel is the flavored gelatin products used in salads and desserts. Gelified humin is called *huminite*. In contrast, lipids do not undergo gelification but rather, with increasing compaction of the sediment, coalesce to form bodies called *liptinites*. This process can be summarized in the scheme shown in Table 3.1.

Table 3.1 Transformations of oxygen- and hydrogen-rich components of type III kerogen through diagenesis and early catagenesis

Form	Process	O-rich	H-rich
Plants		Carbohydrates Lignin Glycosides Proteins	Lipids Resins
	Diagenesis		
Kerogen		Humic acids	Lipids Resins
	Early catagenesis		
Coal		Huminite	Liptinite

As type III kerogen passes through the early stages of catagenesis, two classes of materials are produced: huminites, relatively rich in oxygen and having mainly cyclic structures; and liptinites, with low oxygen but relatively rich in hydrogen and having mainly chain structures. Since coal is a solid, the huminite and liptinite products can not blend or somehow dissolve in each other to form a homogeneous material. Thus we would expect coals to contain two distinct materials, huminites and liptinites, each having a distinct composition, propensity to undergo chemical reactions, and physical properties. In addition, severe decomposition of organic matter caused by forest or swamp fires in the environment while organic matter was accumulating, or by extensive bacterial or fungal attack, leads to carbon-rich material resembling charcoal in appearance. The product of this severe attack on organic matter is sometimes called Type IV kerogen; because it is fairly inert in both the subsequent geochemical reactions and in many of the reactions of the coals in which it is formed, these materials are also called *inertinites*.

We will return to this discussion of the heterogeneity of coal in Chapter 6, where we will examine in more detail these components of coal.

Coalification - the later stages of catagenesis

Catagenesis of Type III kerogen is also called coalification. With the chemical changes in the formation of brown coal, lignite, and sub-bituminous coal being primarily a decrease in the O/C ratio as the result of loss of oxygen-containing functional groups, there is no evident hydrogen redistribution. However, continued transformation of kerogen to graphite must at some point require a significant change in slope of the pathway on the van Krevelen diagram. This transformation can proceed in either a continuous sequence of minor changes or by occasional significant changes in reaction chemistry. Which model - continuous minor changes or significant major changes - is correct is an unresolved issue of organic geochemistry. In a sense the question is analogous to the debate in biology regarding the mechanisms of evolution, that is, whether there is a gradual evolution of organisms or a "punctuated equilibrium" of occasional major changes in species. In kerogen catagenesis, the discontinuous sequence of major changes leading to occasional major changes in chemistry are referred to as *coalification jumps*.

The first coalification jump occurs at a composition of about 80% carbon. This point represents the beginning of the composition range of the *bituminous coals*, and is the point where significant change in slope on the van Krevelen diagram occurs. As the carbon content of the kerogen increases beyond 80%, a new type of reaction chemistry is encountered, in which the H/C ratio begins to decrease. At this point that hydrogen redistribution begins.

Several types of reactions become important at the first coalification jump. These reactions include the dealkylation of aromatic rings,

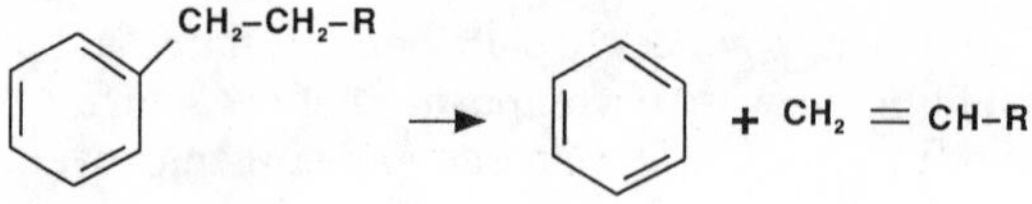

CH$_2$-CH$_2$-R

⟶ + CH$_2$ = CH-R

the dealkylation of resins,

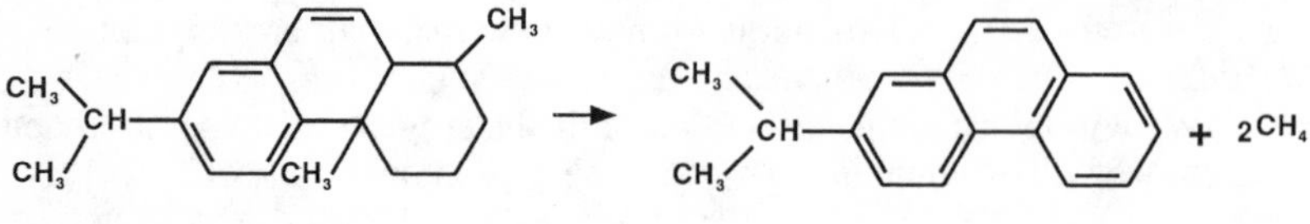

CH$_3$, CH$_3$
CH-
CH$_3$/ CH$_3$

⟶

CH$_3$,
CH-
CH$_3$/ + 2CH$_4$

the partial cleavage of alkyl side chains on aromatic ring systems, and

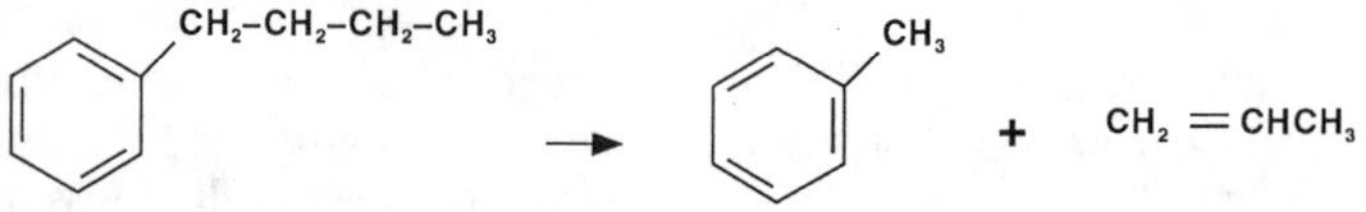

CH$_2$-CH$_2$-CH$_2$-CH$_3$

⟶ CH$_3$ + CH$_2$ =CHCH$_3$

the thermal breakdown of long alkyl chains in lipids (or materials derived from lipids),

$$C_{27}H_{56} \rightarrow C_{13}H_{26} + C_{14}H_{30}$$

Dealkylation of aromatic ring systems is an example of the effects of hydrogen redistribution. For example, dealkylation of propylbenzene (H/C = 1.33) gives products which have H/C ratios of 1.00 (benzene) and 2.00 (propene). One product is hydrogen-rich relative to the starting material and the other is carbon-rich. Hydrogen redistribution is more subtle in the case of a lipid-derived material such as heptacosane, but is nevertheless real. The H/C ratio of heptacosane is 2.07; if the reaction products are taken to be tridecene and tetradecane, the respective H/C ratios of the products are 2.00 and 2.14.

The reactions just discussed occur at temperatures in excess of 60° C (i.e., at depths of 1000 - 2000 m). The compaction resulting from the weight of the overlying sediments results in 7 - 20 m of peat being compressed to about 1 m of bituminous coal. As coalification continues, the huminite takes on a brilliant black, glossy or glass-like appearance. Because of its glassy or vitreous appearance, this component in the bituminous coals is called *vitrinite*.

The second coalification jump occurs in the range of 87% carbon. This composition is well into the range of the bituminous coals. The second coalification jump is accompanied by a marked loss of oxygen, presumably by the destruction of phenolic functional groups, and significant production of methane.When the temperature and corresponding depth of burial exceed about 120° C, thermal breakdown of lipids proceeds to methane. For example,

$$C_6H_{14} \rightarrow CH_4 + C_5H_{10}$$
$$C_5H_{10} \rightarrow CH_4 + C_4H_6$$
$$C_4H_6 \rightarrow CH_4 + C_3H_2$$

It should be apparent that not enough hydrogen exists in the starting material (hexane) to convert all of the carbon to methane. In each step of the process the remaining product is becoming increasingly deficient in hydrogen. Methane has H/C = 4.00. The remaining products of the reactions shown above have H/C ratios of 2.00 in the first step (i.e., pentene), and 1.50 in the second step. As the formation of methane proceeds, the other products become increasingly unsaturated. In this example, the second product would be a butadiene. In addition to becoming increasingly carbon-rich and hydrogen-poor, the highly unsaturated molecules can undergo cyclization. For example, 2,4,6-octatriene could cyclize to produce o-xylene

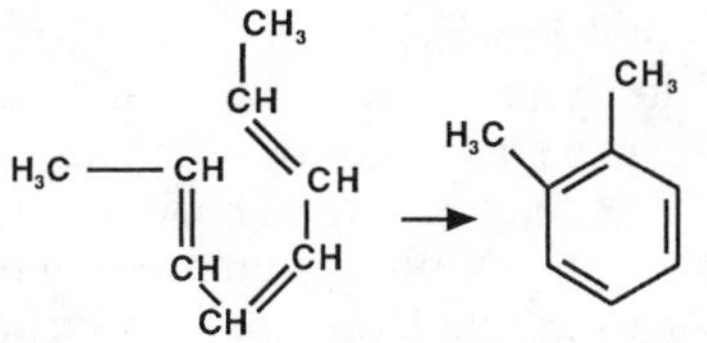

This reaction results in increasing the proportion of aromatic structures in the coal. Compounds such as o-xylene can react further to form larger ring systems, such as 1,2-dimethyltriphenylene:

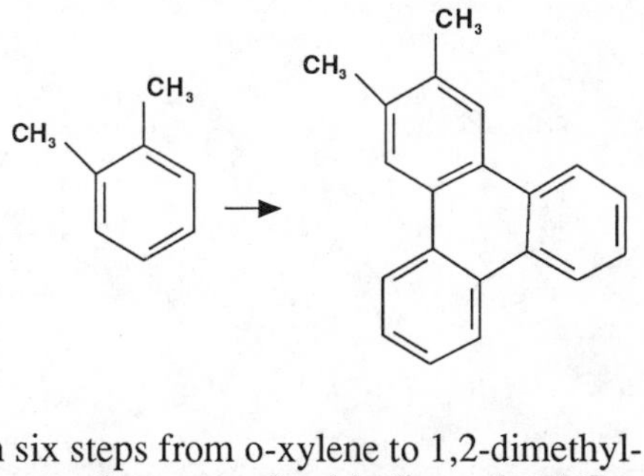

(This reaction proceeds in six steps from o-xylene to 1,2-dimethyl-triphenylene; the details are not important for this discussion.) Consider starting with a molecule of undecane, $C_{11}H_{24}$ (H/C = 2.18). The formation of three molecules of methane from undecane could in principle form as the other product an octatriene, which could aromatize to o-xylene and condense to the 1,2-dimethyltriphenylene. The H/C ratios of the products would be 4.00

51

for methane and 0.80 for the 1,2-dimethyltriphenylene. Again hydrogen redistribution is at work.

In the formation of brown coal, lignite, and subbituminous coal, the principal gaseous product of reaction was carbon dioxide. As coalification proceeds beyond the first coalification jump into the bituminous coal range, the principal gaseous product is now methane. (In fact, measuring the CH_4/CO_2 ratio of gases trapped in coal provides a good indication of the geological maturity of the deposit.) The principal change in composition is now a loss of hydrogen, with relatively little oxygen loss. Thus on a van Krevelen diagram there is a significant drop in H/C and a less pronounced drop in O/C.

The third coalification jump occurs at a composition around 91% carbon. The third coalification jump is the beginning of *anthracite* formation. High temperatures are required for anthracite to be formed; the highest temperatures estimated by some geochemists are in the range of 500° C. In addition, high pressures may also be required. Virtually all commercial deposits of anthracite around the world are found in mountainous regions where extensive folding of the rocks has occurred. The high pressures generated by the rock folding may have contributed to anthracite formation. The reactions accompaning the third coalification jump include extensive formation of methane (up to 200 L per kg of coal) by stripping aliphatic side chains from aromatic structures, further aromatization of ring systems, and condensation of aromatic rings into larger, polycyclic aromatic structures. The formation of coals of high carbon content is also accompanied by the vertical alignment of aromatic ring systems. Cyclophanes are a family of compounds in which vertical stacking of aromatic rings is achieved; a simple example is p-cyclophane:

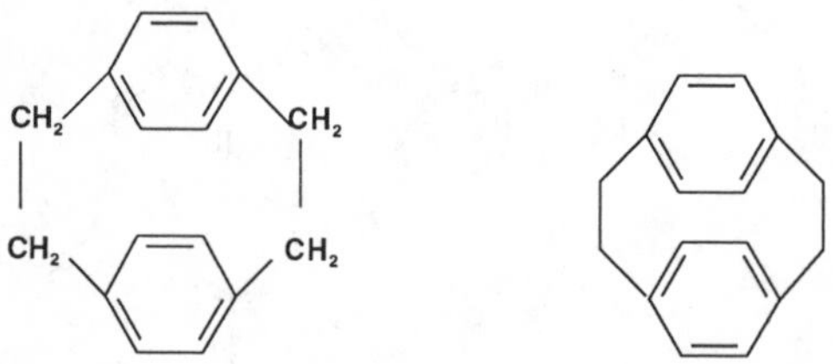

When a polycyclic aromatic compound such as anthracene is heated to high temperatures, condensation reactions of the type shown below lead to the formation of polymeric structures:

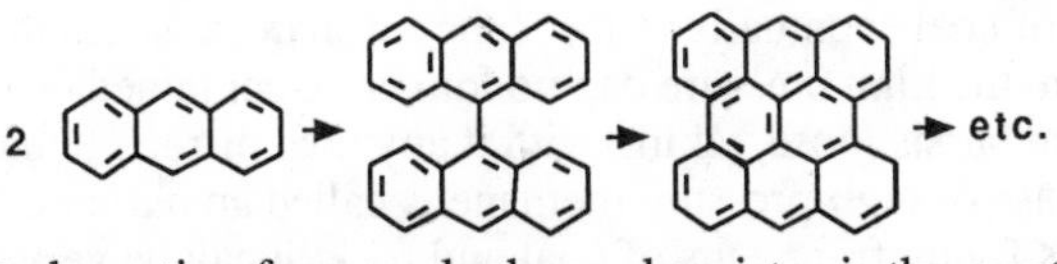

2 → → etc.

A related reaction from cyclophane chemistry is the reaction of a compound containing four "stacked" benzene rings to produce products having much larger ring systems:

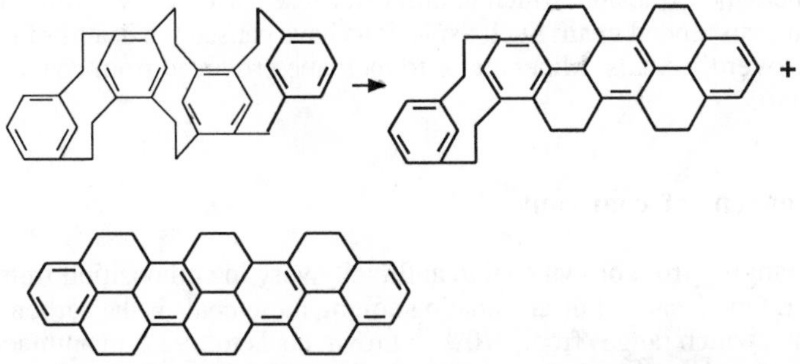

→ +

By the third coalification jump virtually all of the molecular structures which were remnants of the lipids have disappeared, via cracking to methane or cyclization and condensation. Recall that the separate existence of the lipid-derived and lignin-derived materials was responsible for the heterogeneity in the solid coal. The loss of the distinct lipid-derived materials via coalification suggests that the heterogeneity in coal has been eliminated. Indeed, when anthracites are examined under the microscope, it is no longer possible to distinguish the vitrinites and liptinites.

A final coalification jump occurs at carbon contents of about 96%, with the formation of *metaanthracite*. At this stage, the aromatization of cyclic structures and their condensation to a very graphite-like structure is nearly complete.

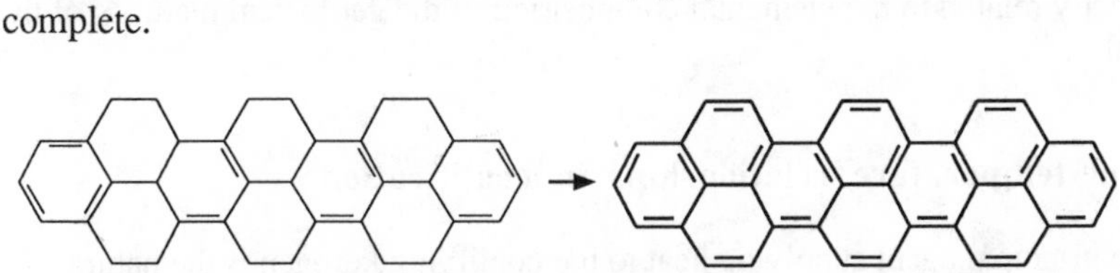

→

Throughout the formation of the bituminous coals and anthracite the principal gaseous product is methane. The suggestion is periodically made that coal beds be tapped to drain off the accumulated methane. This suggestion is motivated not only from a desire to augment supplies of natural gas, of which methane is the princiapl constituent, but also to improve the safety of underground coal mines. Accumulated methane held at high pressure in coal seams can sometimes be released with near explosive violence when the continuous removal of coal eventually reduces the mechanical strength of the seam to a point where the high pressure can no longer be contained. Then the accumulated methane bursts loose, taking with it any coal or rock remaining in the way. This release of high-pressure methane is called an *outburst*. Since an outburst generates flying fragments of coal and rock, it can be very hazardous to miners who happen to be nearby. Even worse, the slow seepage of methane from the coal into the mine can eventually let the concentration of methane in the air reach the explosive limit. Then an inadvertent spark can set off a methane explosion, which in turn can cause a secondary explosion of coal dust suspended in air. Such explosions have caused the deaths of many miners over the years. Miners refer to methane produced from coal as *firedamp*.

The concept of coal rank

Coals ranging from brown coal to anthracite vary in composition from one to another. One means of discriminating among these coals is the carbon contents, which ranges from <70% in brown coals to >92% in anthracites. The progression of coals from brown coal to anthracite also marks a regular transition along the van Krevelen diagram. These facts suggest that it should be possible to classify, or to *rank* coals, either on the basis of carbon content or on the extent to which they have undergone coalification, as indicated by the position on the van Krevelen diagram. In fact, a system of coal ranks indeed exists, with brown coal being the coal of lowest rank and anthracite the coal of highest rank. Brown coals, lignites, and subbituminous coals are sometimes collectively referred to as *low-rank coals* , with bituminous coals and anthracites then being *high-rank coals.*

We will see in Chapter 6 that a formal system of classifying coals by rank exists. That system is based on the behavior of coal as a fuel and is not directly related to the elemental composition or the geological maturity of the coal.

Time-temperature relationships in coalification

The principal agent supplying heat to the coalifying kerogen is the natural thermal gradient in the earth. The deeper a coal is buried, the longer it will be exposed to the elevated temperatures, and the higher will be the temperatures to which the material is exposed. Therefore we should expect that the coal has been converted to a higher rank. Since the process of burial requires long times, it is also reasonable to expect that the deeper a coal is buried, the older it is.

The relationship known as *Hilt's rule* states that the deeper a coal has been buried, the older it is, and the higher the rank it has attained. This three-sided relationship suggests that if we know one of the three properties - age, depth, or rank - we can infer something about the other two. For example, in the United States most lignites are about 70,000,000 years old and tend to lie at shallow depths; most anthracites may be over 300,000,000 years old and were once buried to very great depths.

Because of the great variety of geological environments in which coals were formed, Hilt's rule must be applied with great caution. It is most successful when applied to a single group of coal seams lying in an area where there is no indication of major geological disturbances which may in some way have disturbed the ordering or arrangement of the seams. Its application is tenuous at best when considering coals from widely separated geographical regions and regions which have no geological continuity. For example, some small deposits of anthracites exist in the Rocky Mountains of the United States. These coals are no more than 100,000,000 years old, but have attained their high rank because of extremely high thermal gradients associated with intrusions of magma near the coal deposits. The process may also have been helped by the high pressures associated with the folding of rocks accompanying the formation of the mountains. In contrast, a deposit of brown coal outside Moscow in the Soviet Union is estimated to be over 300,000,000 years old. This coal evidently was never buried very deeply even during all that time, and consequently has never experienced thermal gradients high enough to cause coalification beyond the brown coal stage.

Catagenesis of algal and liptinitic kerogen

This section discusses the conversion of Type I (algal) and Type II (liptinitic) kerogen to fuels. Both are rich in lipids, and their elemental composition is characterized by high H/C and low O/C ratios. The important chemical distinction between algae and plankton on the one hand and higher plants on the other is that lignin is unique to higher plants. Consequently we expect that the chemistry of Types I and II kerogen, where lignin is not present, will be different from that just discussed for Type III kerogen.

The comparison of the three types of kerogen on a van Krevelen diagram is shown in Figure 3.8. The relative positions of these kerogens on the H/C vs O/C diagram follow directly from the nature of the starting materials from which these kerogens were produced. A typical lipid (wax) structure would be

$$CH_3 \ (CH_2)_{22} \ COO \ CH_2 \ (CH_2)_{30} \ CH_3$$

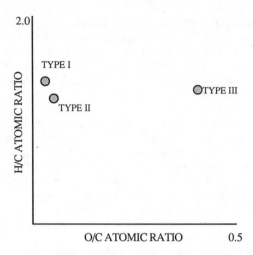

Fig. 3.8 The relative positions of the three types of kerogens shown on a van Krevelen diagram.

A fragment of the lignin structure is shown on the following page. Inspection of these two structures indicates that the O/C ratio of the lipid-derived material will be much lower than that of lignin-derived material.

Figure 3.8 indicates that *roughly* all three kerogens have about the same H/C ratio. However, there is a key distinction - very important for the chemistry of petroleum formation - in the way in which the hydrogen is incorporated. Consider a fragment of a cellulose structure.

Cellulose has a high H/C ratio, ~1.7, but many of the hydrogen atoms are bonded to oxygen atoms. A loss of oxygen from this structure will be

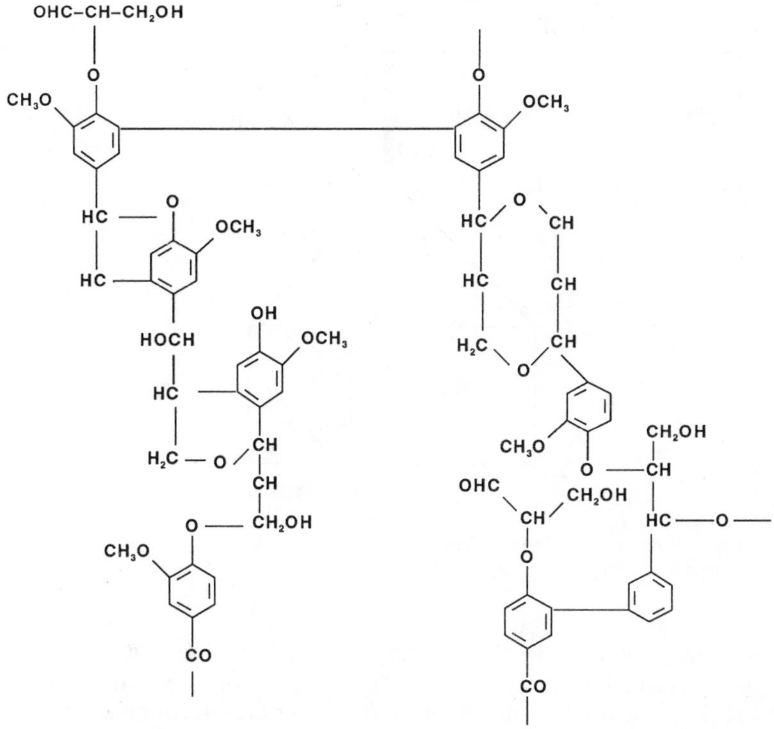

OHC–CH–CH₂OH

CH₃O   OCH₃

HC   O   OCH₃

HC   HC   CH

HOCH   OH   OCH₃   HC   CH

HC   H₂C   CH

H₂C — O   CH   O

CH   CH₃O   CH₂OH

O — CH₂OH   O — CH

CH₃O   OHC   CH₂OH   HC — O —

CO   CH

O

CO

accompanied by a loss of hydrogen at the same time. In comparison, in lipids or their hydrolysis products virtually all of the hydrogen is attached to carbon. Any oxygen lost will likely remove only carbon at the same time. Thus carbohydrates, lignin, and similar highly oxygenated compounds lose oxygen via both carbon dioxide and water, and as a consequence both the O/C and H/C ratios drop. Lipids, on the other hand, lose oxygen mainly as carbon dioxide, so that O/C drops but H/C remains fairly high.

The principal focus of discussion will be sediments in which the total organic matter is about 1%, that is, the oil source rocks. Of all the organic matter accumulated in these rocks, about 10% is soluble in some common organic solvents, such as carbon disulfide or chloroform. This material is *bitumen*. The insoluble material is the kerogen. This situation is shown schematically in Figure 3.9.The formation of oil is essentially the production of bitumen from kerogen.

The reactions of lipids are ones already discussed: saponification of esters, dehydration of alcohols, and decarboxylation of acids and esters. All result in the formation of long chain aliphatic compounds: acids, alcohols, alkanes, and alkenes. This mixture of compounds is sometimes called

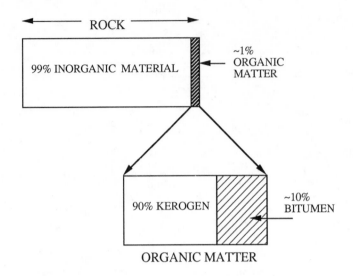

ROCK

99% INORGANIC MATERIAL

~1% ORGANIC MATTER

90% KEROGEN

~10% BITUMEN

ORGANIC MATTER

Fig. 3.9 A schematic relationship among the inorganic material, kerogen, and bitumen in oil source rocks.

protopetroleum. These products tend to be solids at room temperature. For example, eicosane, $C_{20}H_{42}$, has a melting point of 37°C. The smaller molecule dodecane has a melting point of -10°C. The formation of a liquid material - oil - thus requires the production of molecules smaller than the original products of lipid diagenesis. The formation of these smaller molecules requires the breaking of C-C bonds. The key process for accomplishing the reduction in molecular size is *thermal cracking*, the cleavage of bonds at elevated temperatures. Thermal cracking is a very important reaction in the formation of petroleum; we will discuss it again in some operations in petroleum refining and in coal chemistry.

Introduction to free radical reactions

When a covalent chemical bond breaks, two general options exist:

Case I $\quad$ A:B $\rightarrow$ A• + •B
Case II $\quad$ A:B $\rightarrow$ A$^+$ + :B$^-$

Case I is called homolytic cleavage and Case II, heterolytic cleavage. In homolytic bond cleavage the electron pair forming the bond is split so that one electron remains with each of the fragments. The products each have unpaired electrons. These species are called free radicals. Free radicals are electrically neutral, but because of their unpaired electrons, are usually highly reactive.

At high temperatures, many compounds undergo thermally induced breaking of bonds, *cracking*, to produce smaller molecules. In laboratory or industrial processes where time is important, temperatures in the cracking reactions are often 500-700°. Such temperatures never prevail in catagenesis, but because extremely long reaction times are available, extensive reaction is still able to occur. When an alkane (other than methane) undergoes cracking, bond cleavage occurs at a C-C bond rather than at a C-H bond, because of the significant difference in energies required to break the two types of bonds.The high reactivity of free radicals derives from the tendency for the atom bearing the unpaired electron to obtain the "missing" electron and once again possess a pair of electrons. The formation of paired electrons from radicals is a process sometimes called *radical capping*.

The essential feature of thermal cracking is homolytic bond cleavage to produce free radicals. The thermal cracking of ethane, for example, would result in the formation of two methyl radicals

$$C_2H_6 \rightarrow CH_3\bullet + \bullet CH_3$$

A C-C bond and not a C-H bond breaks. (In other words, the cleavage of ethane to an ethyl radical and hydrogen atom is not important at these conditions.)

$$C_2H_6 \text{ -X-> } C_2H_5\bullet + \bullet H$$

The C-C bond cleaves preferentially to the C-H bond because the C-C bond is weaker. For example, in ethane the bond dissociation energy needed to rupture one of the C-H bonds is about 410 kJ/mol, but the bond dissociation energy of the C-C bond is only 352 kJ/mol. Thus if ethane were to be heated, the C-C bond cleavage would occur more readily than C-H bond cleavage. For larger molecules, cleavage of a C-C bond occurs at random positions along the chain. Once a C-C bond has cleaved to form a pair of radicals, the radicals may then react further through a variety of processes.

Hydrogen abstraction
A free radical can remove a hydrogen atom from another molecule (or radical). For example,

$$CH_3\bullet + CH_3CH_2CH_3 \rightarrow CH_4 + CH_3CH\bullet CH_3$$

Notice that in this example the new radical formed is the isopropyl radical rather than the other possibility, the n-propyl radical. The course of the reaction is governed by the stability of the free radicals. A secondary radical, such as isopropyl, is more stable than a primary radical such as n-propyl. The more stable of the two products will be formed in preference. In general the stability of free radicals is in the order

$$3° > 2° > 1° > methyl$$

One way of considering this order of stability is to recognize that the more alkyl groups attached to the carbon which is the radical site, the more stable is the radical. This order of radical stability not only provides a way to predict the preferred product of the reaction, but also shows that the reverse reaction, e.g.,

$$CH_4 + CH_3CH{\cdot}CH_3 \xrightarrow{-X} CH_3{\cdot} + CH_3CH_2CH_3$$

is unlikely to occur. Notice too that the product of the hydrogen abstraction reaction is itself a radical, and is therefore able to participate in subsequent reactions. Hydrogen abstraction is also affected by the variation in the C-H bond dissociation energy with the specific type of C-H bond being cleaved. The C-H bond dissociation energy for a terminal carbon atom, producing a 1° free radical, is about 410 kJ/mol; for a methylene carbon (-CH_2-) in the interior of a chain, the C-H bond dissociation energy to form a 2° radical is 398 kJ/mol; and for a C-H bond at a point at which a carbon chain branches, forming a 3° radical, the bond dissociation energy is 389 kJ/mol. From this information and the order of stability of free radicals shown above, we would expect that the C-H bond dissociation of methane, to form a methyl radical would be even higher than that required for formation of a 1° radical, and in fact it is. The C-H bond dissociation energy in methane is about 440 kJ/mol.

Disproportionation
Disproportionation is a reaction of two radicals in which one abstracts a hydrogen from the other. For example,

$$CH_3CH{\cdot}CH_3 + CH_3CH{\cdot}CH_3 \rightarrow CH_3CH_2CH_3 + CH_3CH=CH_2$$

Here the reactants are two identical radicals. The products are an alkane and an alkene. Because the products are not radicals, disproportionation is one way in which radical processes are terminated.

   In inorganic chemistry disproportionation is the conversion of an element into higher and lower oxidation states, with simultaneous oxidation and reduction. An example is the disproportionation of potassium chlorate used in the formation of potassium perchlorate:

$$4KClO_3 \rightarrow 3KClO_4 + KCl$$

Here the chlorine atom is in the +5 oxidation state in the chlorate, and disproportionates to +7 in perchlorate and -1 in the chloride. Although it is certainly possible to track changes in formal oxidation states of the carbon atoms in the organic free radical reactions, it is more helpful in this context to regard disproportionation as a change in the H/C ratio. In the example of the disproportionation of the isopropyl radical, the H/C ratio changes from 2.33 in the radical to 2.67 and 2.00 in the two products.

β-Bond scission

Next to the initial cracking reaction itself, β-bonds scission is the most important reaction for shortening chains of carbon atoms. An important distinction between these two processes is that the first cracking process starts with an alkane or alkane derivative, but β-bond scission starts with a free radical. β-bond scission involves cleavage of the C-C bond at the β position to the carbon atom bearing the radical. The bond between the carbon with the radical and the immediately adjacent carbon is the α bond; that between the carbon immediately adjacent to the radical and the next carbon atom is the β bond.

$$\overset{\beta}{} \quad \overset{\alpha}{}$$
$$CH_3\text{-}CH_2\text{-}CH_2\text{-}CH_2\text{-}CH_2\bullet$$

Cleavage of the β bond in the pentyl radical would result in the formation of ethylene and a propyl radical:

$$CH_3CH_2CH_2CH_2CH_2\bullet \rightarrow CH_3CH_2CH_2\bullet + CH_2=CH_2$$

Notice that ethylene is the product because the radical undergoing β-bond scission was a 1° radical. Had the reactant been a 2° radical, the unsaturated product would have been a longer 1-alkene. Regardless of the initial reactant, however, the new radical formed in the process is a 1° radical. In principle this new radical can itself undergo β-bond scission to form ethylene and another, newer 1° radical. Indeed the process can continue with the successive generation of ethylene and new, shorter 1° radicals until only an ethyl or methyl radical is left, too short to undergo further β-bond scission. The continual reduction of length of a long chain radical by removing two-carbon fragments as ethene is sometimes called *unzipping*.

Recombination

The recombination process is the direct reaction of two radicals to form a new C-C bond by the pairing of the unpaired "radical" electrons. For example, two methyl radicals could recombine to form ethane. Since the product of recombination is not a free radical, this process also terminates any free radical reaction sequence.

Since energy is required to break a chemical bond and form radicals, the formation of new bonds by the recombination of radicals releases energy, and thus, from an energetic point of view, is a favorable process. Furthermore, most radical recombination reactions are very fast. The ability of radicals to enter into other types of reactions without recombining virtually instantaneously derives from the relative concentrations of radicals and other non-radical species in the reaction. If the concentration of radicals is low, there is a good opportunity for radicals to encounter other, non-radical molecules and participate in, for example, a hydrogen abstraction reaction. On the other hand, if the radical population is very high, then the likelihood of radicals encountering each other and recombining is much greater.

Hydrogen capping

Hydrogen capping is very similar to hydrogen abstraction. A distinction is that the hydrogen in the capping process does not necessarily originate in another hydrocarbon molecule or radical. For example, hydrogen gas could be used to cap radicals in a process such as

$$CH_3\bullet + H_2 \rightarrow CH_4 + H\bullet$$

In fact, we will see later that hydrogen capping by some external source of hydrogen such as hydrogen gas is a crucial process for the successful conversion of coal to synthetic petroleum-like fuels.

Rearrangements

As a rule, free radical species do not undergo structural rearrangement or isomerization by shifting alkyl groups or hydrogens to allow the formation of a more stable radical from one of lower stability. This tendency not to isomerize is an important distinction between the behavior of free radicals and carbocations, and as we will see in Chapter 9, has important implications for the processing of petroleum products. In some cases in which a 1° radical has a very long carbon chain, it may be possible for the chain to twist back on itself to allow the 1° radical to abstract a hydrogen from somewhere in the interior of the chain, capping the 1° radical and creating at the same time the more stable 2° radical.

The multiplicity of products from cracking

Cracking reactions can given rise to a very large number of possible products. To illustrate, consider the cracking of butane. The initial cracking can proceed in two ways, to form a propyl and a methyl radical, or to form two ethyl radicals:

$$CH_3CH_2CH_2CH_3 \rightarrow CH_3CH_2CH_2\bullet + \bullet CH_3$$

and

$$CH_3CH_2CH_2CH_3 \rightarrow CH_3CH_2\bullet + \bullet CH_2CH_3$$

One starting hydrocarbon compound produces, in this case, three different radicals. To determine the final products of reaction, it is necessary to consider the chemical fates of each of these three radicals. These possible reactions are summarized in Table 3.2.

Table 3.2 Reaction products of the radicals produced during thermal cracking of butane

Initial radical	Process	Product
Methyl	Hydrogen abstraction	Methane
	Recombination	Ethane
	ditto, with ethyl	Propane
Ethyl	Hydrogen abstraction	Ethane
	Disproportionation	Ethane and ethylene
	Recombination with methyl	Propane
Propyl	Hydrogen abstraction	Propane
	Disproportionation	Propane and propylene
	β-bond scission	Ethylene and methyl
	Cracking	Methyl and $\bullet CH_2CH_2\bullet$

In this example, reaction products larger than the starting molecule, butane (e.g., pentane from the recombination of an ethyl and propyl radical) were not considered, on the assumption that conditions severe enough to crack butane would also crack larger molecules. Even so, we see from Table 3.2 that one relatively simple hydrocarbon with four carbon atoms gives rise to *five* stable reaction products: methane, ethane, ethylene, propane, and propylene (propene).

Suppose now that instead of the thermal cracking of butane we were to consider the thermal cracking of an alkane typical of lipids, say triacontane, $C_{30}H_{62}$. First, there are 29 possible radicals formed in the initial cracking of triacontane, ranging from the methyl radical to the 1° radical with 29 carbon atoms. Second, each radical can potentially undergo all of the reactions which have just been discussed, and, for those reactions in which a new radical is formed as one of the products, the "second generation" radicals can themselves undergo further reactions. The number of possible products can be very large. The situation is further complicated by the fact that no naturally occurring lipid consists of a single pure compound, such as triacontane, but is rather a mixture of many individual compounds. Some of the compounds in the lipids may have branched chain structures, as for example pristane and phytane, and some may have cyclic structures, as in some of the terpenes.

Thermal cracking of lipids in nature begins with a mixture of compounds containing roughly 16 to 40 carbon atoms, representing straight chain, branched chain, and cyclic structures, and applies all of the radical reactions discussed in this section. The liquid product - petroleum - which eventuates from these reactions is therefore a mixture of the hundreds of possible reaction products arising from the application of a half dozen free radical processes to the components of the natural lipids.

The effects of temperature

Conversion of kerogen to bitumen begins at approximately 60°. (Recall that this is the same temperature as the first coalification jump.) The lipid molecules undergoing cracking are relatively large, having about 16 to 40 carbon atoms. As they begin to break during thermal cracking any of the C-C bonds in the molecule might break. Using eicosane as an example, there are two C-C bonds which could break to form methyl radicals and the primary nonadecyl (C_{19}) radical. However, there are seventeen C-C bonds in the interior of the molecule. There is a greater chance that the molecule will crack somewhere in the interior rather than at the two terminal C-C bonds. Consequently, the products are likely to be medium to large sized molecules.

As the temperature increases, the rate of cracking also increases. More C-C bonds are broken, and, with increased cracking, the size of the product molecules will become smaller. For example, using hexane as a model, we might expect at relatively low temperature as reaction such as

$$CH_3CH_2CH_2CH_2CH_2CH_2CH_3 \rightarrow CH_3CH_2CH_2\cdot + \cdot CH_2CH_2CH_3$$

resulting from the cleavage of one C-C bond, whereas at higher temperatures the molecule might crack into a greater number of smaller fragments

$$CH_3CH_2CH_2CH_2CH_2CH_3 \rightarrow CH_3\cdot + \cdot CH_2CH_2\cdot + \cdot CH_2CH_2\cdot + \cdot CH_3$$

The first significant amounts of methane begin to appear at about 110°. By the time the temperature has reached 170° the cracking is so extensive that methane is essentially the only product. Thus at 170° the formation of a liquid product, petroleum, has ceased and the principal product of reaction is methane.

Although the high temperature thermal cracking tends to drive the system to methane, it is important to recognize that not enough hydrogen is available for complete conversion of all of the carbon to methane. There are two ways of thinking about this situation, both of which lead essentially to the same conclusion. Using heptane as an example, we can consider utilizing all of the hydrogen in the molecule to contribute to methane formation

$$C_7H_{16} \rightarrow 4CH_4 + 3C$$

where the carbon in this equation represents a graphitic solid. Alternatively, we could consider the complete conversion of all of the carbon in the heptane molecule to methane, by a hypothetical process

$$C_7H_{16} + 12(H) \rightarrow 7CH_4$$

Where does the hydrogen come from to drive this process? The available sources are hydrogen from cyclization or aromatization of other molecules in the mixture of compounds undergoing cracking. However, loss of hydrogen via cyclization or aromatization eventually results in the transformation of those molecules losing the hydrogen into graphitic solids. In other words, whether we consider that the cracking of heptane to methane is limited by the hydrogen available in the heptane molecule, or whether we consider that all of the carbon in the heptane is converted to methane by abstracting hydrogen from other molecules, the net effect is the same - the production of methane and a carbon-rich graphitic solid. This is an instance of the hydrogen redistribution process at work. By the time the reaction temperature has reached 225° even the formation of methane ceases and the only product is a graphitic solid.

The entire sequence of diagenesis and catagenesis, and the attendant production of oil and gas, is summarized in the *kerogen maturation diagram* (Fig. 3.10). The vertical axis represents increasing depth, and therefore, also represents increasing temperature. The first product of interest is the methane produced from the biochemical reactions of diagenesis. Oil formation (catagenesis) begins at a depth corresponding to about 60°. The onset of oil formation is sometimes referred to as the opening of the *oil window*. The importance of the elevated temperatures associated with the opening of the oil window is illustrated by data from drilling into the earth in California. Samples from the drill core taken near the surface show about 50 parts per million (ppm) of free hydrocarbons associated with the rock; samples from depths where the ambient temperature is about 140° show a hundred-fold increase in free hydrocarbons, to 5000 ppm.

The first oil to be formed after the opening of the oil window still contains relatively large molecules and is therefore viscous, of relatively high density and high boiling. For example, the values of these properties are shown for octadecane, tridecane, and octane in Table 3.3.

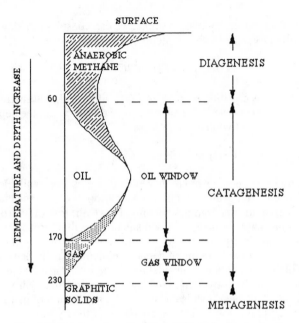

Fig. 3.10 The kerogen maturation diagram for the production of biogenic methane, oil, and gas from kerogen.

Table 3.3 Selected physical properties for three alkanes.

Compound	Boiling point	Density, g/mL	Viscosity, mPa•s
Octadecane	317°	0.777	2.86
Tridecane	234°	0.757	1.55
Octane	126°	0.704	0.54

(The relationship between molecular structure and physical properties will be discussed in detail in Chapter 5) The dense, viscous, high-boiling products formed when the oil window opens are sometimes said to be *heavy oils*. As cracking continues, the molecular components of the oil become progressively smaller, and thereby lower in boiling point, viscosity, and density - in other words, *light oil*. At 170° the oil window closes and the only product is gas. This transition represents the opening of the *gas window*. Finally, at about 225°, even the formation of gas ceases and the final product is a carbonaceous solid. Transformations of the carbonaceous solid above 225° are known as *metagenesis*.

Gas is formed in two different regions - biochemical gas during diagenesis, and gas produced in the gas window during catagenesis. The only hydrocarbon component in the biochemical gas is methane,because the it is the result of biochemical reactions which proceed with the specific formation of methane. The gas formed in the gas window, however, is a mixture of methane and other light hydrocarbons such as ethane, propane, and butane, because the gas is not coming from a biochemical reaction which yields exclusively one gaseous hydrocarbon product, but rather derives from the cracking processes which can produce a variety of products. During cracking, the molecules can fragment to produce many molecular species, and those with five or fewer carbon atoms will be in the gas phase. The more severe the thermal cracking during catagenesis, the greater is the likelihood that the gaseous molecules with two or more carbon atoms will themselves have cracked to methane. Thus a gas in which methane is the predominant, or exclusive, hydrocarbon constituent can also be formed near the bottom of the gas window. As we will see, the distinction between gas for which methane is the principal constituent and a gas which is a mixture of small hydrocarbon molecules has implications for the processing of these gases for fuel use.

The transformation of Type II kerogen can be shown on a van Krevelen diagram as Figure 3.11.

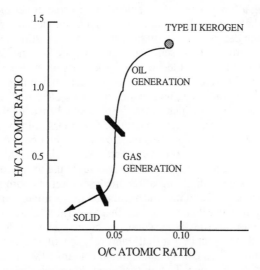

Fig. 3.11 The transformation of Type II kerogen through the oil and gas windows to the formation of a carbonaceous solid.

The curve for Type I kerogen is generally similar but there is usually not as large a gas yield.

Other sapropelic fuels

The discussion on the formation of oil and gas has presumed that the sediment experiencing catagenesis is an oil source rock. According to our previous classification, an oil source rock is a sediment containing about 1% of Type I or Type II kerogen mixed with inorganic sediments. There are two additional categories of fuel precursors in this classification, the oil shales, containing 2 - 50% kerogen, and the coals, which have more than 50% kerogen mixed with the inorganic sediment.

Oil Shales

Virtually all sedimentary rocks contain some amount of organic matter. In fact the average amount of organic matter in sedimentary rocks around the world is 2%. Since many of the important organic compounds which accumulate in sediments contain roughly 50-60% carbon, the average amount of carbon in sedimentary rocks is therefore about 1%. A potentially important future source of fuels are black shales which contain several percent of organic matter. There is no specific upper limit on the amount of organic matter, and in principle there is a continuous gradation from black shales containing a few percent of organic matter to coal or petroleum. A special kind of shale, *oil shale*, contains appreciable amounts of organic matter from which petroleum-like hydrocarbons can be distilled. On the basis of the amount of kerogen in the sediment, oil shales are intermediate between the oil source rocks and coals. Indeed, in a crude sense oil shales can be thought of as being intermediate between oil and coal in nature.

Oil shales are sapropelic; non-marine algae are the principal contributors to the kerogen which forms the oil shale. When algal colonies dry up, the organic material which accumulates is a mass of the waxy, gelatinous cell walls of the algae. This mass of waxy, gelatinous material is called *coorongite*, some deposits of which are found today in Australia. Coorongite can be considered to be the equivalent, in the formation of oil shale, of the peat stage of coal formation. Incorporation of coorongite into inorganic sediments results in the transformation of the mixture to the material we call oil shale. It is unfortunate that the term oil shale has come in to widespread use to describe these sediments, because in many "oil shales" the inorganic material is not actually a shale, and very few oil shales actually contain a liquid oil. A "typical" oil shale might contain 90% inorganic matter, 8% kerogen, and 2% bitumen. The inorganic portion of oil shale may consist of a variety of minerals, such as quartz, feldspars, mica, clays, and pyrite.

The conversion of kerogen to bitumen begins around 60°, marking the onset of thermal decomposition of the kerogen. In contrast, most coals begin to undergo significant thermal decomposition around 350°. In this regard oil shales are like coal, because thermal decomposition to produce a liquid oil, does not occur until temperatures in the vicinity of 350° are reached. Oil shale is also like coal in that it is essentially insoluble in organic solvents and, when heated, does not melt but rather undergoes an irreversible thermal decomposition. The composition of the organic material in oil shale is usually in the range 70 - 80% C, 7 -11% H, 1 -2% N, 1 - 8% S, and 9 - 17% O. In comparison with coal, an oil shale of the same carbon content will have a higher hydrogen content but lower oxygen content. The oxygen content of the organic material in oil shale is generally much higher than that of petroleum of the same carbon content. The liquid hydrocarbons which are produced by heating oil shale are much more petroleum-like than similar to liquids formed in the thermal decomposition of coals.

Because kerogen in nonvolatile and insoluble in all common solvents, the only practical way to remove organic matter from the oil shale for use as a fuel is to heat the shale. This process is called *retorting*. By heating to 350-600° the kerogen undergoes thermal decomposition. The decomposition products vaporize and can then be collected. If we were to assume a 12% concentration of kerogen in an oil shale, the mining of 100 tons of shale would afford only 12 tons of oil (assuming that no losses occurred in retorting). Unfortunately the entire 100 tons of material must be heated, and the 88 tons of useless waste rock must be cooled and disposed of. Most of the energy involved in retorting is expended in heating the rock and is wasted when the heated rock cools. Although the enormous reserves of oil shale make this material a potential future energy source, the difficulties associated with extraction of liquid hydrocarbons are formidable barriers to its commercial exploitation.

Sapropelic coals

At high concentrations of organic matter (Types I or II kerogens), when the organic portion represents greater than 50% of the sediment, the materials formed are coals. To differentiate the coals formed from Type I or Type II kerogen from those formed from Type III, the former are known as *sapropelic coals* and the latter as *humic coals*.

The key distinction between sapropelic and humic coals is their origin. In petroleum formation the importance of having hydrogen bonded to carbon, rather than to oxygen, was mentioned in the context of a loss of oxygen results only in a decrease of O/C ratio during the catagenesis of lipid-derived materials, but involves a decrease in both O/C and H/C ratios during catagenesis of kerogen derived from highly oxygenated compounds. Since sapropelic coals formed from Type I or II kerogen, which have a high proportion of hydrogen bonded directly to carbon rather than via oxygen to carbon, formation of the sapropelic coals retains a greater amount of the hydrogen in the kerogen than does the formation of humic coals.

Two kinds of sapropelic coals are recognized. *Boghead coal* is coal formed primarily from algae (Type I kerogen). Accumulated spores, pollen grains, and similar plant debris (Type II kerogen) gives rise to *cannel coal*. Compared with humic coals, the sapropelic coals are very rich in hydrogen content. In fact, some sapropelic coals can be ignited simply with a match, burning with a luminous, smoky flame. The term "cannel" is a corruption of the Scots word for candle.

A typical cannel coal has about 8% hydrogen, and a boghead coal, about 11 -12%. Since petroleum contains about 12 - 14% hydrogen, the conversion of a sapropelic coal to a synthetic, petroleum-like liquid fuel should be much simpler than the conversion of a humic coal, most of which have 4 - 6% hydrogen. Thus the sapropelic coals would be very desirable feedstocks for the production of synthetic liquid fuels. Unfortunately, sapropelic coals are very scarce and very few deposits of sapropelic coals with commercial potential exist. Because few deposits of sapropelic coals exist, all of the discussion of coals throughout this book will pertain to humic coals unless specifically stated otherwise.

Impure cannel or boghead coals are sometimes found associated with oil shales. *Cannel shales* are oil shales which are composed mainly of mineral grains which are completely encased in organic matter, and the organic matter itself is primarily derived from algae. Cannel shales are therefore very similar to cannel coals, except for the former having a higher inorganic content. Cannel shales are also sometimes referred to as *torbanites*. The torbanites derive from the alga *Botryococcus*, which is a freshwater organism. A related material derived from marine organisms is *tasmanite*, which occurs in two locations almost literally poles apart: Tasmania and the Brooks Range of northern Alaska. The tasmanites derive from the alga *Tasmanites*.

Summary

The two preceding sections have discussed the chemical processes involved in the transformations of kerogen into the principal hydrocarbon fuels. Although the discussions were presented separately, because in much of the remainder of this book it is convenient to consider coals on the one hand and oil and gas on the other, it is important to recognize that all of the transformations can be summarized collectively on a single van Krevelen diagram, Figure 3.12. All kerogens initially show a region of carbon dioxide and water formation via decarboxylation and dehydration. This reaction does not occur for the inertinites, because the easily displaced functional groups which are the source of carbon dioxide and water were lost when the inertinites were formed.

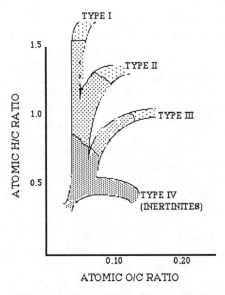

Fig. 3.12 Summary van Krevelen diagram of the transformations of the kerogens into fuels.

Type I and Type II kerogen show a long region of oil formation. Oil forms as a result of thermal cracking and, as it forms, it usually migrates out of the source rock, and essentially is lost from further reaction. The reaction pathways shown on the van Krevelen diagram in Figure 3.12 represent the changes in composition of the remaining organic material; the oil (or gas) which has migrated away does not appear in this diagram. It is possible that some oil is formed from Type III kerogen, although these kerogens are not usually thought of as sources of oil. (However, there is a similarity in temperature between the first coalification jump and the opening of the oil window, and there is interest among some geochemists in the possibility of coals acting as sources of oil.)

The formation of gas is possible from any kerogen. The gas-forming reactions include dealkylation to remove -CH₃ side chains from cyclic structures and and extensive thermal cracking to CH₃• radicals. When extensive gas formation is occurring, severe decomposition of the organic material is taking place. A point is reached at which it is no longer possible to distinguish, either chemically or optically, between virtinites and liptinites, or between vitrinites and inertinites. Thus as gas production continues, the solid becomes increasingly homogeneous and increasingly graphite-like.

The occurrence of oil and gas together is not uncommon, and would be expected for oils formed in the lower half of the oil window, where some gas formation is occurring at the same time. Indeed most oil accumulations have

71

at least some gaseous hydrocarbons associated with them (discounting the biogenic methane produced by bacteria decomposition of organic matter in early diagenesis). Gas is also formed during coalification, accumulating in coal seams as the hazardous firedamp. To address the possible accumulation of coal, oil, and gas in the same geographic region, it is instructive to recall the hydrogen redistribution scheme of Figure 3.1. As kerogen matures to an increasingly graphite-like solid, more and more methane production should be expected. Increasing the rank of coal represents a progression of the carbonaceous solid toward graphite, with a significant reduction in the H/C ratio. Reduction of the H/C ratio is accomplished by greater and greater degradation of molecular structures. In the formation of oil and gas, methane production from cracking begins approximately in the middle of the oil window. With increasing temperature, cracking increases and methane becomes more and more prevalent as the oil molecules are cracked into smaller and smaller fragments. Eventually the temperature increases to a point at which the oil window closes and the gas window opens. Under such severe reaction conditions oil formation is no longer possible, and hydrogen redistribution again drives the system toward methane and graphite.

Provided that the appropriate organic source materials were available coal and oil could occur in the same geographic region. As the combined effects of temperature and time increase, the rank of coal increases. As temperature increases, oil cracking to progressively lighter oils increases. Thus as the rank of coal increases, any oil which might be associated with it becomes lighter. At very high ranks of coal, associated with high reaction temperatures or long reaction times,.we would expect no oil in the same region as the coal but rather only gas and coal together. The relationship between the carbon content of coal and the associated petroleum and gas is shown in Table 3.4.

Table 3.4 Relationships among the carbon content of coals and the occurrences of petroleum and gas

% Carbon in coal	Petroleum	Gas
>83	Rare	Rare
79-83	Very light	Frequent, but small quantity
75-79	Light	Major accumulations
71-75	Medium	Rare
<71	Heavy	Rare

We have seen that rank increases with increasing carbon content of coal, and that increasing rank implies, via Hilt's rule, that the organic material has been exposed to higher temperatures. At the same time, we have seem in the kerogen maturation diagram that deeper burial, which again implies higher temperatures, results in a progressive change from heavy oil through medium and light oils to gas. Thus the qualitative correlations of Table 3.4 reflect a common effect of temperature as applied to the different types of kerogens.

As examples of the relationships indicated by Table 3.4, in Pennsylvania the rank of coal decreases heading westward across the state. Anthracites are found in the northeastern portion of the state, and there are no accumulations of oil or gas. Bituminous coals, light oils, and gas are all found in the western part of Pennsylvania. The Williston Basin, a broad area of western North Dakota, eastern Montana, and southern Saskatchewan contains enormous energy reserves. The coal is lignite, and the oils are heavy. An excellent quantitative relationship exists between the abundance of oil and the abundance of coal in the Oficina Formation of Venezuela. This oil probably derived directly from the coal. Oil has been found associated with coal in many of the coal fields of England; the oil may have been formed along with the coal and has not been subject to migration. The large deposit of natural gas in the Slochteren gas field of the Netherlands has arisen from the catagenesis of deeply buried coals which underlie the gas deposits.

To conclude this chapter, we return to the hydrogen redistribution diagram, now amended to show the effects of temperature (Fig. 3.13).

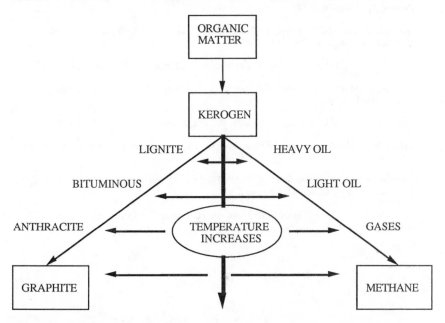

Fig. 3.13 Summary hydrogen redistribution diagram. As temperatureincreases, so does redistribution of hydrogen. The temperature effects on hydrogen redistribution help explain why lignites tend to be associated with heavy oils, bituminous coals with light oils, and anthracites with gases.

Further reading

Barker, C. (1979). *Organic Geochemistry in Petroleum Exploration*. Tulsa: American Association of Petroleum Geologists; Chapters 4,6,7.

Berkowitz, N. (1979). *An Introduction to Coal Technology*. New York: Academic Press; Chapters 1,2,3.

Boggs, S. Jr. (1987). *Principles of Sedimentology and Stratigraphy*. Columbus, OH: Merrill; Chapter 8.

Bustin, R. M., Cameron, A. R., Grieve, D. A., and Kalkreuth, W. D. (1985). *Coal Petrology: Its Principles, Methods, and Applications*. St. John's, Newfoundland: Geological Association of Canada.

Fieser, L. F. and Fieser, M. (1961). *Advanced Organic Chemistry*. New York: Reinhold Chapter 4.

Gates, B. C., Katzer, J. R., and Schuit, G. C. A. (1979). *Chemistry of Catalytic Processes*. New York: McGraw-Hill; Chapter 1.

Haun, J. D. (ed.) (1976). *Origin of Petroleum. I*. Tulsa: American Association of Petroleum Geologists; Chapters 2,11.

Haun, J. D. (ed.) (1974). *Origin of Petroleum. II*. Tulsa: American Association of Petroleum Geologists; Chapter 6.

Levoersen, A. I. (1954). *Geology of Petroleum*. San Francisco: Freeman; Chapter 11.

Loudon, G. M. (1988). *Organic Chemistry,* 2d edn. Menlo Park, CA: Benjamin/ Cummings; Chapter 5.

North, F. K. (1985). *Petroleum Geology*. Boston: Allen and Unwin; Chapters 5,6,9.

Selley, R. C. (1985). *Elements of Petroleum Geology*. San Francisco: Freeman; Chapters 5,9.

Streitwieser, A., Jr. and Heathcock, C. H. (1985). *Introduction to Organic Chemistry*, 3rd edn., New York: Macmillan; Chapter 6.

van Krevelen, D. W. (1961). *Coal : Typology - Chemistry - Physics - Constitution*. Amsterdam: Elsevier; Chapters III and IV.

Chapter 4

Natural gas

This chapter deals with the composition and properties of natural gas. Concepts of the classification of natural gas as wet or dry gas and sour gas are developed. The properties of natural gas affecting its use as a fuel are discussed. The important issue of the relationship of molecular structure to volatility is introduced.

This chapter begins the second major unit of the book, on the composition and properties of fuels. Two topics are of particular concern in this unit. The first is the relationship between fuel composition and schemes for classifying these complex materials. The second is the very important issue of the relationships between the molecular structures of fuel components and their physical properties.

Natural gas is defined as a mixture of hydrocarbons with varying quantities of non-hydrocarbons, that exists either in the gas phase or in solution with crude oil in natural underground reservoirs. The principal hydrocarbon component of natural gas is methane. Methane arises in two general ways: diagenesis of accumulated organic matter, and catagenesis of kerogen via hydrogen abstraction or hydrogen capping methyl radicals produced in thermal cracking.

Gas produced during catagenesis usually migrates through the porous rock in the earth until it encounters a formation of non-porous rock which prevents further migration and acts as a *reservoir*. Three classifications of gas are recognized, depending upon how it is found in the earth. *Associated gas* is gas found in conjunction with oil. Two kinds of associated gas are *gas cap gas*, in which the gas forms a separate phase over the accumulated oil (Fig. 4.1) or *dissolved gas*, in which the gas is in solution in the oil. *Non-associated gas* is gas found without oil. Non-associated gas could arise from gas migrating to a location separate from that to which oil has migrated, or could result from the

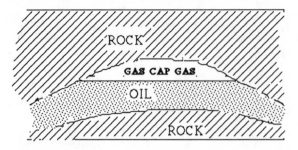

Fig. 4.1 The association of gas cap gas with oil in a reservoir inside the earth. Potentially this formation could also contain some dissolved gas, held in solution in the oil.

formation of gas without oil in the gas window. About 40% of the world's supply of natural gas is associated gas, the remaining 60% being non-associated.

Wet and dry gas

The principal hydrocarbon component of natural gas is methane. Often, however, the gas also contains the other light alkanes and a variety of inorganic compounds, including hydrogen sulfide, carbon dioxide, helium, and hydrogen. Normally the only hydrocarbon component of biochemical gas is methane, because in this case the methane is the product of specific biochemical reactions which do not yield ethane or other small alkanes. On the other hand, thermal cracking during catagenesis can give rise to an enormous variety of possible products, including most of the small alkanes and alkenes. (Recall the example of butane, forming five smaller molecules during cracking.)

In terms of the hydrocarbon components, we may encounter two types of natural gas: one in which methane is the only hydrocarbon component, produced either during diagenesis or in the gas window; and one which contains methane and other small alkanes and alkenes, as might be expected for gas formed in the oil window. Among the first few members of the alkanes, there are very significant differences in the boiling points of the compounds as the size of the molecules increases (Table 4.1). These differences in volatility make it easy to condense the gases heavier than methane. The ability to condense hydrocarbons other than methane from natural gas provides a classification of the gas as *dry gas*, which by definition contains less than .013 L/m^3, and *wet gas*, which contains greater than 0.040 L/m^3 of the condensible hydrocarbons. It is important to recognize that the terms wet and dry used in this context have nothing to do with moisture

Table 4.1 Boiling points of the smaller alkanes

Compound	Boiling point
Methane	-162°
Ethane	-89°
Propane	-42°
n-Butane	-1°
i-Butane	-10°
n-Pentane	36°
i-Pentane	28°

content - they refer only to the presence or absence of condensible hydrocarbons.

Condensing liquid from a wet gas and subjecting the condensate to fractional distillation (which must be performed at elevated pressure, to keep some of the components in the liquid phase) makes it possible to produce ethane and propane as pure fractions. The ethane is valued as a starting material for the production of ethylene, which is a feedstock of enormous importance for the petrochemical industry. Propane, which boils at -42° at atmospheric pressure, is liquefied fairly easily and is the principal component of the useful fuel *liquefied petroleum gas*, which is almost universally known by its acronym, *LPG*. The hydrocarbons having three to eight carbon atoms condensed from natural gas compose a material known as *natural gasoline*. A typical composition of natural gasoline is shown in Table 4.2.

Table 4.2 A typical composition of natural gasoline

Component	Percentage in mixture
C_{3-4}	20
C_5	30
C_6	24
C_7	20
C_8	4
$>C_8$	2

The butane and 2-methylpropane (isobutane) are separated and referred to as the *C_4 cut*. These gases are easily liquefied and are commonly used as a liquid fuel. For example, the liquid seen in transparent cigarette lighters is butane. Because butane boils at -1° at atmospheric pressure, only a slightly elevated pressure is required to keep it in the liquid phase at ambient temperatures in the range of 20 - 30°. The pentane, 2-methylbutane (isopentane), which occur in roughly equal amounts, and a small amount of 2,2-dimethylpropane (neopentane) comprise the *C_5 cut*. As we will see, natural gasoline does not have the combustion characteristics desirable in a gasoline for modern, high-compression automobile engines. However, natural gasoline is a useful feedstock for the petrochemical industry.

The composition of a natural gas as it comes from the well varies from one location to another. An example of the variability of gas composition is provided by the data in Table 4.3, for three sources of gas.

Table 4.3 Composition of natural gas from different sources

Component (volume percent)	Algeria	North Sea	Libya
CH_4	83.0	85.9	66.9
C_2H_6	7.2	8.1	19.4
C_3H_8	2.3	2.7	9.1
C_4H_{10}	1.0	0.9	3.5
C_5H_{12} and larger	0.3	0.3	3.5
N_2	5.8	0.5	-
CO_2	0.2	1.6	-
He	0.2	-	-

Sour gas

A natural gas containing hydrogen sulfide is called *sour gas*. The hydrogen sulfide in natural gas has several possible sources. One is the decomposition of proteins. Some amino acids contain the thiol functional group, -SH. The anaerobic decay of sulfur-containing proteins or their thermal decomposition at mild conditions could liberate the sulfur as H_2S. If sulfur from proteins or from the sulfate ion (supplied by brackish or marine water to the accumulating organic matter) survives diagenesis, it could become incorporated into a variety of organic compounds. The sulfur-containing functional groups can undergo reactions during catagenesis which are similar to the reactions of analogous oxygen-containing compounds discussed in Chapter 3. The C-S bond is weaker than the C-O bond; consequently sulfur functional groups are even more likely to undergo reactions. For example,

$$R\text{-}CH_2\text{-}CH_2\text{-}SH \rightarrow R\text{-}CH=CH_2 + H_2S$$

A third potential source of hydrogen sulfide is the anaerobic reduction of the sulfate ion, without its first being converted into organic sulfur compounds:

$$3\ SO_4^{-2} + C_6H_{12}O_6 \rightarrow 6\ HCO_3^- + 3\ H_2S$$

Sour gas is undesirable for several reasons. The hydrogen sulfide itself has an unpleasant smell, being the compound responsible for the unforgettable odor of rotten eggs. Hydrogen sulfide can dissolve in water to form a mildly acidic solution, corrosive to components of a gas handling and storage system, such as valves and pipes. If hydrogen sulfide is present in the gas when it is burned, sulfur oxides are formed as products of combustion:

$$2 H_2S + 3 O_2 \rightarrow 2 H_2O + 2 SO_2$$
$$H_2S + 2 O_2 \rightarrow H_2O + SO_3$$

Not only do the sulfur oxides themselves have irritating odors, but they also dissolve in water to form sulfurous and sulfuric acids, which are very corrosive. The oxides of sulfur are readily soluble in water. The release of sulfur oxides to the atmosphere will, after a few days to a few weeks, result in their being washed out of the atmosphere by dissolution into rain. The pH of the rain is lowered resulting in the serious environmental problem of *acid rain*. Finally, hydrogen sulfide and its combustion products the sulfur oxides are injurious to health.

When sour gas contains small quantities of hydrogen sulfide, the gas is purified by removal of hydrogen sulfide in a process known as *sweetening* (Chapter 7). If a gas contains very large quantities of H_2S (an extreme case being a gas deposit discovered near Emory, Texas which contained 42% H_2S) the hydrogen sulfide is converted to sulfur, which is then sold to the chemical industry.

Other components of natural gas

Some sources of natural gas contain the noble gases, particularly helium. In some cases, natural gases containing as much as 8% He have been found. In the United States, the principal helium-containing gas deposits are in the Texas panhandle, the gas having about 2% He. Helium is recovered from the gas from Texas, and this natural gas is the principal source of most helium used commercially in the United States. The helium derives from the radioactive decay of uranium, thorium, and radium in granitic rocks deep inside the earth. Helium is recovered by cooling the natural gas to condense hydrocarbons, leaving a gaseous mixture predominantly nitrogen and helium. Further cooling allows condensation of nitrogen as a liquid. The remaining helium (which is still gaseous because of its extremely low boiling point of - 269°) is purified by passing through activated carbon adsorbents at the temperature of liquid nitrogen, - 196°. Smaller quantities of the other noble gases, such as argon and radon, may also occur in some natural gases. They are also products of radioactive decay; argon, for example, is a decay product of ^{40}K.

Hydrogen is a rare component in some natural gases. Some hydrogen may be produced during the late stages of catagenesis, as for example, during aromatization of cyclic terpene molecules. The occurrences of hydrogen are rare because it is quite mobile and easily escapes from the confining rock strata.

Carbon dioxide is usually a minor constituent of natural gas. A variety of processes can lead to the formation of carbon dioxide, including anaerobic decay reactions and decarboxylation reactions discussed previously. In addition, carbon dioxide can be produced by the oxidation of organic compounds if highly oxygenated surface waters percolate through a geological formation in which the organic compounds are trapped, and can be formed by the thermal decomposition of carbonate rocks if they are heated by contact with magma.

Introduction to volatility relationships

The data in Table 4.2 show that the alkanes differ in volatility. The volatility differences can be taken advantage of to effect a separation of the alkanes in natural gas. This section addresses the question of why natural gas is a gas; that is, why some of the smaller alkanes are gaseous at ordinary conditions of temperature and pressure whereas others are liquids.

For matter to exist in a condensed state at any temperature there must be an interaction of some kind between the constituent atoms or molecules. In an ionic substance, such as sodium chloride, the interaction is a Coulombic attraction between the ions of different charge. In a covalent compound having a permanent dipole moment, such as chloromethane, a Coulombic interaction occurs between the dipoles. Alkanes are neither ionic nor possess permanent dipole moments. Attraction between alkane molecules is based on electron correlation. The intermolecular attractions arising from electron correlation are called *London forces*, and are an example of the class of attractions known as van der Waals forces. London forces arise from the existence of very short lived dipoles resulting from the movement of electrons within a molecule. If we were able to examine the movement of electrons in a methane molecule, for example, we might find a situation such as shown in Fig. 4.2

As two molecules, each possessing temporary, short-lived dipole moments, approach each other, the movement of electrons in one molecule correlates with that of the second molecule (Fig. 4.3).

Any n-alkane molecule, other than methane, could be viewed as a nearly cylindrical array of hydrogen atoms or as a nearly cylindrical electron cloud.

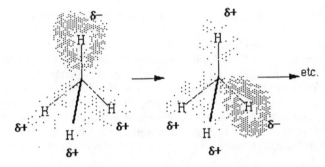

Fig. 4.2 The formation of temporary dipoles in methane, by the movement of electrons creating regions of low electron density (lightly shaded) and high electron density (heavily shaded). This movement of electrons creates short-lived regions having partial positive (δ+) and partial negative (δ-) charges.

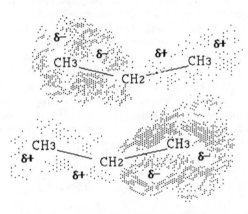

Fig. 4.3 As a propane molecule with temporary dipoles approaches a second molecules, the movements of electrons in the two molecules begin to correlate, producing a weak electrostatic interaction between the molecules.

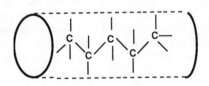

As the size of the molecule increases (that is, as the number of carbon atoms increases) two effects become important. First, the surface area of the molecule increases. As surface area increases, so do the London forces between molecules. Second, the weight of the molecule increases. A greater weight requires in turn a greater amount of kinetic energy either to remain in the gas phase or to escape the liquid phase. The combined effects of greater intermolecular London forces and greater weight result in the boiling point of alkanes increasing as the molecular size or molecular weight increases. Thus while very low temperatures are required to liquefy methane, the fifth compound in the family, pentane, is a liquid at normal room temperatures.

Natural gas as a premium fuel

Among the fuels discussed in this book, natural gas is, by several criteria, the best. Natural gas liberates by far the most heat when an equal weight of various hydrocarbon fuels is burned. Because it is a gas, natural gas leaves no residue of ashes when it is burned. Sour gas can be sweetened by straightforward processing operations, eliminating the problem of sulfur emissions during combustion. Fluids are generally easy to handle, meter, and regulate in combustion systems, especially in comparison to labor-intensive drudgery such as shovelling coal. However, because of their very low density at ordinary conditions, storage of a useful weight of gaseous fuel requires either high pressure storage, or liquefaction of the gas at low temperatures. Since gas is invisible, it cannot be seen if it happens to be leaking (in contrast to an oil leak or coal spill) and the potential exists for a potentially hazardous accumulation of gas to build up without its being noticed.

The energy liberated in the combustion of a fuel is known as the *calorific value*. For natural gas, assuming that methane is the only combustible component, the combustion reaction in an excess of oxygen or air is

$$CH_4 + 2\,O_2 \rightarrow CO_2 + 2\,H_2O$$

The calorific value is the enthalpy change associated with this reaction. This enthalpy change is also called the *heat of combustion* or heating value. (In this book, the term "calorific value" will be used for fuels, and "heat of combustion" for pure compounds.) For methane, the enthalpy change for the formation of gaseous carbon dioxide and liquid water is -892 kJ/mol.

A comparison of the calorific values of natural gas, petroleum products, and coal must recognize that each exists in a different physical state and that in common practice the calorific values might be expressed per unit of gas volume, or of liquid volume, or of mass. The calorific value expressed per unit mass is a convenient basis of comparison. Natural gas has a calorific value of 55,600 kJ/kg, whereas petroleum-derived fuels typically have calorific values around 46,000 kJ/kg and a good quality coal might have a calorific value of 35,000 kJ/kg.

We now introduce one of the factors relating to the composition of the fuels which is responsible for this difference in calorific values. For methane, the H/C ratio is 4. Many petroleum products have H/C of ~2, and coals have H/C <1. For illustration we can consider decane, $C_{10}H_{22}$, to be a typical component of petroleum, and assume that the macromolecular structure of a coal could be represented by a formula such as $C_{100}H_{80}$. We further assume that each of these fuels will be burned with sufficient oxygen to form carbon dioxide and liquid water. The total enthalpy liberated in the experiment will be the enthalpy contributed by the formation of carbon dioxide from the carbon in the fuel and the enthalpy contributed by the formation of water from the hydrogen. (For the purposes of this qualitative argument, we neglect the heat of formation of the CH_4, $C_{10}H_{22}$, and $C_{100}H_{80}$.) The heats of formation of carbon dioxide and water are -393 kJ/mol and -285 kJ/mol, respectively. Again there is a need to find a common basis for comparing the fuels, since there are different numbers of carbon and hydrogen atoms in each molecule. A convenient approach is to make the comparison on the basis of an equal number of carbon atoms or equal moles of carbon. Doing so provides the data summarized in Table 4.3.

Table 4.3 Comparison of the contributions of the heats of formation (kJ/mol) of carbon dioxide and water to the observed calorific values of fuels, per mole of carbon.

Fuel	CO_2 contribution	H_2O contribution	Total
CH_4	-393	2 x (-285)	-963
$C_{10}H_{22}$	-393	1.1 x (-285)	-707
$C_{100}H_{80}$	-393	0.4 x (-285)	-507

If the fuels are compared on an equivalent number of moles of carbon - that is, are compared as CH_4, $CH_{2.2}$, and $CH_{0.8}$, the contribution of the formation of carbon dioxide to the calorific value is the same for each. The contribution from the formation of water decreases as the amount of hydrogen decreases. Even though the -393 kJ contributed by the carbon dioxide remains constant from fuel to fuel, the contribution from the formation of water decreases, and, consequently, the sum of these contributions must also decrease.

The analysis shown in Table 4.3 leads to a useful rule-of-thumb for comparing hydrocarbon fuels: among a group of fuels compared on an equal amount of carbon, the calorific value decreases as the H/C atomic ratio decreases. It is very important to recognize that this rule is a qualitative argument useful only for a quick, approximate comparison of fuels. As noted above, the analysis is not rigorous because it does not take into account the heats of formation of the different fuels. Furthermore, we will see in Chapter 6 that the argument is made more complex by the presence of abundant oxygen and aromatic structures in some coals. This rough rule of thumb is reliable if the comparisons are confined to materials which are chemically similar, as in the case of a group of alkanes (Table 4.4, for example).

Table 4.4 An example of the application of the relationship of calorific value to atomic H/C ratio for typical alkanes

Compound	H/C	Heats of combustion, kJ/mol	
		kJ per mole of compound	kJ per mole of carbon
Methane	4.00	- 882	-882
Ethane	3.00	-1542	-771
Propane	2.67	-2223	-741
Pentane	2.40	-3489	-698
Hexane	2.33	-4144	-691
Heptane	2.29	-4810	-687
Octane	2.25	-5454	-682
Decane	2.20	-6740	-674

Comparing these alkanes on the basis of a mole of carbon shows a distinct drop in the heat of combustion accompaning the decrease in H/C ratio.

Further reading

Fieser, L. F., and Fieser, M. (1961). *Advanced Organic Chemistry*. New York: Reinhold; Chapter 4.

Loudon, G. M. (1988). *Organic Chemistry*, 2d edn., Menlo Park, CA: Benjamin/ Cummings; Chapter 3.

North, F. K. (1985). *Petroleum Geology*. Boston: Allen and Unwin; Chapter 5.

Streitwieser, A., Jr., and Heathcock, C. H. (1985). *Introduction to Organic Chemistry*, 3rd. edn., New York: Macmillan; Chapter 5.

Chapter 5

Petroleum

Natural gas, when purified for commercial use, is almost entirely a single chemical compound, methane. Petroleum, however, consists of thousands of individual compounds. The classification of petroleum, and the relationship of physical properties to composition, are much more complex than for natural gas. The important topics discussed in this chapter are systems for classifying petroleum and for predicting properties on the basis of classification; and an expansion of the discussion introduced in Chapter 4 on the relationships between molecular composition and structure and the physical properties of compounds.

The components of petroleum

Analysis of crude oils from around the world shows that the elemental composition varies over a narrow range: 82 - 87% carbon, 12 - 15% hydrogen, the balance being oxygen, nitrogen, and sulfur. However, if these same samples are analyzed to determine the specific compounds present, any particular sample will be found to contain hundreds to thousands of individual compounds. In other words, on an elemental basis most crude oils have about the same composition, but on a molecular basis it is likely that no two samples are exactly alike. The reason for this seeming contradiction is that most components of petroleum belong to homologous series of compounds for which the composition, expressed on a weight percent basis, actually varies very little even over a long range of the series. For example, consider pentane, C_5H_{12}, and pentadecane, $C_{15}H_{32}$. The composition of pentane is 83.3% C and 16.7% H; that of pentadecane is 84.9% C and 15.1% H. A list of all of the possible isomers of all the alkanes between pentane and pentadecane would amount to several thousand compounds, yet across this range the elemental composition on a weight basis changes by less than two percentage units.

The components of petroleum fall into four classes of compounds: the alkanes, the cycloalkanes, aromatic compounds, and compounds containing the heteroatoms oxygen, nitrogen, or sulfur. In the informal nomenclature of petroleum technology, the alkanes are called *paraffins* and the cycloalkanes are called *naphthenes*. The compounds containing heteroatoms are collectively known as the *NSO's*, the name deriving from the symbols of the three principal heteroatomic elements.

Paraffins

The paraffins are formed primarily by the cracking of lipids, including the terpenes. Among the n-alkanes, compounds of less than five carbon atoms are gases at ordinary temperatures and pressures. The largest compound which is liquid at room temperature is heptadecane, $C_{17}H_{36}$, with a melting point of 22°. The next n-alkane, octadecane, has a melting point of 28°. Octadecane and larger alkanes can exist in solution in the liquid crude oil. The largest alkane ever to be reported in crude oil is $C_{78}H_{158}$.

In addition to the normal alkanes, branched chain alkanes also occur in petroleum. These compounds arise from the cracking of terpenes. The cracking of a terpene molecule can give rise to a radical containing a branched structure

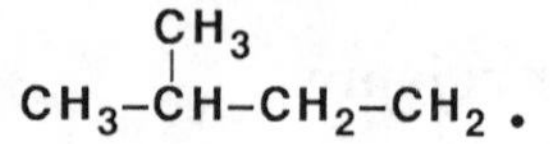

$$CH_3-\overset{\overset{\displaystyle CH_3}{|}}{C}H-CH_2-CH_2 \cdot$$

which can undergo a recombination reaction with a straight chain radical

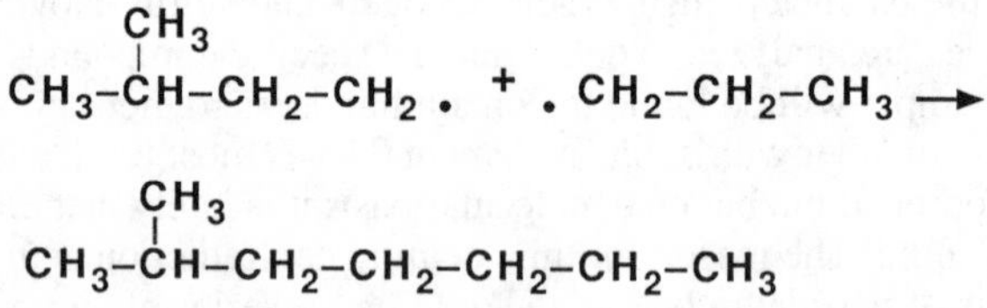

$$CH_3-\overset{\overset{\displaystyle CH_3}{|}}{C}H-CH_2-CH_2 \cdot \ + \ \cdot CH_2-CH_2-CH_3 \longrightarrow$$

$$CH_3-\overset{\overset{\displaystyle CH_3}{|}}{C}H-CH_2-CH_2-CH_2-CH_2-CH_3$$

or with another branched chain radical (next page).

The original branched chain radical can also undergo the other radical reactions discussed in Chapter 3, such as hydrogen abstraction. The result of these cracking and recombination reactions is the existence of a variety of branched chain alkanes, some having a single branch and others multiply branched.

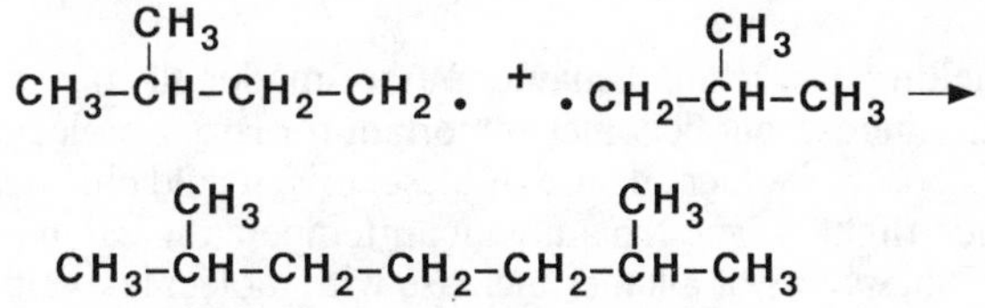

$$\begin{array}{ccccc}
& \overset{\displaystyle CH_3}{|} & & & \overset{\displaystyle CH_3}{|} \\
CH_3-CH-CH_2-CH_2 \cdot & & + & \cdot CH_2-CH-CH_3 & \longrightarrow
\end{array}$$

$$\begin{array}{cc}
\overset{\displaystyle CH_3}{|} & \overset{\displaystyle CH_3}{|} \\
CH_3-CH-CH_2-CH_2-CH_2-CH-CH_3
\end{array}$$

Examples of branched chain compounds which have been isolated from crude oil are 2,6,10,14-tetramethylpentadecane (pristane) and 3,5,11,15-tetramethylhexadecane (phytane). These compounds, with their abundant methyl groups, could produce a variety of branched-chain radicals by cracking anywhere along the carbon chain.

As a rule the branched chain alkanes are more volatile than the straight chain compound of the same number of carbon atoms. For a series of isomers, it is generally true that the compound having the most highly branched structure is the most volatile. This difference in volatility arises from reduced London forces between branched chain molecules, because the surface area of a branched chain molecule is smaller than that of the corresponding straight chain compound

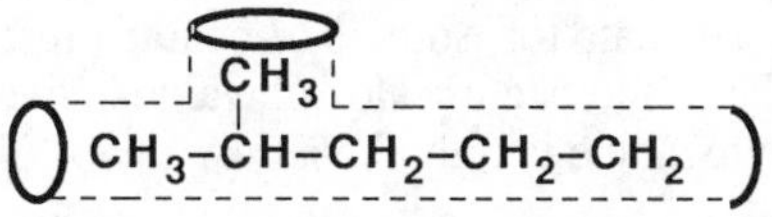

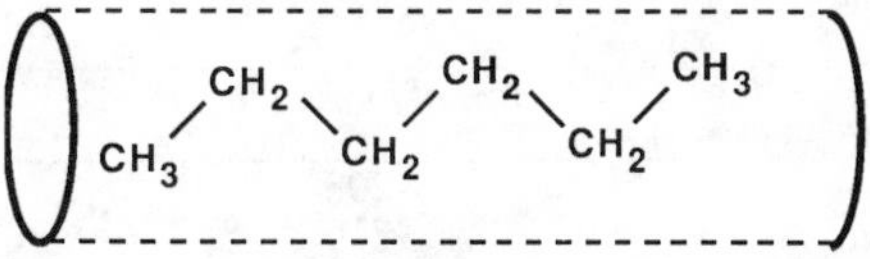

We have seen in Chapter 4 that London force interactions are proportional to surface area. (If the comparison is restricted to compounds having the same number of carbon atoms, there is no difference in molecular mass and thus there are no effects of kinetic energy to be considered in the comparison.) This relationship is illustrated in Table 5.1 for isomers of hexane.

Table 5.1 Effects of chain branching on volatility of the isomers of hexane

Compound	Boiling point, °C
Hexane	69
2-Methylpentane	60
2,3-Dimethylbutane	58
2,2-Dimethylbutane	50

The melting or freezing behavior of the smaller alkanes is generally of little practical interest, but becomes important for larger molecules. Examples of concern include the use of diesel oils in cold climates or jet fuels in high altitude flight, where the ambient air temperature can be very low. As a rule the melting points of alkanes increase with molecular weight. Melting points also increase with increasing symmetry in the structure, so some very symmetrical branched-chain structures have melting points much higher than less symmetrical isomers. A remarkable example is afforded by alkanes of eight carbon atoms: the linear compound, octane, has a melting point of -56.5°, whereas the highly symmetrical branched isomer, 2,2,3,3-tetramethylbutane, melts at +104°!

Branched chain compounds are slightly more stable thermodynamically than the corresponding straight chain isomer. That is, the heats of formation of the branched chain compounds are slightly more exothermic than for the straight chain compound. Because the heats of formation of the branched chain compounds are more negative, the heat of combustion of the straight chain compound will therefore be slightly more exothermic. In other words, the straight chain isomer has is a slightly better fuel in terms of the available calorific value. However, the differences are so small as to be of negligible effect in practice. The data for isomers of octane presented in Table 5.2 illustrate these points. The difference in calorific value between octane and its isomer 2,2,4-trimethylpentane is only 0.3%.

Table 5.2 Heats of formation and heats of combustion (kJ/mol) for selected isomers of octane

Compound	Heat of formation	Heat of combustion
Octane	-208	-5117
2-Methylheptane	-216	-5110
2,2,4-Trimethylpentane	-224	-5101

We will see later that considerations other than the calorific value are also important in the practical combustion of fuels, with these isomers of octane being a particularly good example.

Among the paraffins, the heat of combustion per mole increases with molecular weight. For example, in the n-alkanes the increase is about 653 kJ for each additional $-CH_2-$ group. When the heats of combustion are compared on a weight basis, methane has the highest value among the alkanes. The value falls off with increasing molecular size, and becomes almost constant for molecules larger than octane. This behavior reflects the importance of hydrogen in determining heats of combustion. The elements themselves have heats of combustion of 142 kJ/g for hydrogen and 34 kJ/g for carbon. Thus the compounds having the highest hydrogen content have a clear advantage in heats of combustion per gram. As the atomic H/C ratio drops, so does the heat of combustion per gram. For paraffins, the heat of combustion becomes nearly constant for compounds larger than octane because the H/C ratio is nearly constant; *e.g.*, compare decane, $C_{10}H_{22}$, H/C = 2.20; and the much larger molecule triacontane, $C_{30}H_{62}$, H/C = 2.07. (Also see the data in Table 4.4.) In practical use, however, it is not necessarily the mass of available fuel which is a determining factor in fuel selection. If heat of combustion per mass were the most important consideration, then clearly combustion systems should be designed to utilize liquid methane or liquid ethane. In many systems, such as vehicles, it is the volume of available fuel storage rather than weight that is most important. Compounds of low molecular weight, such as methane and ethane, also are of low density. With the larger alkanes, the greater density compensates, at least in part, for the lower heat of combustion per gram relative to methane and ethane. Thus in the diesel oil range, say C_{16} to C_{20}, the alkanes have a heat of combustion per volume about 57% greater than methane. Thus when volume is limited, as on a fuel tank in a road vehicle or airplane, the vehicle can utilize a greater fuel load if higher alkanes are used, rather than small compounds such as methane and ethane. (There is of course an additional factor governing fuel selection - the ignition and combustion characteristics of the fuel. This issue, specifically with regard to gasoline and diesel oil, will be discussed in Chapter 8.) The comapison of the heats of combustion on a mass and volume basis is illustrated for selected alkanes in Table 5.3.

Table 5.3 Comparative heats of combustion for n-alkanes, expressed on a mass and volume basis.

Compound	Heat of combustion:	kJ/g	kJ/mL
Methane		55.7	23.4
Ethane		51.9	28.0
Propane		51.0	28.9
Pentane		48.6	30.6
Hexane		48.6	31.8
Heptane		48.1	32.6
Octane		48.1	33.5
Decane		47.7	34.7
Hexadecane		47.3	36.8

Naphthenes

Almost all of the cycloalkanes present in crude oil are derivatives of cyclopentane and cyclohexane. This fact reflects the lack of internal bond strain in these two rings and the ease of synthesis of five- or six-membered rings. In principle cyclic compounds can exist with any number of carbon atoms greater than two. However, derivatives of the first two cycloalkanes (*i.e.*, cyclopropane and cyclobutane) are very rarely encountered in nature because of propensity of the rings to undergo reactions to relieve the internal strain. In methane, the four bonds from the carbon atom are oriented toward the vertices of tetrahedron, so that the H-C-H bond angle is 109°.

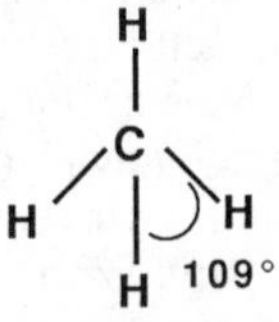

In cyclopropane, the triangular shape of the molecule forces the C-C-C bond angles to be 60°, a very significant deviation from the preferred 109°. Consequently the C-C bonds are severely strained, and cyclopropane and its derivatives readily undergo reactions which will relieve the strain by opening the ring. The internal angle of a square, as might be expected for the structure of cyclobutane, is 90°, which is not as bad as cyclopropane but is still a significant difference from the preferred tetrahedral angle of 109°. However, with more than three carbon atoms it is not necessary that they all lie in the same plane, and the cyclobutane molecule can achieve a small degree of strain relief by having one of the four carbon atoms slightly out of the plane of the other three.

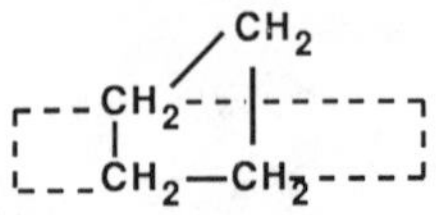

Despite adopting a non-planar structure, cyclobutane is still strained and, like cyclopropane, undergoes a variety of ring-opening reactions.

The internal angle of a pentagon is 108°. Cyclopentane can easily achieve the desired tetrahedral angle of 109° by folding one of the carbon atoms slightly out of the plane of the other four. This process can be visualized as the folding of the flap of an envelope, and this non-planar configuration of cyclopentane is called the envelope configuration. Since most of the strain in cyclopentane is relieved, this cyclic structure is about as stable as most non-cyclic alkanes. Cyclohexanes would have a hexagonal array of carbon atoms. The internal angle of a hexagon is 120°, so at first sight it might seem that cyclohexanes would also be strained molecules, with the C-C-C bond angles forced outward to 120° from the preferred 109°. However, the ability to attain a non-planar configuration allows cyclohexane to adopt configurations which provide the 109° bond angle and eliminate strain among the C-C bonds. In the case of cyclohexane, two configurations are adopted, the chair and the boat

The chair configuration is preferred.

Another way of considering the stability of cyclohexane is in terms of heats of formation. As a rule, a single -CH₂- group contributes -20.5 kJ/mol to the heat of formation of an n-alkane. (In other words, the heat of formation of a given n-alkane differs by -20.5 kJ/mol from that of the n-alkane with one less -CH₂- group.) This value, -20.5 kJ/mol, is nearly the same as the ΔH°_f per -CH₂- group for cyclohexane. In other words, cyclohexane is as stable as a typical n-alkane.

Any cycloalkane larger than cyclohexane can also use the same strategy of adopting non-planar configurations to eliminate C-C bond strain. In this respect, therefore, there is no reason why cycloalkanes containing seven or more carbon atoms in a single ring could not exist. Although such compounds are known in the laboratory, they tend to be rare in nature. To see why this should be the case, it is helpful to consider the origin of the cycloalkanes.

Most of the naturally occurring cycloalkanes derive from the cyclic terpenes, an example of which is menthane

$$CH_3$$

$$CH$$
$$CH_3 \quad CH_3$$

The terpenes in turn are the products of biosynthesis, via reaction sequences which at some point must involve a ring closure reaction to form the cyclic structure. Ring closures to form cyclopropanes or cyclobutanes are difficult to carry out because of the large bond strain in the resulting structures. Bond strain is no longer a significant issue in the cyclopentanes and cyclohexanes, so that compounds based on these structures are widespread in nature. Although bond strain is important in cycloheptane and larger cyclic compounds, a second problem arises. Too illustrate, consider the formation of a cyclic structure by the recombination of a diradical of the type ·CH₂(CH₂)ₙCH₂·. As the length of the carbon chain increases, the likelihood diminishes that one end of the chain will find the other before some competing radical reaction occurs. For chains longer than seven carbon atoms, ring closure by bringing the ends of the two chains together becomes very difficult. In fact, when large cycloalkanes are synthesized in the laboratory, the reactions in some cases must be carried out in extremely dilute solution to maximize the chances for intramolecular ring closure and minimize the chances for unwanted intermolecular reactions. The graph shown as Figure 5.1 emphasizes the compromise between internal bond strain and probablility of ring closure, indicating that the cycloalkanes with five or six atoms in the ring are the most prevalent because of, on the one hand, the lack of bond strain, and, on the other hand, the probability of formation via ring closure reactions.

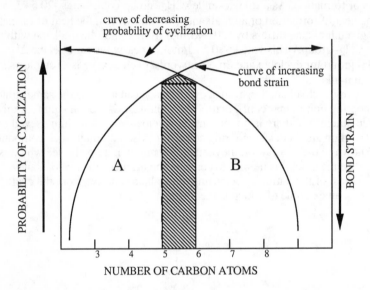

Fig. 5.1 As the number of carbon atoms increases, the bond strain in the cycloalkane decreases, but the probability of cyclization also decreases. The optimum situation is a low internal bond strain and high probability of cyclization. This situation is achieved for cycloalkanes of five or six carbon atoms, indicated by the region of shading.

In the region A in Fig. 5.1 high bond strain makes the compounds highly reactive and therefore unlikely to exist because of their high reactivity. In region B, the difficulty of ring closure also makes the compounds unlikely to exist because of the low probability of cyclization.

In comparison with alkanes, cycloalkanes are less volatile. For example, hexane boils at 69° while cyclohexane boils at 81°. The reduced volatility of the cycloalkanes is due to greater London force interactions, since the cyclic structure presents a much larger effective surface area than the chain structure. In addition, the cycloalkanes are more dense: the density of hexane is 0.66 g/mL while that of cyclohexane is 0.78 g/mL. The density difference arises from a more efficient filling of a given volume by the cyclic structures, which tend to favor a single molecular configuration (e.g., the chair form of cyclohexane), relative to a relatively "floppy" chain molecule which can potentially adopt a variety of configurations in the liquid.

Alkanes have slightly greater heat of combustion than the cycloalkanes of similar carbon numbers: 4144 kJ/mol for hexane compared with 3926 kJ/mol for cyclohexane. Note that this comparison conforms to the rule of thumb relating heats of combustion to H/C ratios. However, when considering the practical use of a fuel in situations in which the volume of the fuel storage is a limiting factor, the greater density of the cycloalkanes means that a naphthene-rich fuel will have a greater calorific value *per unit volume* than will a paraffin-rich fuel. Continuing the comparison of hexane and cyclohexane, the heats of combustion per unit volume are 36.4 kJ/mL for cyclohexane, but only 31.8 kJ/mL for hexane, giving approximately a 15% higher value for the cyclic compound. An example of a situation in which the calorific value per volume is of interest is in jet aircraft, where the amount of fuel available is constrained by the volume of the fuel tanks. A greater calorific value for a given volume of jet fuel provides an advantage for increased range or performance of the aircraft. Fuels rich in naphthenes are sometimes called high-density fuels.

Aromatics

The principal aromatic components of crude oils are benzene, toluene, the isomers of xylene, ethylbenzene, and 1,3,5-trimethylbenzene, which has the common name mesitylene. Aromatic compounds having condensed ring systems, such as naphthalene, anthracene, and phenanthrene, are solids at room temperature, but may be present in crude oils in small quantities in solution. Some alkylated derivatives of the larger aromatic compounds, such as the methylnaphthalenes, may also be present.

Some aromatic compounds occur in crude oil because they have persisted from the original organic matter. For example, amino acids such as phenylalanine

contain aromatic structures and, in this case, could be a component of the proteins which formed Type I or II kerogen. Other aromatic compounds may have formed as a result of aromatization reactions during catagenesis.

Aromatics are more dense than the corresponding cycloalkanes: 0.88 g/mL for benzene compared with 0.78 g/mL for cyclohexane. The benzene ring is completely planar, and may therefore be even more efficient at packing molecules into a given volume than is cyclohexane. The heats of combustion of aromatics are much lower than the cycloalkanes: 3273 kJ/mol for benzene vs. 3926 kJ/mol for cyclohexane. Although benzene has a much lower H/C ratio than cyclohexane, the resonance stabilization energy of benzene also plays a very significant role in reducing the heat of combustion of benzene relative to cyclohexane. Volatility relationships are difficult to generalize. Benzene and cyclohexane have almost identical boiling points: 80° and 81°, respectively. The same is true for some of the alkylated derivatives. For example, the respective boiling points of isopropylbenzene and isopropylcyclohexane are 152° and 154°. In other cases there are significant differences between aromatic and cycloalkane compounds, as in the case of the 1,3-dimethyl- and 1,4-dimethyl- derivatives, where in each instance the benzene compound boils 18° higher than the corresponding cyclohexane.

Aromatic compounds containing more than one ring tend to exist in asphalts. Examples of such compounds are naphthalene, anthracene, and phenanthrene

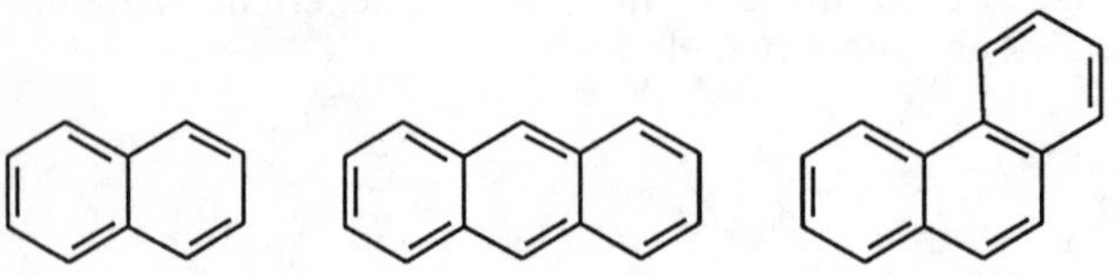

The larger aromatic hydrocarbons in crude oil are divided into two classes on the basis of their solubility in pentane. Those components of asphalt which are soluble are classified as *resins*; those which are insoluble are called *asphaltenes*. (As used in this context, "resin" is simply an operational definition describing a class of compounds on the basis of their solubility behavior, without specific regard to molecular structure. These resins are not the same as the plant constiteunts introduced in Chapter 1.)

NSO's

NSO's are the compounds containing the heteroatoms nitrogen, sulfur, or oxygen. In principle, NSO's could include compounds containing any functional group of these elements. However, the heterocyclic compounds are of special concern. These compounds are exemplified by furan, pyrrole, and thiophene:

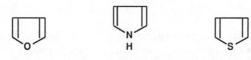

Nitrogen and sulfur are undesirable constituents of fuels. We have discussed some of the concerns with sulfur in the context of sour gas. Nitrogen compounds can form oxides of nitrogen during combustion. Like the oxides of sulfur, nitrogen oxides can dissolve in water to form corrosive, acidic solutions, of which acid rain is the most notorious example. Nitrogen oxides also react in the atmosphere with hydrocarbon molecules to generate the form of air pollution known as smog. To help protect the environment, efforts are made to remove nitrogen and sulfur from fuels in refining or processing steps before the fuels are actually used. The concern for the heterocycles is that often these compounds tend to be much less reactive than compounds containing the heteroatoms in other types of functional groups, such as mercaptans or amines. Thus the heterocycles provide a particular challenge to fuel refiners.

Because of the very large diversity of NSO compounds, it is not possible to develop rule of thumb generalizations comparing the NSO's with the other three families of compounds. We will be discussing some properties of specific NSO compounds on a case-by-case basis as these compounds are introduced.

Some additional properties of petroleum

This section introduces some properties of petroleum which will be helpful in the discussion of systems for classifying crude oils. The classification systems will be introduced in the section which follow this one.

API gravity

Oils are sometimes qualitatively classified as *heavy* or *light oils*. Heavy oils are formed at the top of the oil window. As a rule, heavy oils have a relatively high density and high viscosity. In comparison, light oils have low densities and viscosities. The "heavy" and "light" classifications care often expressed in terms of the *API gravity*. (The API is the American Petroleum Institute.)

The API gravity is an unusual unit, but is introduced here and will be used in the text because it has widespread use in the petroleum industry. The API gravity is defined by the equation

$$\text{API gravity} = [141.5 / (\text{specific gravity})] - 131.5$$

API is customarily expressed in degrees; for example, one might speak of an oil as having an API gravity of 56°. Note that the equation uses the *specific gravity* of the liquid rather than the density. (For purposes of determining the API gravity, the specific gravity is determined by measuring the density at 15.6° (60° F) and comparing it to the density of water at the same temperature.) Specific gravity and density are sometimes informally used as synonyms, but it is important to recognize that the specific gravity of a fluid is a dimensionless number relating the density of the fluid of interest to the density of a reference fluid. For liquids, the standard of comparison of water, which happens to have a density of 1 g/mL. Therefore when using metric units the specific gravity of a liquid happens is essentially numerically equal to its density, but density must be expressed in units of mass per volume while specific gravity is dimensionless. Note also that the API gravity is inversely proportional to specific gravity, so that the higher the specific gravity, the lower the API gravity, and *vice versa*. Water, which has a specific gravity of 1.00, will have an API gravity of 10°. Liquids more dense than water will have a low (possibly even negative) API gravity, whereas liquids of low density, such as the smaller alkanes, will have high API gravities. For petroleum, those materials having API gravities over 40° are said to be *light oils*, whereas those with API gravities less than 10° are said to be *heavy oils*.

To examine the effects of molecular structure on API gravity, consider the compounds in Table 5.3:

Table 5.3 API gravities of simple hydrocarbon compounds

Compound	Gravity
Hexane	83°
Heptane	75°
Hexadecane	51°
Eicosane	50°
Hexane	83°
Cyclohexane	50°
Benzene	30°
Benzene	30°
Naphthalene	-8°

Comparison of the first set of compounds shows that API gravity decreases with increasing chain length of alkanes. The effect is more pronounced for the smaller alkanes, and as the chain length increases beyond fifteen carbon atoms changes in gravity with incremental addition of an extra carbon atom become quite small. For compounds with the same number of carbon atoms,

API gravity decreases in the order alkanes > cycloalkanes > aromatics. Finally, API gravity decreases with increased ring condensation in aromatics. It is important to recognize that these rules of thumb, like our previous discussion of the effects of H/C ratio on calorific value, are useful when approximate comparisons need to be made and no experimental data are readily at hand. No rule of thumb should ever be substituted for actual experimental data, provided of course that such data are available.

The effect of heteroatoms on API gravity is also an important concern. Sulfur compounds are of special interest, for reasons similar to those discussed previously in connection with sour gas: odors which range from the obnoxious to the downright revolting, formation of the pollutants sulfur oxides during combustion, and corrosion of metal surfaces. In addition, some sulfur compounds "poison" catalysts used in refining operations. There is a very marked effect of sulfur on API gravity, and a measurement of the API gravity can in some cases be used as a rough-and-ready estimate of the sulfur content of an oil.

Consider the compounds heptane and dipropyl sulfide:

$$CH_3CH_2CH_2CH_2CH_2CH_2CH_3 \qquad CH_3CH_2CH_2SCH_2CH_2CH_3$$

The structures of these compounds can be considered to be related by the replacement of the central methylene group in heptane with a sulfur atom. (This is not meant to imply that dipropyl sulfide is synthesized by reacting sulfur with heptane.) The C-C bond length is .154 nm; the C-S bond length is .208 nm. If we consider for simplicity that the two molecules are essentially cylinders, the replacement of a $-CH_2-$ group by $-S-$ increases the molecular volume by about 5%. However, the difference in atomic mass, 14 vs 32 respectively, means that the molecular weight of dipropyl sulfide is 18% greater than that of heptane (118 vs 100). The introduction of a sulfur atom into a hydrocarbon results in the mass increasing faster than the volume - 18% vs 5% in this example. As a result, we should expect sulfur compounds of this type to have densities higher than the related hydrocarbon compound. If the density is increased, the API gravity must be decreased. Although the detailed example has treated an alkane, the relationship holds for cycloalkanes and aromatics as well, as shown by the examples in Table 5.4.

Table 5.4 Comparative API gravities of sulfur containing compounds and analogous hydrocarbons

Hydrocarbon	API gravity	S-Compound	API gravity
Heptane	75°	Dipropyl sulfide	42°
Cyclohexane	50°	Thiacyclohexane	12°
Indene	10°	Benzothiophene	-10°

We have seen that as the API gravity decreases, the boiling point generally decreases also. Thus the amount of desirable fuel components of crude oil obtained by distillation (Chapter 8), especially so-called distillate fuels such as gasoline and kerosene, that can be derived from crude oil drops as the API gravity decreases. A useful relationship is summarized in Table 5.5 for various naturally occurring hydrocarbons.

Table 5.5 Relationships among API gravity, asphaltene content, distillate fuel yield, and desirability for naturally occurring hydrocarbons

Hydrocarbon source	API gravity	Asphaltenes	Distillate	Desirability
Crude oil	>16°	↓	↑	↓
Heavy crudes	10 - 16°			
Tar sands	6 - 12°	Increasing	Increasing	Decreasing
Native asphalts	-5 - 12°			
Asphaltite	-15 - -5°	↓	↑	↓

Viscosity

Viscosity is a measure of the resistance to flow. For fluids in which the shear rate is proportional to the shear stress, viscosity is the constant of proportionality. Thus

$$\tau = \eta(dV/dy)$$

where dV/dy is the shear rate and τ is the shear stress. Suppose that a fluid is confined between two planes (which are essentially infinite in size, so that the effects occurring at the edges of planes may be ignored), one of which is moving with a constant velocity, v, and the other is stationary. The layer of fluid molecules immediately adjacent to the stationary plane will also be stationary; similarly, the layer of molecules immediately adjacent to the moving plane will be moving with the velocity of the plane. Between these two layers of molecules, other molecules will be moving with velocities intermediate between 0 (*i.e.*, the velocity of the stationary plane) and v. This situation establishes a velocity gradient between the two planes, and which is therefore perpendicular to the direction in which the fluid is moving. If we assume that the direction of flow is the x direction in a coordinate system and that the velocity gradient is in the y direction, the force which is required to move one plane of fluid molecules relative to the next higher or lower one (in the y direction) can be written as

$$F = \eta \, (dv_x / dy)$$

The unit of viscosity is the Pascal second, Pa•s. The Pascal is equal to 1 N/m². Suppose that a force of 1 N (1 kg•m / s²) is needed to produce a velocity, v_x, of 1 m/s in a plane of area 1 m² relative to a second plane 1 m away in the y direction. In that case the viscosity will be 1 Pa•s.

Viscosity is an important physical property of petroleum and its derivatives. The viscosity (among other properties) relates to such operations as pumping a crude oil from the well, the flow of fuels through fuel lines, and the lubricating properties of oils. For the alkanes, viscosity increases dramatically with increasing chain length, as shown in Figure 5.2

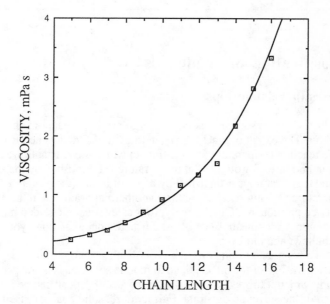

Fig. 5.2 Variation of viscosity with length of carbon chain for n-alkanes.

If we consider viscosity as a measure of the ability of molecules to flow past each other, viscosity will increase with chain length for two reasons. First, as chain length increases, the surface area of the molecules also increases. Increased surface areas provide increased London forces, which in turn result in increased intermolecular attractions. Second, as the length of the carbon chains increases, so do opportunities for the chains to become entangled with one another. As entanglement of chains increases, so too does the difficulty of moving the molecules past each other. As a crude example, consider the difficulty of pushing cooked spaghetti through a pipe.

An illustration of the relationship among London forces, volatility, and viscosity is provided by the data in Table 5.6.

Table 5.6 Comparative boiling points and viscosities for hydrocarbons with six carbon atoms

Compound	Boiling point	Viscosity
Cyclohexane	80°	1.02
Hexane	69°	0.33
2-Methylpentane	60°	0.30

Classification of crude oils

Age-depth relationships

Oil liberated from source rock can undergo cracking reactions in the reservoir. The extent of cracking is determined - as in any chemical process - by the reaction temperature and the time to which the reactants are exposed to that temperature. For oil, time at temperature is established by the age of the oil, and temperature is established by depth of burial (the elevated temperatures arising from the natural geothermal gradients in the earth, as discussed previously). Qualitatively, four categories of age-depth (i.e., time-temperature) relationships can be established: young-shallow, young-deep, old-shallow, and old-deep.

To begin classifying oils in an age-depth system, we must begin by assessing the age of the oil. Coal, an immobile solid, stays in the geologic strata in which it formed, so that the age of a coal deposit can be inferred from the age of the surrounding strata. Furthermore, solid coal can itself retain fossilized plant remains which provide useful information for determining the age of the coal. Because oil often migrates away from the source rock, the age of the surrounding reservoir rocks may not necessarily be the age of the oil trapped in the reservoir. A liquid cannot retain the imprints of fossils. Therefore we need to consider a way of dating oil based on the chemical nature of the oil, rather than evidence provided by surrounding rocks or fossils.

Fatty acids are components of fats, oils and waxes, which are major constituents of lipids. In most organisms the biosynthesis of fatty acids begins with an acetic acid molecule and increases the chain length by two carbon atoms at a time. Therefore most naturally occurring fatty acids contain an even number of carbon atoms in their chains. Consider the thermal decarboxylation of a fatty acid, as in the case of the eighteen-carbon acid stearic acid:

$$CH_3(CH_2)_{15}CH_2COOH \rightarrow CH_3(CH_2)_{15}CH_3 + CO_2$$

The loss of one carbon atom as carbon dioxide shortens the carbon chain, so that the paraffins produced directly from fatty acids have one less carbon atom than the parent acids. If fatty acids with an even number of carbon atoms predominate in lipids, the alkanes made from them will predominantly have an odd number of carbon atoms. As thermal cracking begins, C-C bond cleavage will occur at random positions along the chain. The radical fragments produced are as likely to have an even number of carbon atoms as an odd number of carbon atoms. As these radicals become stabilized by hydrogen capping, recombination, or other reactions, there is gradually a "scrambling" of the chain lengths, so that the predominance of odd-number carbon chains relative to even-number carbon chains is diminished.

The *Carbon Preference Index* (CPI) is a method of indicating the predominance of odd-number carbon chains among the alkanes in a sample of oil. The CPI is calculated from

$$CPI = (1/2) \cdot [(O/E_1) + (O/E_2)]$$

where O is the sum of the concentrations of all alkanes having *odd numbers* of carbon atoms, in the range between C_{17} and C_{31}; E_1 is the sum of concentrations of all alkanes having *even numbers* of carbon atoms in the range between C_{16} and C_{30}; and E_2 is defined similarly, except that the range over which the even-carbon-number alkane concentrations are summed is C_{18} - C_{32}. Since the odd-number chains decrease in prominence as age increases, the CPI provides an indication of the age of the oil. Hydrocarbons extracted from recently deposited sediments have CPI values in the range of 4 to 5, whereas hydrocarbons from ancient bitumens have CPI values of about 1. The quantitative determination of n-alkanes is easily performed by by modern instrumental analytical methods such as gas chromatography; calculation of the CPI from the analytical data then provides an estimate of the age of the oil.

Young-shallow oils
At the top of the oil window some biochemical action is still occurring. Anaerobic bacteria can generate hydrogen sulfide by such reactions as

$$C_6H_{12}O_6 + 3 SO_4^{-2} \rightarrow 3 H_2S + 6 HCO_3^-$$

Therefore we expect young-shallow crudes to be sour. In addition, sulfur can be incorporated into a variety of organic sulfur compounds, such as mercaptans, thiophenes, and sulfides. A young-shallow crude is also likely to be a high-sulfur crude. (Strictly speaking, there is a distinction between the sweet/sour description, which is an indication of the presence of absence of *hydrogen sulfide*, and the high/low sulfur description, which refers to the presence or absence of *organosulfur compounds*. In some of the literature these two descriptors are used interchangeably.)

The shallow depth indicates that the oil has experienced relatively low temperatures, and the young age indicates that little time (on a geological scale) has been available for reactions to occur. Therefore the alkane molecules in the oil will still be relatively large. Because of the large size of these molecules, we expect the oil to be viscous, to have a high boiling range, and to have a low API gravity.

When cracking occurs, the decomposition of long chain alkanes to shorter chain alkanes cannot be the only process occurring. For example, consider the hypothetical reaction of eicosane cracking to tetradecane and hexane:

$$C_{20}H_{42} \text{ -X-> } C_{14}H_{30} + C_6H_{14}$$

As written, this equation is not balanced; there are two "extra" hydrogen atoms on the right side of the equation. The conversion of large alkanes into smaller alkane molecules requires some source of hydrogen. The hydrogen is derived by the concurrent formation of some aromatic compounds as cracking proceeds. We have seen previously that the formation of compounds with several double bonds via hydrogen abstraction can lead to the formation of aromatics, as in the cyclization of 2,4,6-octatriene to o-xylene:

To sum up, young-shallow crudes are often sour and are characterized by CPI values of ~4, low API gravity, high aromatic content, high boiling range, high viscosity, and high sulfur content. A high viscosity means the oil could be difficult to pump. The high boiling range indicates that the amount of low-boiling, valuable products, such as gasoline, will be low. A sour, high sulfur crude will require removal of sulfur. In short, young-shallow crudes are not particularly desirable feedstocks for the petroleum industry,

Young-deep oils

An oil buried deeply will be exposed to higher temperatures from the thermal gradients in the earth. With higher temperatures, more cracking reactions are likely to occur. Since cracking shortens carbon chains, a young-deep crude will have a lower API gravity, a lower viscosity, and a lower boiling range than will a young-shallow crude. Cracking also decreases the predominance of odd number carbon chains, and as a result the CPI of a young-deep crude will be lower than that of a young-shallow crude.

During cracking C-C bonds tend to break in preference to C-H bonds because the former are weaker bonds (~350 kJ/mol compared to ~410 kJ/mol, respectively). C-S bonds are even weaker than C-C bonds (~270 kJ/mol vs. 350 kJ/mol), so that as cracking proceeds the sulfur content of the oil decreases.

As cracking proceeds there is also an increased need for hydrogen to supply the radical capping processes. The hydrogen is obtained from further aromatization of unsaturated molecules, with the production of increasingly larger aromatic molecules. As large aromatic molecules are formed, it is increasingly difficult for such species to remain in solution in the mixture of alkanes and cycloalkanes. The large aromatic molecules precipitate from the oil as a separate solid phase, the *asphaltenes*. Therefore the amount of aromatic molecules in the liquid phase decreases.

Young, deep crudes are formed in the middle or bottom of the oil window. They are generally out of the region of biochemical H_2S production, and therefore are not as sour as the young-shallow crudes. However, some organic sulfur compounds may be present in the oil.

Young-deep crudes tend to have moderate values of API gravity, boiling range, viscosity, and content of aromatic compounds and sulfur.

Old-shallow oils

In many geological processes there is a rough equivalence of time and temperature, in the sense that a long time at relatively low temperatures can eventually accomplish the same transformations as a shorter time at higher temperatures. In many respects an old-shallow crude has properties which are similar to those of a young-deep crude. Old-shallow crudes are more likely to be sour than young-deep crudes.

Old-deep oils

An old-deep oil has experienced the longest times at the highest temperatures. The cracking reactions have had the longest time to proceed, and the oil has cracked about as far as possible without its being completely converted to wet gas. Mercaptans and other sulfur compounds will crack almost completely to H_2S. Because of the diminished solubility of gases in liquids at high temperatures, the H_2S is driven out. The hydrogen necessary for the generation of H_2S from mercaptans or other functional groups comes from

103

internal redistribution of hydrogen in the oil. Old-deep crudes tend to be sweet crudes and have the lowest sulfur content.

Since cracking produces short alkane molecules, old-deep crudes have high API gravity, low viscosity, and a low boiling range. Old-deep crudes also have low contents of aromatic molecules. The CPI of an old-deep crude is about 1. A low boiling range is indicative of the potential of obtaining a high yield of desirable products such as gasoline. The low viscosity provides easy handling of the liquid. A low sulfur content minimizes or eliminates operations necessary for sulfur removal. As a result, old-deep crudes are highly desirable feedstocks for the oil industry. The standard crude oil, by which the quality of others is judged, is Pennsylvania crude. A good Pennsylvania crude has an API gravity of 45 -50°, an extremely low sulfur content (about 0.07%), a low viscosity, and large amounts of alkanes.

Unfortunately, by 1980 the amount of Pennsylvania crude and other crude oils of this quality amounted to about 2% of the world's oil supply. As a result, the oil industry must deal with crudes of poorer quality, and, as the total oil supply dwindles, must also consider the prospect of producing synthetic petroleum-like fuels from other hydrocarbon sources. Processes for treating crude oils and producing synthetic liquid fuels will be a major topic of Unit 3.

Composition relationships

The age-depth system represents an approach to classifying crude oils from a geological perspective. An alternative approach is to use a system which classifies oils on the basis of their composition. Such an approach derives from the perspective of chemistry or chemical engineering, rather than from geochemistry, and represents a greater focus on the end use of the fuel rather than on its origins.

Classification of crude oil in terms of composition is based on the amounts of alkanes, cycloalkanes, and aromatic compounds in the oil. Using the terminology of petroleum technology, these are the paraffins, naphthenes, and aromatics, respectively. The *aromatic* NSO's are lumped with the aromatic hydrocarbon compounds for purposes of this classification.

The classification system is based on the ternary diagram shown in Figure 5.3. Six classes of crude oils are defined on the basis of this classification system. *Paraffinic* crudes are those oils having a total of paraffins plus naphthenes greater than 50%, the paraffins greater than the naphthenes, and paraffins greater than 40%. *Naphthenic* crudes also have the total paraffins plus naphthenes greater than 50%, but in the naphthenic crudes the naphthenes are greater than 40%, and the naphthenes are greater than paraffins. In *paraffinic-naphthenic* crudes the total paraffins and naphthenes are again greater than 50%, but neither the paraffins nor the naphthenes

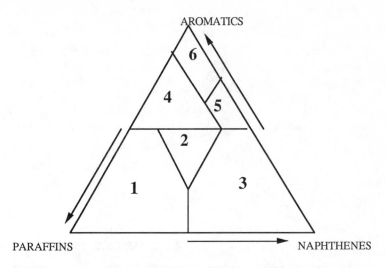

Fig. 5.3. Ternary classification of crude oils based on composition. The principal classifications are **1**, paraffinic; **2**, paraffinic-naphthenic; **3**, naphthenic; **4**, aromatic-intermediate; **5**, aromatic-naphthenic; and **6**, aromatic-asphaltic.

exceed 40% each. *Aromatic-naphthenic* crudes have an aromatic content exceeding 50%, naphthenes greater than 25%, and paraffins greater than 10%. In all four of these classes of oil the sulfur content is generally less than 1%.*Aromatic-intermediate* crudes have aromatic contents over 50% and paraffin contents over 10%. *Aromatic-asphaltic* crudes also have greater than 50% aromatics, paraffins less than 10%, and naphthenes are less than 25%. The aromatic-intermediate and the aromatic-asphaltic crudes usually have sulfur contents greater than 1%. The limits of composition of these six classes of crude oil can be derived from the positions of the boundaries in Fig. 5.3.

As a rule the relative proportions of paraffins, naphthenes, and aromatics vary from one crude oil to another. This is true even for oils found in the same geographic area. Although the axes in the ternary diagram of Fig. 5.3 are scaled from 0 - 100%, implying the existence of any possible composition, in fact most of the crude oils in the world have compositions clustering in a relatively narrow band as shown in Figure 5.4.

In general, young-shallow crudes tend to have moderately high aromatic contents, while old-deep crudes tend to be paraffinic. The young-deep and old-shallow crudes are intermediate between these extremes. Thus there is a relationship between the age-depth classification and position on the paraffin-naphthene-aromatic ternary diagram, as shown in Figure 5.5.

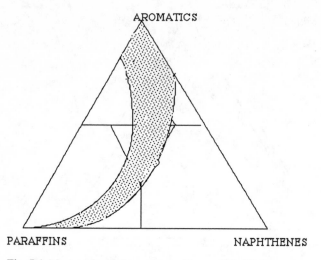

Fig. 5.4. Most crude oils have compositions which fall in the region defined by the shaded band.

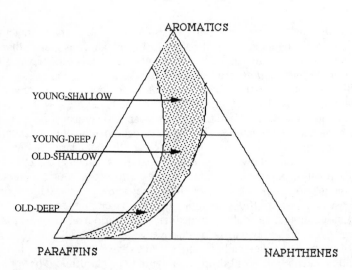

Fig. 5.5. The relationship between age-depth classification and composition of crude oils.

We will see in some detail in Unit 3 how crude oil is separated into useful products on the basis of boiling ranges. For example, gasolines typically have boiling ranges of 25 - 150°, kerosenes of 205 - 260°, and diesel fuels 260 -

315°. As increased cracking occurs smaller molecules, having lower boiling points, are produced. Therefore the extent to which any of the crude oil fractions (that is, gasoline, kerosene, etc.) is the major product of distillation of an oil will depend on the composition of the oil, particularly the sizes of the molecules being distilled. The age and depth of burial help to establish the range of molecular sizes, which in turn helps to establish the boiling ranges and therefore the useful products obtained on distillation. This *rough* relationship is illustrated in Figure 5.6.

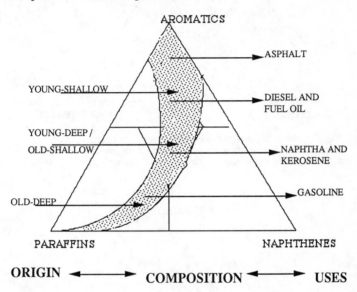

Fig. 5.6 The relationship between the fuel origin, as indicated by the age-depth classification, fuel composition, and the end uses of petroleum.

Fig. 5.6 illustrates the three major foundations of fuel science - how the origin of a fuel establishes its composition and properties, and how in turn the composition and properties of the fuel relate to its use.

As cracking proceeds sulfur content is reduced and API gravity increases. Thus we find relationships of the types illustrated in Figures 5.7 and 5.8.

The similarity of trends shown in Figs. 5.7 and 5.8 for sulfur and API gravity suggests that there must also be a relationship between these two fuel properties. Such a relationship is illustrated in Fig. 5.9.

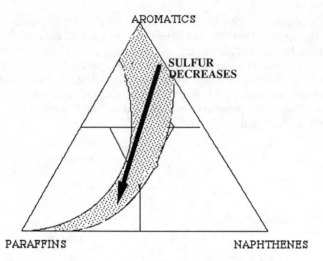

Fig. 5.7 The sulfur content of crude oils tends to decrease in a progression through the composition band from aromatic-asphaltic and aromatic-naphthenic oils to the paraffininc oils.

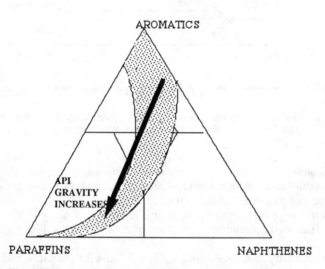

Fig. 5.8 The API gravity increases (*i.e.*, density decreases) on progressing along the composition band from aromatic to paraffinic crudes.

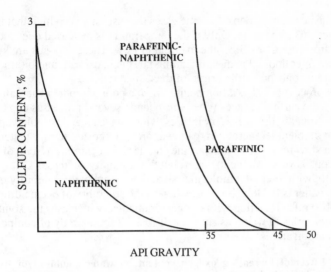

Fig. 5.9 Relationships among the sulfur content, hydrocarbon composition, and API gravity for various crude oils. As a rule, the sulfur content increases as API gravity decreases.

The trends shown in Fig. 5.9 should not be surprising, based on the relationships between composition and density (and, by extension, API gravity) developed previously. We have seen how the introduction of sulfur into a molecule increases the density relative to structurally analogous hydrocarbon compounds. Furthermore, the densities of alkanes (paraffins) are generally lower than those of the corresponding cycloalkanes (naphthenes). Thus we should expect an inverse relationship between sulfur content and API gravity, and that for a given sulfur content the gravity should increase in the order naphthenic < paraffinic-naphthenic < paraffinic.

In the past the highly aromatic crudes were valued as a source of aromatic compounds for chemicals production. For example, some crude oils from Borneo contain 40% aromatics. During World War I the Borneo crudes were in great demand as a source of toluene, which is the parent compound of the high explosive 2,4,6-trinitrotoluene (TNT).

Asphalts and tar sands

To complete the discussion of petroleum, we will focus briefly on the materials found at the very top of the ternary classification diagram of Fig. 5.3. There are, in many places around the world, deposits of extremely viscous, plastic hydrocarbons. In some cases the viscosity is so high that the material appears to be solid. Only if the temperature is increased does the viscosity drop enough to allow the material to flow. These deposits are known by a variety of colloquial names, including tar, pitch, asphalt, and bitumen. There are two concepts of the origin of these materials.

The process of *inspissation* begins when a pool of crude oil actually migrates to the earth's surface. When this occurs, several processes can then take place. Small hydrocarbon molecules, which have an appreciable vapor pressure at ambient temperatures, can evaporate. Exposure to rain water results in a slow removal of some of the smaller hydrocarbons. (None of the hydrocarbons is very soluble in water, but the solubility decreases significantly with increased molecular size. For example, the solubility of pentane in water is 0.036 g/100 mL, whereas the solubility of octane drops to 0.0015 g/100 mL.) Reaction with oxygen dramatically increases the solubility of small molecules in water, as shown in Table 5.7 for a family of compounds with five carbon atoms.

Table 5.7 Effect of increasing oxygen content on water solubility of five-carbon-atom hydrocarbons

Compound	Formula	Solubility, g/100 mL
Pentane	$CH_3CH_2CH_2CH_2CH_3$	0.036
1-Pentanol	$CH_3CH_2CH_2CH_2CH_2OH$	2.7
Pentanoic acid	$CH_3CH_2CH_2CH_2COOH$	3.7
Glutaric acid	$HOOCCH_2CH_2CH_2COOH$	64

Compared to the parent hydrocarbon pentane, the oxidized compound glutaric acid has a solubility in water almost 2000 times greater. As the interaction with atmospheric oxygen proceeds, the solubility of small, oxidized molecules in rain water or streams increases greatly. Larger molecules also incorporate some oxygen, but the increase in water solubility is not nearly so dramatic as in the case of the small compounds. For example, octane has, as noted, a solubility of 0.0015 g/100 mL in water. 1,8-Octanedioic acid (suberic acid) is soluble to the extent of 0.14 g/100 mL, and increase of only 100 times, rather than the factor of 2000 increase for the five-carbon-atom compounds. Therefore some oxidized species derived from larger hydrocarbons are likely to remain in the deposit. The net effect of inspissation is that we are left with a material containing relatively large molecules, partially oxidized, with very low volatility, low API gravity, and a high viscosity.

Inspissation of paraffinic crudes leads to plastic, waxy materials of a yellowish to brown color, known as *ozokerite*. Typical ozokerites contain about 81% paraffins plus naphthenes, 10% aromatics, and 9% NSO's. The solid, waxy material contains hydrocarbons in the C_{22} to C_{29} range. The composition is 84-86% carbon, 14-16% hydrogen, and small amounts of the heteroatoms.

Inspissation of naphthenic crudes leads to the formation of *asphalts*, soft, sticky materials which are semi-solid. Typical naturally occurring asphalts have 18% paraffins plus naphthenes, 48% aromatics, and 34% NSO's. The elemental composition is in the range 80-87% carbon, 9-13% hydrogen with small amounts of nitrogen and oxygen but, in extreme cases, up to 8% sulfur. In comparison with the ozokerites, asphalts may have 18% paraffins and naphthenes, 48% aromatics, and 34% NSO's. Asphalts melt in the range 65 - 95°. The API gravity is -8° or lower. Further natural decomposition of asphalts leads first to the formation of *asphaltites* and then to *asphaltic pyrobitumens*. The API gravity of asphaltites is less than -14° and that of pyrobitumens, less than -18°. Asphaltites are harder than asphalts and melt at higher temperatures, typically 120-320°. An important naturally occurring asphaltite is *gilsonite*. The commercial uses of gilsonite include the manufacture of waterproof coatings, road paving, and wire insulation. Gilsonite is highly aromatic, while ozokerite is highly aliphatic. As a rule, the elemental composition of these materials is close to that of the more familiar liquid petroleum, with the amounts of carbon and hydrogen being slightly lower than the liquid petroleum, and the amounts of the heteroatoms being slightly higher. The pyrobitumens do not melt on heating, but rather swell and then decompose. Ozokerite has the highest H/C ratio among these materials, while albertite, a pyrobitumen, has the lowest H/C ratio, nearly as low as some coals.

The second view of the origin of these heavy hydrocarbons is that the deposits were not inspissated; rather, they are materials that never experienced extensive cracking. In a sense, they represent "very young - very shallow" materials. Some asphaltic materials contain porphyrin complexes derived from the organisms which produced the accumulated organic matter. The preservation of these materials suggests that the sediment has not experienced severe reaction conditions. Regardless of whether the inspissation or very young - very shallow origin is correct, it is important to recognize that the amount of free liquid hydrocarbons (*i.e.,* petroleum) increases with depth in the earth, at least to the bottom of the oil window, whereas the amount of asphaltic and similar materials, and the amount of unconverted organic matter, is inversely proportional to depth. Increasing depth of burial, or increasing age, is reflected in increasing conversion to liquid petroleum.

Some very highly asphaltic materials occur in nature. They are extremely viscous, very dense (API gravities of 5 - 15°) materials associated with highly porous sands, and generally referred to as *tar sands* or oil sands. Some tar sand deposits may have formed from inspissation, and others may be the very young - very shallow deposits. The highly paraffinic, high sulfur tar sands are probably very young - very shallow deposits, while the highly aromatic, low sulfur materials are highly degraded, very old oils. The sulfur content of tar sands is highly variable, from 0.5 to 8%. An example is the La Brea deposit in Trinidad, in which the tar sand is 8% sulfur and has an API gravity of 1°.

The interest in tar sands derives from their potential as hydrocarbon sources equivalent to enormous deposits of oil. The La Brea deposit, for example, covers only 125 acres, yet contains hydrocarbons equivalent to 60 million barrels of oil. Heavy oils can be extracted from tar sands provided that we can get the sand away from the oil, or *vice versa*. The methods of extraction take advantage of the property of liquids that viscosity decreases with increasing temperature. An example of this property is provided by eicosane, which has a melting point of 38°. Normally eicosane is a waxy solid which would melt only on exceptionally hot days. However, by increasing the temperature to 150° eicosane becomes a liquid with a viscosity of 0.8 mPa•s, a value which is lower than that of water at room temperature. By treating tar sands with superheated steam, or raising the temperature by partial combustion of some of the hydrocarbons *in situ*, the organic components become mobile enough to be separated from the associated sands.

Further reading

Boggs, S., Jr. (1987). *Principles of Sedimentology and Stratigraphy*. Columbus, OH: Merrill; Chapter 8

Fieser, L. F., and Fieser, M. (1961). *Advanced Organic Chemistry*. New York: Reinhold; Chapter 4.

Haun, J. D. (ed.) (1976). *Origin of Petroleum. I*. Tulsa: American Association of Petroleum Geologists; Chapters 1,7,11.

Haun, J. D. (ed.) (1974). *Origin of Petroleum II*. Tulsa: American Association of Petroleum Geologists; Chapter 2.

Loudon, G. M. (1988). *Organic Chemistry,* 2d edn. Menlo Park, CA: Benjamin/Cummings; Chapter 7.

Selley, R. C. (1985). *Elements of Petroleum Geology*. San Francisco: Freeman; Chapters 1,9.

Chapter 6

Coal

This chapter continues the theme established in Chapters 4 and 5, presenting a discussion of the composition, structure, and physical properties of coals, and how these properties are used to help classify coals. The consideration of coals adds another level of complexity compared to petroleum. Coal does not have a unique molecular structure, but rather has a macromolecular structure of varying composition. Because coal is a solid, some water and mineral components are associated with the carbonaceous material as it is extracted from the earth. In addition, because coal is a solid, it is important to investigate some of the properties which are unique to it by virtue of being a solid, properties such as porosity and surface area.

Coal compared with gas and oil

We have seen that natural gas may contain various amounts of impurities, which led to classifications of gas as sweet or sour and wet or dry. However, once natural gas has been purified, over 90% of the gas is a single chemical compound, methane, of which the structure and properties are well known. Since all gases are infinitely miscible, natural gas is a completely homogeneous fluid. It contains no inorganic impurities that might leave a combustion residue of ash.

Petroleum is a homogeneous liquid. In terms of elemental composition, petroleum has a fairly narrow range, about 82-87% carbon, 12-15% hydrogen, and the balance nitrogen, sulfur, and oxygen. The H/C ratio is about 2. On a molecular level, petroleum consists of hundreds to thousands of individual compounds in four classes: paraffins, aromatics, naphthenes, and NSO's. Although it would serve no practical purpose, one could in principle separate and identify every one of the constituents of petroleum, and assign a unique molecular structure to each. Inorganic impurities in petroleum that would leave an ash on combustion are typically less than 0.1% and may be in the range of 0.01%

Petroleum contains an enormous number of individual chemical compounds, which have come from a variety of sources. For example, chlorophyll may have decomposed to pristane and phytane, which in turn crack to short, branched-chain alkanes; proteins decompose to heterocompounds; and plant waxes yield long chain alkanes. Despite this diversity, these compounds are usually mutually soluble, so that a crude oil is fairly or completely homogeneous. Gases are infinitely miscible, so that even a very sour, very wet gas is homogeneous. Since coal is a solid, the coalified plant residues can not mix with each other, and consequently a piece of coal will contain distinct regions representing different coalified plant residues.

In contrast to petroleum and natural gas, coals have an extremely wide range of composition: 65-95% carbon, 2-6% hydrogen (with even higher hydrogen contents in the sapropelic coals), 2-30% oxygen, 1-13% sulfur, and up to about 2% nitrogen. The chemical components of coal cannot be separated by classical laboratory procedures. Coal cannot be distilled reversibly and is not completely soluble in any solvent. Coal is generally believed to have a macromolecular structure, but the exact structure is not known and varies from one coal to another. Coal is distinctly heterogeneous because, as a solid, there is no way for the individual components to mix with one another. Coal contains a variable, but appreciable, amount of inorganic material that forms ash during combustion. The amount of this material ranges from about 1 to over 25%. In addition, coal as mined also contains a variable but appreciable amount of water. The as-mined moisture content of coal ranges from about 3% to nearly 70%.

The heterogeneity of coal

We begin by considering the van Krevelen diagram of Figure 6.1. When kerogen transforms to gas or oil, the products are homogeneous fluids. On a macroscopic or microscopic level there is no way to observe a distinction in the appearance of the fuel based on the original kerogen source material. Because there is no physical mixing of the components of coal, for virtually any coal (except the anthracites) it is possible to discern microscopic regimes in the coal which are clearly related to the original sources. A common rock, such as granite, provides a useful analogy. When a piece of granite is examined, it is possible to see with the microscope - and sometimes with the unaided eye - the individual mineral grains which make up the rock. In the same way, when coal is examined under a microscope it is possible to see"grains" of individual components. By analogy with granite, we can consider coal to be an organic rock. In an inorganic rock, the individual .

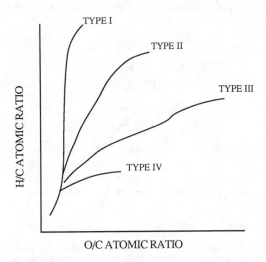

Fig. 6.1 A summary van Krevelen diagram showing the transformations of kerogen; for greater detail, refer to Figure 3.12

components are the minerals. In the organic rock coal the individual components are called *macerals*.

Macerals are described and classified on the basis of their optical appearance and evident relationship to botanical structures. The study of coal macerals is the field of *coal petrography*. The foundations of the discipline of coal petrography were established in 1921 by the British scientist Marie Stopes. The importance of petrography in coal research and utilization is this: The different macerals come from different components of the original plant material which eventually formed the coal. Different plant components have different molecular structures (for example, compare lignin and a wax). Substances having different molecular structures undergo different kinds of chemical reactions at a given set of conditions or participate in a given chemical reaction in different ways. Although plant components are altered chemically during coalification, the macerals should still reflect some of the chemical differences inherent in the original plant components. Consequently, we expect the various macerals to show differences in their chemical behavior. Thus by knowing the relative proportions of the different macerals in a coal sample it should be possible to predict something about the chemical behavior and reactivity of that sample.

The successful application of coal petrography to assess their suitability for utilization is a major example of the three cornered relationship of the foundations of fuel science:

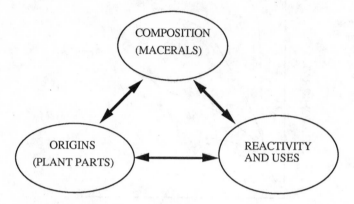

Fig. 6.2 The "fuel science triad" illustrating the relationships among the origins of the components of coal, the petrographic composition, and reactivity or use.

The individual macerals in coal are classified into maceral groups. The properties of the maceral groups are summarized in Table 6.1.

Table 6.1 Properties of the maceral groups

Maceral group	Appearance	Composition	Plant origins
Vitrinite	Shiny, glass-like	Oxygen rich Moderate hydrogen Moderately aromatic	Wood, bark
Liptinite	Waxy or resinous	Hydrogen rich Highly aliphatic	Algae, plant debris: such as spores, leaf cuticles, resins, pollen
Inertinite	Dull, charcoal-like	Carbon rich Highly aromatic	Degraded, carbonized wood, bark, and cell protoplasm

Each maceral group includes one or more individual macerals. However, for purposes of this book the macerals will almost always be referred to by group names, such as the vitrinites. Individual macerals will be discussed on a case-by-case basis as the need arises.

In most of the coals of the northern hemisphere, the vitrinites are the predominant macerals. Vitrinites are, in many reactions of coals, the most

116

reactive macerals. For these reasons many fundamental studies of coal composition and reactivity focus on vitrinite. Liptinites (sometimes also called exinites) form from plant debris such as resin, spores or pollen, leaf cuticles, or algae. The principal liptinite macerals are resinite, sporinite, alginite, and cutinite, the names deriving from the plant structures from which the maceral formed. Some coals in the United States (Utah) contain large blocks of resinite which can be picked by hand from the coal. Cutinite is rare except in the unusual "paper coals" of the United States (Indiana) and the Soviet Union, which contain much cutinite. Alginite is also rare, except in the torbanites of Scotland and the cooronginte from Tasmania, which are largely algal coals. Inertinites appear to have derived from the oxidation or extensive degradation of wood or bark tissue. Inertinite macerals include fusinite and semifusinite, which form from oxidized wood. The fusinite in coals of the northern hemisphere appears to be a fossil charcoal. Macrinite and micrinite are thought to have arisen from cell protoplasm and sclerotinite arises from fungal attack on wood.

As an alternative to Table 5.1, macerals can also be classified by botanic origin. Macerals arising from wood or bark are vitrinite, fusinite, and semifusinite; macerals deriving from specialized plant parts include liptinites and scelrotinite; and those deriving from cell contents are macrinite and micrinite. Humic coals may contain any of these macerals in varying proportions. Of the sapropelic coals, boghead coals are predominantly alginite, and the cannel coals are mainly sporinite and micrinite.

The classification of coal by rank

In the discussion of coalification, the names of various kinds of coals were introduced informally as a progression along the van Krevelen diagram, following an increase in carbon content (Fig. 6.3). This progression provides an approach to classifying coals such that each class implies a higher carbon content than the preceding, e.g. bituminous coals have a greater carbon content than subbituminous coals. Consequently this provides one approach to ranking coals; that is, we might say that a bituminous coal is of higher rank (higher carbon content) than a subbituminous coal.

In the United States the formal system for ranking coals is not based on carbon content or on the position on the van Krevelen diagram, but rather on the behavior of coal as a fuel. There are two considerations for evaluating coal as a fuel. How does the coal behave during combustion? How much heat is liberated when the coal is burned?

The *calorific value* is measured by burning a weighed sample of coal in a high pressure oxygen atmosphere and measuring the heat liberated into a surrounding water jacket. In the United States the calorific value is expressed in British units of Btu/lb.

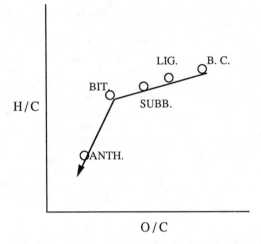

Fig. 6.3 A van Krevelen diagram showing changes in composition with coalification: BC = brown coal; LIG = lignite; SUBB = subbituminous coal; BIT = bituminous coal; and ANTH = anthracite.

When coal is heated in an inert atmosphere to about 105°, a weight loss occurs. The weight loss a result of water being driven off; the weight loss is used to calculate the *moisture content* of the sample. If the inert atmosphere is maintained but the temperature is increased substantially, to 950°, a second weight loss is observed. Under these conditions a variety of materials, including carbon dioxide, carbon monoxide, and an assortment of hydrocarbons, is evolved. The compounds emitted during this experiment are never determined individually. Rather, they are lumped together under the term *volatile matter*. The weight loss observed at 950° thus provides a measure of the volatile matter associated with the coal sample. At the end of the volatile matter test, a black, carbonaceous solid still remains. It contains carbon which was not emitted during the volatile matter determination. In the early days of chemistry substances which did not readily undergo changes of state were referred to as "fixed." If the material remaining from the volatile matter test is heated in air it burns, leaving behind an incombustible inorganic residue, *ash*. The ash is collected and weighed. Three components - moisture, volatile matter, and ash - are determined directly. The *fixed carbon* is calculated indirectly as

$$\%FC = 100 - (\%M + \%VM + \%A)$$

118

In none of these analyses are the actual components of the volatile matter, ash, or fixed carbon determined. In other words, the carbon dioxide evolved in the volatile matter determination is never collected, determined, and reported as some percentage of CO_2. In analytical chemistry the practice of lumping a variety of components and reporting them as a single entity is called *proximate analysis*. The proximate analysis of coal is therefore the determination of moisture, volatile matter, fixed carbon, and ash. The name proximate is unfortunate in a sense because of its sounding much like the word approximate. A proximate analysis *is not* approximate analysis! The procedures for the proximate analysis are rigorously established by the American Society for Testing and Materials (widely known by its initials, ASTM), along with standards for the acceptable levels of error within one laboratory and between different laboratories.

The portion of coal that actually burns during combustion is the volatile matter and the fixed carbon. The moisture and ash vary widely from coal to coal, the moisture of United States coals used commercially being <5 to >40% and ash <5 to ~28%. Because moisture and ash do not contribute to the combustion, and because they vary widely, the behavior of coals is often better compared in terms only of the amounts of volatile matter and fixed carbon, without the influence of moisture or ash. To make this conversion, proximate analysis or calorific value data can be converted to a moisture-free or moisture-and-ash-free basis by straightforward algebra. To convert a property X (volatile matter, fixed carbon, or ash) to from the as-received value determined in the laboratory to a moisture-free basis

$$\%X_{mf} = 100 \cdot \%X_{ar} / (100 - \%M_{ar})$$

The conversion to an ash-free basis would be done in an analogous manner. Similarly, the conversion to a moisture-and-ash-free basis is

$$\%X_{maf} = 100 \cdot \%X_{ar} / [100 - (\%M_{ar} + \%A_{ar})]$$

The concern for the values of volatile matter and fixed carbon is that these are the properties that determine the combustion characteristics of the coal. Coals with a high volatile matter are usually easy to ignite; burn with a large, often smoky flame; and burn quickly. Coals with high fixed carbon are hard to ignite, but burn slowly with a short, clean flame. The *fuel ratio* of a coal is defined as the ratio of fixed carbon to volatile matter.

It is important to recognize that, strictly speaking, there is no ash in coal. The incombustible residue, ash, remaining after the combustion of coal, is actually the product of high temperature reactions of inorganic components, termed *mineral matter*, originally present in the coal. As will be explained later in this chapter, the amount of ash is not necessarily equaly to the amount of original mineral matter. Since the isolation and weighing of ash is much easier than obtaining unaltered mineral matter from coal, an algebraic method, introduced later, is usually used to calculate the mineral matter content of a coal from the ash determination. For the most precise work, coals are compared on a mineral-matter-free basis, rather than an ash-free basis. Provided that one has a value for the mineral matter in a coal sample, the algebraic conversion to a mineral-matter-free basis is analogous to the methods just discussed.

The ASTM rank classification of coals is based on the volatile matter and fixed carbon reported on a moisture-and-mineral-matter-free basis, and on the calorific value reported on a moist, mineral-matter-free basis. (Notice that the calorific value is corrected only for mineral matter; the moisture is *not* removed algebraically.) The rank classification is shown in Table 6.2.

Table 6.2 ASTM rank classification of coals

Class	Group	%FC_{mmmf}	%VM_{mmmf}	Calorific value Btu/lb, m,mmf
Anthracitic				
	Metaanthracite	>98	<2	
	Anthracite	92-98	2-8	
	Semianthracite	86-92	8-14	
Bituminous				
	Low volatile	78-86	14-22	
	Medium volatile	69-78	22-31	
	High volatile A	<69	>31	>14,000
	High volatile B			13,000 - 14,000
	High volatile C			10,500 - 13,000
Subbituminous				
	Subbituminous A			10,500 - 11,500
	Subbituminous B			9,500 - 10,500
	Subbituminous C			8,300 - 9,500
Lignitic				
	Lignite A			6,300 - 8,300
	Lignite B			<6,300

Often an informal nomenclature is used which is a combination of class and group names; for example, lignitic coals are often referred to simply as lignites without a distinction between lignite A and lignite B.

The variation of coal composition with rank

The determination of the principal elements in coal - carbon, hydrogen, oxygen, nitrogen, and sulfur - is called the *ultimate analysis*. It is not "ultimate" in the sense of determining completely the elemental composition of coal, because the careful analysis of a coal sample, including the mineral matter, to the trace level would show that coal contains virtually every element in the periodic table except the rare gases and the man-made, highly unstable elements. Throughout this book, ultimate analysis data will always be discussed on a moisture-and-ash-free basis unless explicitly stated otherwise.

The variation of composition with rank

Despite a wide variation in carbon content over the range of coal ranks, the hydrogen content tends to be fairly constant at ~5.5±0.5% (for the humic coals) up to a carbon content of ~88%. The variation of hydrogen content as a function of carbon content is represented as the *Seyler chart* or Seyler coal band (Fig. 6.4).

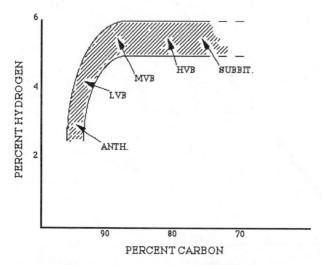

Fig. 6.4 The Seyler coal band. Most of the humic coals in the world have carbon and hydrogen contents which lie in the shaded region of the band. SUBBIT = subbituminous; HVB = high volatile bituminous; MVB = medium volatile bituminous; LVB = low volatile bituminous; ANTH = anthracite.

It should be noted that the abscissa is "backwards" in the sense that the origin is at the point (100% C, 0% H).

The Seyler chart as originally constructed ends with coals of ~75% carbon, because the not data for coals of lower rank were available to Seyler. The dashed lines represent an extrapolation of the original data to the lignitic region. The near-zero slope of the H vs. C band in the region below 88%

carbon indicates that coalification has occurred almost entirely as a result of loss of oxygen. Above 88% carbon, coalification involves loss of both oxygen and hydrogen. The dramatic change of slope at 88% carbon indicates a significant change not only in the reactions occurring during coalification, but also in many properties of the coals. We will see later that when many properties of coal are plotted as a function of carbon content, the data pass through a maximum, minimum, or charge of slope at 88% carbon.

The variation of oxygen content with rank is illustrated in Figure 6.5. (Note again that the abscissa is "backwards".)

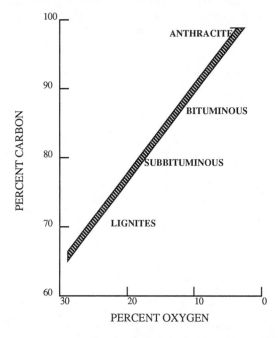

Fig. 6.5 The variation of oxygen content with carbon content of coal. The oxygen content drops steadily as rank increases.

There is very large variation in oxygen content as a function of carbon content, from nearly 30% in the brown coals through ~2% in the anthracites. Since the hydrogen content of coals are reasonably constant through 88% carbon, *i.e.* the medium volatile bituminous range, the increase in carbon content with increasing rank must be accompanied by a decrease in oxygen.

Aside from the fact that, on a weight basis, oxygen is generally the second most important element in coal, the oxygen content of coal has several practical implications. The presence of oxygen detracts from the calorific value. As a rule, as the oxygen content increases and hydrogen decreases, for

122

a given amount of carbon, the calorific value will drop. For example, methane (H/C = 4, O/C =0) has a heat of combustion of 883 kJ/mol, whereas for formaldehyde (H/C = 2, O/C = 1) the heat of combustion has dropped to 548 kJ/mol. Some additional data in support of this is shown later in Table 6.3. Furthermore, the oxygen-containing structures represent functional groups, the sites where chemical reactions can occur. The principal oxygen functional groups in coals are carboxylic acids, phenols, ketones or quinones, and ethers. Other oxygen functional groups of little importance or absent from coals are esters, aliphatic alcohols, aldehydes, and peroxides. The variation of oxygen functional groups as a function of carbon content in vitrinites is shown in Figure 6.6.

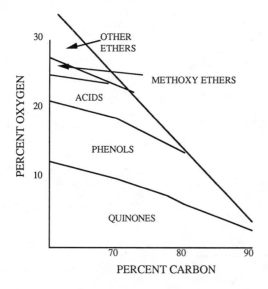

Fig. 6.6 The variation of the principal oxygen functional groups with carbon content in vitrinites.

Methoxy groups are important only in coals of <72% carbon. These groups are derived from lignin, and their loss with increasing coalification suggests that lignin structures have been completely coalified by the time that the coal has reached subbituminous rank. Carboxyl groups are not important for coals of >80% carbon, *i.e.,* at about the subbituminous A - high volatile C bituminous boundary. Phenols and quinones are the main oxygen groups in the high rank coals, although some ethers may persist into high ranks.

123

The variation of calorific value with rank

We have already established the rule-of-thumb that heat of combustion or calorific value decreases as the H/C ratio drops. The calorific value of natural gas is about 56 MJ/kg, and that of many petroleum products is in the range of 46 MJ/kg. If we assume that H/C is 2 for petroleum and is 0.75 for a typical coal, a linear extrapolation allows us to forecast a calorific value of about 44 MJ/kg for coal (Figure 6.7).

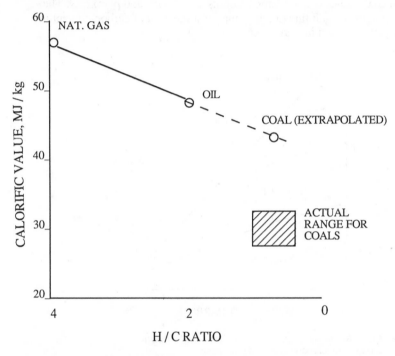

Fig. 6.7 Extrapolation of the calorific values of natural gas and oil, as a function of atomic H/C ratio, suggests that coals should have calorific values of ~45 MJ/kg. In fact, the actual calorific values of coals are much lower, generally in the range indicated by the shaded box.

In fact, a typical coal having 0.75 H/C is likely to have a calorific value in the range of 28 - 32 MJ/kg on a moisture-and-ash-free basis, and even the very best quality coals seldom have calorific values in excess of 36 MJ/kg. (If this example serves no other purpose, it should point out the dangers of extrapolating results outside the range for which data exists.)

Why is there such a discrepancy between the calorific value predicted on the basis of the rule-of-thumb relationship with H/C and the actual value?

To answer this question we must consider two very important ways in which coals differ from natural gas and petroleum. First, most coals contain appreciable amounts of oxygen, nearly 30% in the lowest ranked coal. Second, the macromolecular structure of coal is based mainly on aromatic, rather than aliphatic, carbon structures.

The incorporation of oxygen into an organic molecule means that, in a sense, a portion of the carbon is already oxidized. Therefore the heat liberated when the compound burns is less than when a structurally similar compound containing no oxygen is burned. An extreme example is the case of carbon and carbon monoxide. The reaction

$$C + O_2 \rightarrow CO_2$$

has an enthalpy change of -393 KJ/mol, whereas the reaction

$$CO + 0.5 \, O_2 \rightarrow CO_2$$

has an enthalpy change of only -283 kJ/mol. If we examine the effect of oxygen on the heats of combustion of simple organic molecules we find trends like those of Table 6.3.

Table 6.3 Effect of increasing oxygen content on heat of combustion of four-carbon-atom compounds

Compound	Formula	$-\Delta H$, kJ/mol
Butane	$CH_3CH_2CH_2CH_3$	2880
1-Butanol	$CH_3CH_2CH_2CH_2OH$	2675
2-Butanone	$CH_3CH_2COCH_3$	2436
Butanoic acid	$CH_3CH_2CH_2COOH$	2193
Butanedioic acid	$HOOCCH_2CH_2COOH$	2156

Other factors being kept as comparable as possible, then when the oxygen content increases, the heat of combustion drops.

Why do we believe that the macromolecular structure of coal is based mainly on aromatic carbon? First, the low H/C ratios of coals are typical of aromatic compounds. For example, the H/C ratios of octane and cyclohexane are 2.25 and 2.00, respectively; but for simple aromatic compounds benzene is 1.00; naphthalene, 0.80; and phenanthrene, 0.71. Second, humic coals, which are the predominant kind found throughout the world, derive mainly from Type III kerogen. This kerogen originates from higher land plants, which are distinguished from other types of organisms by the presence of lignin. Recall that the basic carbon framework of lignin is based on phenylpropane structures, which are mainly aromatic.

Two parameters are used to characterize the aromatic structure of coal. The *aromaticity*, f_a, is the fraction of the total amount of carbon incorporated in aromatic structures. The *ring condensation*, R, is the average number of fused aromatic rings in a structure. For example, the f_a of phenylpropane is 0.67 (six of the nine carbons are aromatic); the ring condensation of 2-benzylnaphthalene,

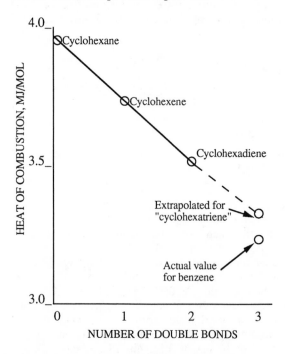

is 1.5 ((1+2) / 2). For coals of a given rank, both f_a and R vary in the order inertinites > vitrinites > liptinites.

To consider the effect of aromatic structure on calorific value, we can first consider the variation of the heats of combustion of the cyclic six-carbon-atom compounds, Figure 6.8.

Fig. 6.8 The variation in heat of combustion of cyclic six-carbon-atom compounds, as a function of the number of double bonds in the molecule.

The discrepancy between the expected heat of combustion for the hypothetical cyclohexatriene molecule and the actual heat of combustion of benzene indicates that benzene is more stable than would be anticipated from the extrapolation of data for cyclohexane, cyclohexene, and cyclohexadiene. This additional stability of benzene is the resonance stabilization energy, which arises from the delocalization of electrons throughout the π bonding system in the benzene molecule. Increasing aromaticity offers an opportunity for increased resonance stabilization, which in turn results in a decrease in the heat of combustion. An example is provided by decahydronaphthalene (decalin), 1,2,3,4-tetrahydronaphthalene (tetralin), and naphthalene:

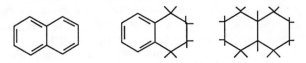

Here the heat of combustion of naphthalene, with f_a of 1.00, is 18% less than that of decahydronaphthalene, with f_a of 0.00.

The variation of the calorific value of coals with rank is expressed in the Mott chart, Figure 6.9

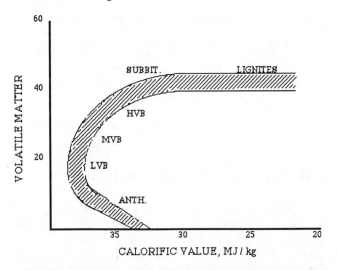

Fig. 6.9 The Mott chart, which illustrates the relationship between calorific value and volatile matter, expressed on a moisture-and-mineral-matter-free basis. Data for most humic coals falls within the shaded band. SUBBIT = subbituminous; HVB = high volatile bituminous; MVB = medium volatile bituminous; LVB = low volatile bituminous; and ANTH = anthracite.

The volatile matter and calorific value are both plotted on a moisture-and-mineral-matter-free basis. The variation expressed in the Mott chart is not monotonic; even though anthracite is the highest rank on the ASTM classification system, it does not have the highest calorific value.

The explanation of the behavior of anthracite relative to the bituminous coals is found in the variation of composition with rank. In the series of coals from lignite to low volatile bituminous, the oxygen content drops from 22-25% to about 4%, while the ring condensation increases from ~1 to ~5. Two contradictory effects are at work: the decreasing oxygen content should raise the calorific value, while the increasing ring condensation should lower the calorific value. Here the very significant drop in oxygen content prevails over the increase in ring condensation. On the other hand, the anthracites have oxygen contents of ~2%, compared with ~4% in low volatile bituminous coal; R however, increases from ~5 to >30. Consequently, the tremendous increase in ring condensation, with its attendant lowering of the calorific value, overwhelms any advantage obtained by the relatively slight drop in oxygen content.

The emphasis on ring condensation in this discussion derives from the indication that the resonance stabilization energy *per carbon atom* increases with increasing ring condensation, as shown in Table 6.4.

Table 6.4 Variation of resonance stabilization energy per carbon atom with ring condensation

Compound	Ring condensation	Stabilization, kJ/mol
Benzene	1	25.1
Naphthalene	2	31.4
Phenanthrene	3	33.1
Triphenylene	4	34.7

The macromolecular structures of coal

Two important features that differentiate coal from petroleum are that coal is a solid, and that the chemistry of coal is based mainly on aromatic, rather than aliphatic structures. Both f_a and R increase as the rank of coal increases, as does the percentage of carbon in the coal. Since graphite is 100% carbon with $f_a = 1.00$ and $R = \infty$, as the rank of coal increases the structure is becoming increasingly graphite-like.

It is generally agreed that coal does not have a unique molecular structure, and indeed that even a single maceral of a particular rank of coal does not have a unique structure. This feature distinguishes coals from polymers, such as polystyrene for example, which have regular, repeating structures; and from biological macromolecules such as DNA, which might have very large and complex structures but structures which, once elucidated, are then known unequivocally. Coal scientists studying the "structure" of coal usually work with so-called *average structures*, which agree with the ultimate analysis and contain the principal carbon structures and functional groups, but which should not be believed to represent *the structure* of coal. Given the considerable power of modern instrumentation for organic structural analysis, it is conceivable that someone could carefully separate a single maceral from a single sample of coal and actually determine the structure. However, aside from being a *tour de force* of organic structural determination, it is questionable how much practical information would be generated from such an endeavor.

Low-rank coals

A consideration of the macromolecular structure of coal, and how the structure varies with rank, must begin with lignin, the principal structural component of woody plants and a major contributor to Type III kerogen. As the rank of coal increases we would expect an evolution of the structure from being lignin-like in the lignites to graphite-like in the anthracites. The structure of lignin is not known with certainty, and may vary from one species of plants to another. However, lignin is considered to be a biopolymer of three monomers:

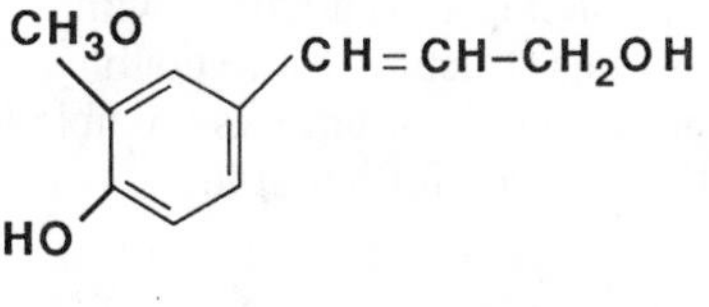

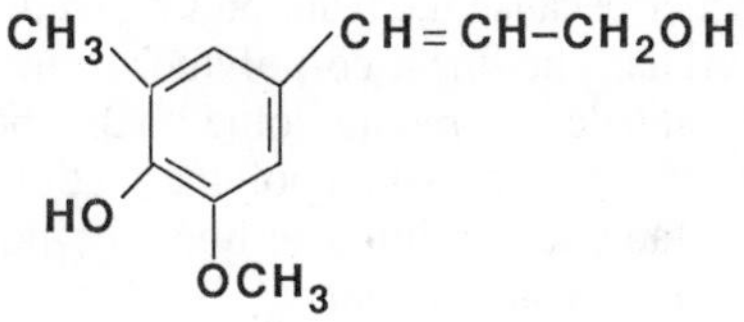

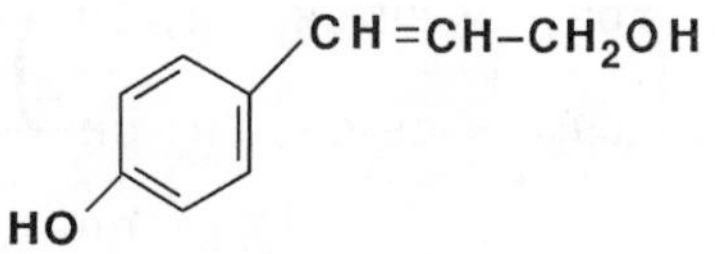

In general, we might expect these monomers to react in one of two ways: polymerization across the double bonds in the side chain

129

$$\underset{\underset{\emptyset}{|}}{\overset{\overset{CH_2OH}{|}}{CH}}=\underset{}{\overset{}{CH}} \quad + \quad \underset{\underset{\emptyset}{|}}{\overset{\overset{CH_2OH}{|}}{CH}}=\underset{}{\overset{}{CH}} \quad \longrightarrow \quad -\underset{\underset{\emptyset}{|}}{\overset{\overset{CH_2OH}{|}}{CH}}-\underset{}{\overset{}{CH}}-\underset{\underset{\emptyset}{|}}{\overset{\overset{CH_2OH}{|}}{CH}}-\underset{}{\overset{}{CH}}-$$

or by addition of the alcohol group to the double bond

$$\underset{\underset{\emptyset}{|}}{\overset{\overset{CH_2OH}{|}}{CH}}=\overset{}{CH} \quad + \quad \underset{\underset{HOCH_2}{|}}{\overset{\overset{\emptyset}{|}}{CH}}=CH \quad \longrightarrow \quad \emptyset-CH-O-CH_2-CH=CH-\emptyset$$
$$\underset{\underset{CH_2OH}{|}}{\overset{\overset{}{|}}{CH_2}}$$

In these equations the symbol $\emptyset$ is used to represent any of the aromatic rings without concern for its specific identity.

Before proceeding with the discussion of the lignin structure, it is helpful to digress to introduce a useful concept from polymer science. Consider the reaction of 1,3-butadiene with acrylonitrile

$$CH_2=CH-CH=CH_2 + CH_2=CH-CN \longrightarrow \underset{\underset{CN}{|}}{+CH_2-CH=CH-CH_2-CH_2-CH+}_x$$

The product of this reaction is linear co-polymer, linear because the polymerization has essentially formed a line of carbon atoms, and a co-polymer because the reaction involved two different monomers as the starting materials. The linear co-polymer formed by reacting 1,3-butadiene with acrylonitrile still retains some double bonds in its structure. Since double bonds are sites at which polymerization can potentially occur, it is possible for the butadiene-acrylonitrile linear co-polymer to react further at the remaining double bonds

$$\underset{\underset{CN}{|}}{+CH_2-CH=CH-CH_2-CH_2-CH+}_x$$
$$+$$
$$\underset{\underset{CN}{|}}{+CH_2-CH=CH-CH_2-CH_2-CH+}_x$$
$$\longrightarrow \left(\begin{array}{c} \underset{\underset{CN}{|}}{-CH_2-CH-CH-CH_2-CH_2-CH-} \\ -CH_2-CH-CH-CH_2-CH_2-CH- \\ \underset{CN}{|} \end{array} \right)_x$$

The resulting product is a three-dimensional, crosslinked polymer. (Crosslinked polymers are also sometimes called network polymers.) In fact this particular example is nitrile rubber, used commercially, in for example, the hoses on gasoline dispensing pumps at service stations. A good example of

130

the effect of crosslinking on polymer properties is provided by the copolymer of divinylbenzene and styrene. The divinylbenzene molecule

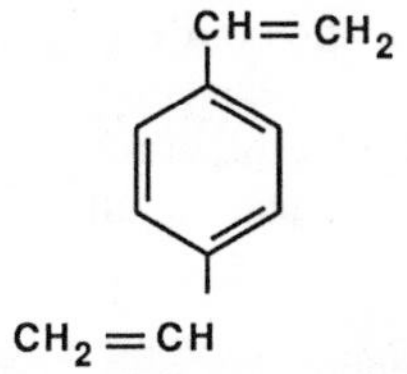

$$CH= CH_2$$

$$CH_2 = CH$$

has two functional groups at which polymerization can occur; each vinyl group can participate polymerization in a different polystyrene chain, thus establishing the crosslinks and forming the three-dimensional network. Because crosslinking restricts the mobility of the polymer chains, it can have a drastic effect on the observed physical properties of a polymer. Ordinary polystyrene, for example, easily dissolves in many common organic laboratory solvents, such as benzene or carbon tetrachloride. However, if 0.1% of divinylbenzene is incorporated into the polymer, the extent of crosslinking is enough to destroy the solubility, so that the 99.9 styrene : 0.1 divinylbenzene copolymer only swells in benzene or carbon tetrachloride. Crosslinking is crucial in the performance of the latex derived from the rubber tree. Natural rubber is a polyterpene of structure

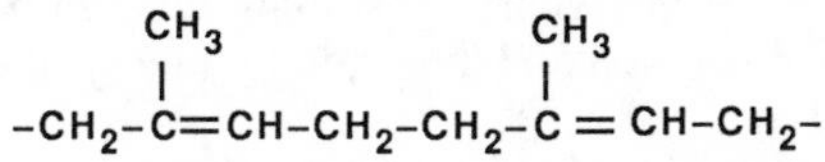

$$CH_3 \qquad\qquad CH_3$$
$$-CH_2-C=CH-CH_2-CH_2-C = CH-CH_2-$$

This linear polymeric structure has virtually no practical utility. However, when the polyterpene is vulcanized by reaction with sulfur, the sulfur atoms participate in crosslinking by addition across the double bonds:

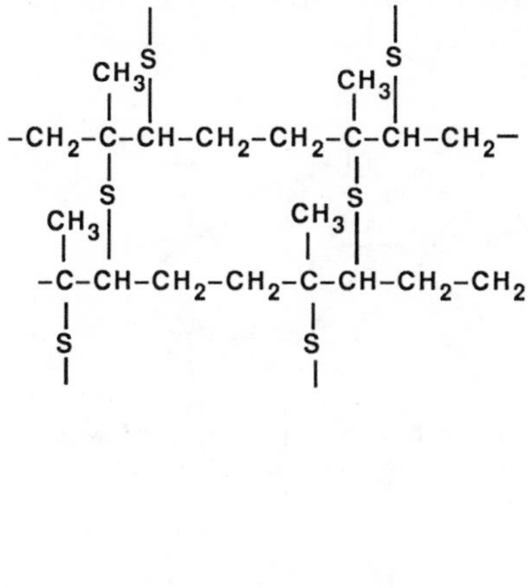

$$-CH_2-\overset{\overset{\displaystyle S}{|}}{\underset{\underset{\displaystyle S}{|}}{C}}-CH-CH_2-CH_2-\overset{\overset{\displaystyle S}{|}}{\underset{\underset{\displaystyle S}{|}}{C}}-CH-CH_2-$$

131

The crosslinked material has the desirable elastic properties we associate with rubber materials.

Lignin can be considered to be a three-dimensional cross-linked "terpolymer" of p-coumaryl, sinapyl, and coniferyl alcohols. Because lignin is of the plant components which is fairly resistant to degradation during the early stages of coalification, we expect that the vitrinites in lignite should have a composition resembling lignin. This resemblance is demonstrated by the data in Table 6.5.

Table 6.5 Comparison of the composition of a "typical" lignite with the lignin monomers

Substance	%C	%H	%O	f_a	R
Coniferyl alcohol	66.7	6.7	26.7	.60	1
Sinapyl alcohol	62.9	6.7	30.5	.55	1
p-Coumaryl alcohol	72.0	6.7	21.3	.67	1
Lignite	70	5.5	22	.6	1.5

The preservation of the lignin structure into the lignites then means that the macromolecular structure of lignites would consist of small aromatic units (mainly single rings) joined by crosslinks of aliphatic (methylene) chains or aliphatic ethers. If the polymerization is assumed to be random, with crosslinks heading off in all directions, the structure can be represented as seen in Fig. 6.10 (next page) where the straight line symbol represents an aromatic ring viewed edge-on and the squiggly line symbol represents an aliphatic crosslink.

This type of structure was deduced independently of considerations of polymerization or of lignin structure from x-ray diffraction analysis of coals. The structure shown above is known as the *open structure*. We will see that the kind of macromolecular structure a coal has influences the physical and chemical properties of the coal. Before examining the relationship between structure and properties, the other general types of macromolecular structures will be introduced.

Fig. 6.10 A sketch of the "open structure" with extensive crosslinking andsmall aromatic ring systems.

132

Bituminous coals

Compared to lignites, bituminous coals have higher carbon content, lower oxygen content, and higher values of f_a and R. The progression of changes accompanying the increase in coal rank is an evolution of the structure toward that of graphite. The structure of graphite itself (100% C, $f_a = 1$, $R = \infty$) is

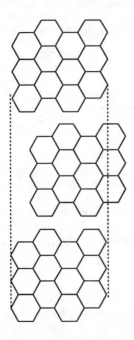

Viewed edge-on, using the convention introduced for depicting the open structure, graphite would be represented as

Fig. 6.11 A "sideways" view of graphite, showing the perfectly stacked aromatic planes.

where the hexagonal layers are perfectly stacked and aligned.

133

Since graphite is a crystalline substance, it produces a characteristic x-ray diffraction pattern which derives from the interatomic and inter-planar distances in the structure. Most coals, in contrast, are nearly amorphous and do not produce sharply-defined x-ray diffraction patterns as does graphite. However, when bituminous coals are examined by x-ray diffraction, it is possible to detect weak graphite-like signals emerging from the amorphous background. This information indicates that in the bituminous coals the aromatic ring systems are beginning to grow and to become aligned.

In the bituminous coal range, roughly between 85 and 91% carbon, the structure can be represented as

Fig. 6.12 The "liquid structure" of bituminous coals, with reduced crosslinking but increased size of aromatic units relative to the open structure.

This is called the *liquid structure*. Compared to the open structure shown previously, the aromatic units are larger (R may range from 2 to 6), the crosslinks are both shorter and fewer in number, and some vertical stacking of the aromatic units is evident.

Anthracites

Anthracites have carbon contents over 91%, and f_a values over 0.9. The ring condensation is probably much greater than 10, and may be on the order of 100. The structure of anthracites is approaching that of graphite. Indeed, x-ray diffraction data shows increased alignment of the aromatic rings with little contribution from aliphatic carbon

Fig. 6.13 The "anthracite structure", with large, fairly well aligned aromatic units and minimal crosslinking.

This is the *anthracite structure*.

Implications of structural variations in coals

Solvent swelling

Although it is not, strictly speaking, correct to say that coals are polymers, the consensus among coal scientists is that coals do have a macromolecular structure, however ill-defined. A polymer dissolves in a solvent in a two-step process: first, solvent molecules diffuse into the polymer to produce a swollen gel, and then the gel dissolves in the solvent to produce a true solution. If the intramolecular forces in the polymer are very high (e.g., as might be the case with covalently bonded crosslinks or strong hydrogen bonds) the solvent-polymer interactions may not be strong enough to overcome the intramolecular forces. In that case the process of dissolution stops at the end of the first step, leaving a swollen gel. A characteristic of crosslinked polymers is that they do not dissolve but only swell (in cases where they interact with the solvent at all). An example of the application of this property of polymers is the highly crosslinked nitrile rubber discussed previously, used to make hoses for handling gasoline. Gasoline can be an excellent solvent for many types of aliphatic materials, but extensive crosslinking in the nitrile rubber makes it resist even swelling in this solvent.

Since coals are crosslinked macromolecules, they do not dissolve in organic solvents, but only swell. The insolubility of coals is a major stumbling block both in the study of their molecular structures in the laboratory and in industrial processes for their conversion to other fuel forms or chemical products. Lignites are highly crosslinked and show very little swelling in solvents. With fewer crosslinks than lignites, bituminous coals swell to a much greater extent. Although there are few covalent crosslinks in anthracites, there are strong π-π interactions between the ring systems. Furthermore, the very large ring systems in anthracite are virtually impossible to solvate. Thus anthracites also show little interaction with solvents.

Porosity and surface area

When the total surface area (the external surface of the particles plus any internal surface on the exposed walls of pores) of lignites is measured by carbon dioxide adsorption, they are found to have the highest values of any coals, on the order of 200 m^2/g. (To put this number in context, a good quality activated carbon used, for example, to adsorb contaminants in water purification, might have a surface area of 1000 m^2/g.) About 90% of the surface area of lignites is in the internal surface of pores. The highly porous nature of lignite is implicit in the name open structure, and the sketch below illustrates the porosity by highlighting the large open spaces between the aromatic groups

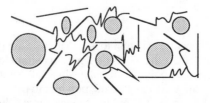

Fig. 6.14 The open structure with abundant, large pores (depicted by the shaded circles.)

One consequence of a highly porous structure is that the pores can fill with water. It is reasonable to expect a high moisture content for lignites. In fact, the moisture content determined in the proximate analysis is often in the range 35 - 45%. The moisture content of a coal includes not only the moisture in the pores but water on the surface and water held in cracks or fissures in the specimen. The actual moisture content measured in the proximate analysis will reflect the amount of handling a sample has received before it arrives in the laboratory and the conditions (especially temperature and humidity) under which it has been stored. Since the preliminary handling of the sample is out of the control of the analyst (and indeed often not even known to the analyst), information on porosity of coals cannot be derived with much confidence from such data.

The capacity of a coal for holding moisture in its pores could be determined from a measurement of the moisture content if it were certain that the pores were full and no extraneous moisture were present on the surface. Such a determination can be made by saturating a coal with water and storing the wet coal in a chamber at 100% relative humidity until an equilibrium has been attained. At that point it is presumed that the moisture in the coal is held entirely in the pore system, and the measured moisture content is referred to as the *equilibrium moisture* or sometimes the *capacity moisture*.The equilibrium moisture can vary from coal to coal even among coals of the same rank. However, as a group lignites have the highest equilibrium moisture contents of any rank of coal, in the range of 20 - 40%.

Because coal does not dissolve completely in any solvent, all reactions of coal are heterogeneous reactions between the solid coal and a liquid or gaseous reagent. In a heterogeneous reaction the rate and extent of reaction will often be limited by the total amount of reacting surface that can be attacked by the reagents. The reactions may also be limited by the ability of reagents to diffuse into pores or of the reaction products to diffuse out.

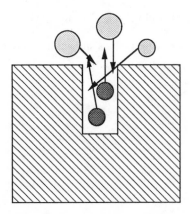

Fig. 6.15 A sketch of a coal particle undergoing a reaction, with reactants (lightly shaded) attempting to diffuse into a pore and products (darkly shaded) attempting to diffuse out.

With the highest surface area and highest porosity, lignites are generally very reactive coals. (Of course another factor accounting for the high reactivity of lignites is their abundance of functional groups which provide the potential for a variety of chemical reactions to occur.)

A qualitative comparison of the liquid structure with the open structure shows that the liquid structure is less porous

Fig. 6.16 The liquid structure with pores; pores are smaller and fewer than in the open structure (compare Fig. 6.14).

In fact bituminous coals do have lower porosities and lower total surface areas than lignites. The lower pore volume results in a lower equilibrium moisture content and generally a lower reactivity (when comparing identical reagents and reaction conditions) than lignites.

Despite the increase in ordering in the anthracite structure relative to the liquid structure, anthracites are slightly more porous than the bituminous coals. The bituminous coals consist of relatively small aromatic units (R = 2 - 6) connected by aliphatic crosslinks. The crosslinks can exist in various conformations which may allow for closer filling of space than can be attained

137

with the very large (r = 10 - 100) and very rigid aromatic units connected by shorter and less flexible crosslinks.

The porosity of coals varies with rank as shown in Figure 6.17.

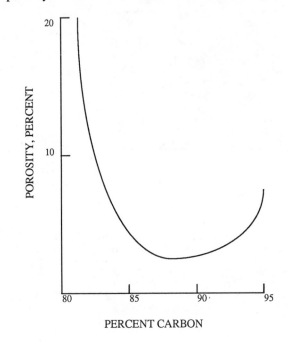

PERCENT CARBON

Fig. 6.17 The variation of porosity of coals with carbon content. A minimum is reached in the region 87-88% carbon.

The variation of surface area with rank is a curve of approximately the same shape. The trend in equilibrium moisture with rank is shown in Figure 6.18. Since the porosity of anthracite is greater than that of bituminous coal, we might expect anthracite to be more reactive. In fact, anthracites tend to be unreactive in most organic reactions of coals. The low reactivity of anthracite derives from the stabilization energy of the large aromatic systems, and from the fact that many reactions of coals occur either at functional groups or at aliphatic crosslinks, both of which are scarce in anthracites.

For many applications it is more important to know the surface area rather than the total porosity, since it is the surface area that controls the extent and rate of reactions. By knowing the surface area and pore volume, and by making an assumption about pore geometry, average pore radius can be calculated, to obtain an idea of the sizes of molecules which can penetrateor leave the pore system. The most commonly used approach to measuring surface area is gas adsorption. If a coal sample is exposed to a gas

138

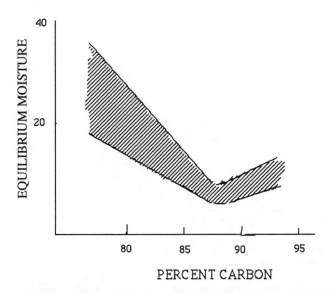

Fig. 6.18 The variation of equilibrium moisture with carbon content. Consistent with Fig. 6.17, a minimum is attained around 87-88% carbon.

held at constant pressure and temperature, it will be seen that the volume of the gas decreases. This observation suggests that some of the gas has penetrated the pores of the coal where it has somehow becoming held, indeed, almost as if gas were condensing inside the coal. The gas is being adsorbed onto the surface of the coal, and the adsorption process can be considered to be, in essence, a condensation. This process is by no means limited to coal, and can take place for any porous solid. The gas being used in the experiment is referred to as the *adsorptive*, the solid on which adsorption is taking place is the *adsorbent,* and the layer of adsorbed gas is the *adsorbate*. The density of the adsorbate is nearly equal to the density of the pure material in its liquid state.

If the pressure of the gas is then increased, the total volume of gas adsorbed also increases. By repeating the measurement several times, enough data is obtained to plot an *adsorption isotherm*. Typically an adsorption isotherm is plotted as V_a, the volume of gas adsorbed, as a function of $P/P°$, where P is the actual pressure of the gas and $P°$ is the vapor pressure of the gas over its liquid at the temperature of the experiment. An adsorption isotherm is shown as Figure 6.19.

Several approaches have been developed to calculate surface areas from adsorption isotherms. In essence the volume of gas adsorbed can be used to calculate the number of molecules adsorbed. Knowing the size of one

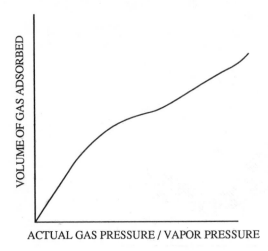

VOLUME OF GAS ADSORBED

ACTUAL GAS PRESSURE / VAPOR PRESSURE

Fig. 6.19 A hypothetical adsorption isotherm for the adsorption of gas into the pore system of a coal.

molecule, and the number of molecules adsorbed, one can then calculate the area covered by the adsorbed gas, that is, the surface area of the sample. The gases most commonly used for coal characterization are carbon dioxide and nitrogen. Surface areas measured by carbon dioxide adsorption are usually 100-200 m^2/g, whereas those measured by nitrogen adsorption are typically <20 m^2/g. This large discrepancy is attributed to the fact that diffusion of gases into pores is an activated (*i.e.* temperature-dependent). Carbon dioxide adsorption experiments are performed at temperatures up to 298 K, whereas nitrogen adsorption might be done at ~90 K. The temperature difference could result in a more rapid diffusion and equilibration of carbon dioxide relative to nitrogen. In fact the activated diffusion would be especially important for the so-called ink bottle pores (Fig. 6.20), for which nitrogen diffusion through the restricted opening would be very difficult at 90 K, but diffusion of carbon dioxide at 298 K would be much easier.

For a hypothetical coal containing a mixture of cylindrical and ink bottle pores, the difference obtained from nitrogen and carbon dioxide adsorption is illustrated in Figure 6.21.

Because the value of the measured surface area depends on the gas being used as the adsorptive, there is no uniquely defined surface area of a porous solid such as coal. In this sense, surface area is also an operational definition because it depends on the way the material (*i.e.,* coal) behaves in the experiment and on the experimental conditions. Reporting the value of surface area without indicating the adsorptive used is meaningless.

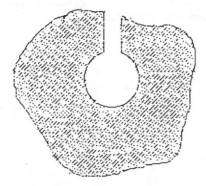

Fig. 6.20 A coal particle containing an "ink-bottle" pore, with a narrow opening but large cavity.

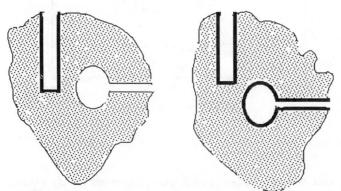

Fig. 6.21 A comparison of gas adsorption in coal particles. The particle on the left adsorbs nitrogen in cyclindrical pores (the dark line indicating a pore surface covered with adsorbed gas) but not in ink-bottle pores. The particle on the right adsorbs carbon dioxide onto pore surfaces in both cylindrical and ink-bottle pores. The calculated surface area for carbon dioxide adsorption would be higher than that calculated from nitrogen adsorption.

Density

As with most solids, the density of coal is measured by the displacement of a fluid. Because coals are porous, the *apparent density* will depend on the fluidused for the measurement, and specifically on the ability of fluid molecules to penetrate the pore system. The apparent density is useful in engineering applications to calculate the size of reactor vessels needed to contain a given quantity of coal. The variation of apparent density of coal in methanol as a function of rank is shown in Fig. 6.22.

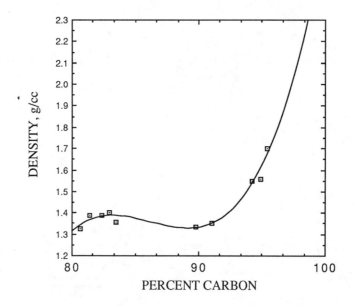

Fig. 6.22 The variation in density of coals, as measured by displacement of methanol, as a function of carbon content.

Other liquids, such as water, hexane, or benzene, give curves having a similar shape to that of Fig. 6.22, but in which the line is displaced vertically. In all cases a broad minimum occurs between 85-90% carbon. For coals of the same rank, the apparent density of macerals varies as inertinites > vitrinites > liptinites.

The *true density* of coals is measured by displacement of helium, on the assumption that helium should be able to penetrate even the smallest of pores. (This measurement is also referred to as the *helium density*.) The helium density varies as a function of carbon content as shown in Figure 6.23. The variation in helium density reflects, at carbon contents below 88%, the effect of decreasing oxygen content. Above 88% the increase in density is a reflection of increasing structural order. These trends are illustrated for some simple aromatic molecules in Table 6.6.

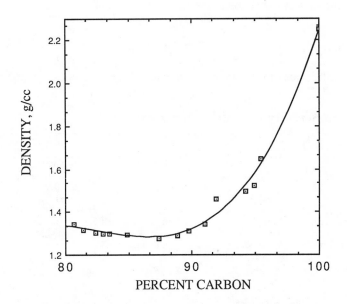

Fig. 6.23 Variation of coal density measured by helium displacement as a function of carbon content.

Table 6.6 Densities of simple aromatic molecules, indicating increased density as structure becomes more aromatic and compact.

Compound	f_a	Density, g/mL
2-Ethylnaphthalene	0.83	1.008
2-Methylnaphthalene	0.91	1.029
Naphthalene	1.00	1.145
9-Ethylanthracene	0.88	1.041
9-Methylanthracene	0.93	1.066
Anthracene	1.00	1.250
Graphite	1.00	2.25

The *bulk density* of coal is the volume occupied by a mass of assorted lumps of coal. The bulk density will vary with the particle size and packing efficiency of the lumps. Bulk density conveys almost no fundamental information about the structure or properties of coal, but can be a useful quantity for calculating the volume needed to store, ship, or handle a given amount of coal, as in sizing a storage bin, for example.

Summary

The key features of macromolecular structures of coals are summarized in Table 6.7.

Table 6.7 Summary of structural features of coals

Feature	Lignite	Bituminous	Anthracite
Structure	Open Lignin-like	Liquid	Anthracite Graphite-like
f_a	~0.6	~0.8	>0.9
R	~1.5	2 - 6	>10
Porosity, surface area, equilibrium moist.	high	minimum at 88%	increased from minimum
Helium density	moderate, decreasing	minimum at 88%	high
Reactivity	high	medium	very low

The summary in Table 6.7, along with the Mott chart (Fig. 6.9) and the Seyler chart (Fig. 6.4) provide the key facts of the properties and composition of coal necessary for the discussions in Unit 3 on coal utilization.

The caking behavior of bituminous coals

Coal is heated in an inert atmosphere decomposes with the evolution of a variety of volatile materials. If the decomposition is monitored as function of temperature (instead of the drastic exposure of coal to a temperature of 950° in the volatile matter determination), thermal decomposition is found to proceed in three basic stages. *Stage 1* occurs at temperatures below 200°. It is usually a slow reaction, and the principal volatile products are water (from thermal dehydration of functional groups), carbon dioxide, carbon monoxide, and hydrogen sulfide. These products arise from a loss of functional groups or through condensation reactions. *Stage 2* occurs between about 350° and 550°. These reactions tend to be fast. The principal products are light hydrocarbon gases as well as a variety of organic compounds which condense at room temperature to a tarry mixture. *Stage 3* begins above 550°. A variety of small gaseous molecules are formed, including water, carbon monoxide, carbon dioxide, hydrogen, methane, ethane, ethene, ethyne (acetylene), and ammonia. The other product is a graphite-like solid of high carbon content.

During Stage 2, some coals pass through a plastic state in which the coal appears to soften, swell, and then resolidify into a porous mass. Coals which exhibit this behavior are called *caking coals*. As a rule, the caking coals are coals of high volatile A, medium volatile, and low volatile bituminous rank. The medium volatile bituminous coals are particularly good caking coals.

The caking behavior of a coal is measured, in United States practice, by the *free swelling index*, often known simply as the FSI. In this test a known quantity of coal is heated in a standard sized crucible under standardized conditions (820° for 2.5 minutes). The FSI is determined by comparing the size of the resulting solid "button" with a series of standard outlines, and assigning a value from 1 to 9. (An FSI of 0 is assigned to a material which does not form a coherent mass but rather falls apart when removed from the crucible.)

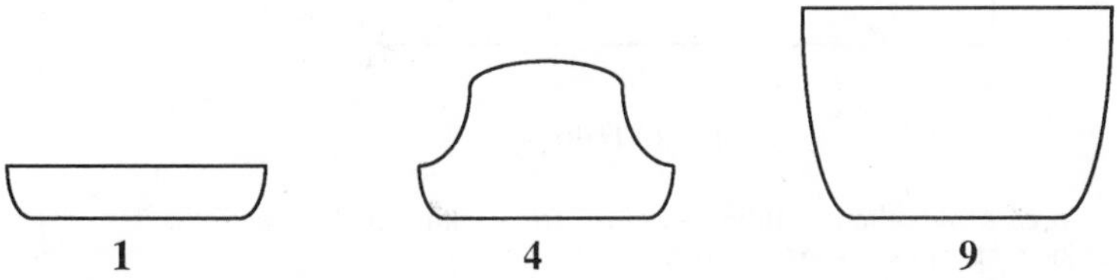

Fig. 6.24 Outlines of the "buttons" of carbonaceous solid remaining at the end of a free swelling index test, for coals of free swelling indices of 1,4, and 9. Coals of other free swelling indices would have buttons intermediate in shape between these.

Coals that are weakly caking have FSI values of 2 to 3.

The behavior of the coal during its plastic stage is measured by the fluidity of the plastic mass and the change in volume which occurs. The fluidity is measured in a Gieseler plastometer. A stirrer is immersed in the coal and a constant torque is applied. The fluidity is measured by the speed of revolution of the stirrer on a scale as the coal is being heated at 3°/min. The fluidity data are expressed in units of dial divisions per minute. A fluidity curve is shown in Figure 6.25. Three temperatures are read from the fluidity curve: the softening temperature T_s, the temperature of maximum fluidity T_m, and the resolidification temperature T_r. (In Fig. 6.25 these values are 400°, 450, and 475°, respectively.) The *plastic range* of the coal is determined from $T_r - T_s$.

The change in volume accompanying the plastic stage is measured in a dilatometer. Typically a coal sample is compressed into a rod or "pencil" and loaded into the dilatometer beneath a piston with scale indicator. Changes in the volume of the coal during heating can then be monitored by noting the position of the scale indicator. The volume changes are characterized by three

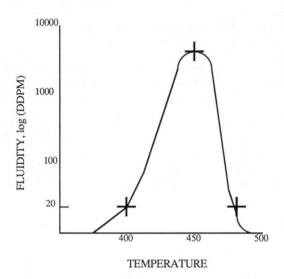

Fig. 6.25 A hypothetical fluididity curve for a caking coal. Notice that the vertical axis is logarithmic.

parameters: contraction, swelling, and dilatation. These parameters are illustrated in Figure 6.26.

Dilatometry experiments are used to describe four classes of plastic behavior, illustrated in Figure 6.27.

Many factors affect the observed plastic behavior of coal; the role of each is at best poorly understood. Increasing the heating rate increases T_m, the maximum fluidity, and the swelling. Heating the coal in an inert atmosphere at ~200° prior to going to higher temperatures reduces fluidity and swelling. Oxidation of the coal (as might occur in prolonged exposure to the atmosphere during long term storage) reduces the plastic range, the maximum fluidity, and the FSI; in contrast, reacting the coal with hydrogen generally increases the plastic range, fluidity, and swelling. Fine grinding of the coal decreases the plastic behavior; removing or reducing ash improves it. These diverse phenomena make it difficult to develop a single mechanism of plastic behavior which accounts for all observations.

It is generally agreed that the thermal decomposition of parts of the coal structure produces a fluid material variously called *metaplast*, thermobitumen, or primary tar. Thus in an initial step coal decomposes to metaplast and a solid material

$$COAL \rightarrow METAPLAST + SOLID\ 1$$

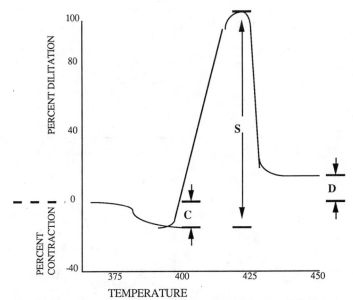

Fig. 6.26 The results of a dilatometry measurement for a coal. **C**, the contraction; **S**, the swelling, and **D**, the dilitation.

The metaplast acts as a plasticizer which allows the structural components of Solid 1 (which may be highly aromatic) to rearrange in the fluid state and undergo condensation reactions of the type.

147

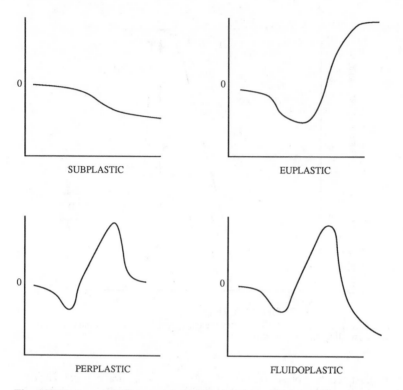

SUBPLASTIC

EUPLASTIC

PERPLASTIC

FLUIDOPLASTIC

Fig. 6.27 The types of dilatometry behavior of coals are illustrated. The axes in each case are percent contraction of dilitation as a function of temperature. Some coals show a net reduction in volume (subplastic and fluidoplastic) while others show a net increase (euplastic).

The metaplast gradually evaporates or cracks to produce volatile compounds. The metaplast vapors or cracking products bubble through the plastic mass and cause it to swell. When most of the metaplast has evaporated or cracked the residue resolidifies. Thus

$$METAPLAST \rightarrow GAS + SOLID\ 2$$

When the plastic behavior is such that the resulting solid product is a very strong, hard, porous mass, the product is called *coke* ,and the original coal is called a *coking coal*. Coke is an important fuel in the metallurgical industry, especially in the iron industry, and supplies of coking coals can be very valuable. (All coking coals are caking coals, but not all caking coals yield a commercially desirable coke, so not all caking coals are coking coals.)

There is a relationship between coking behavior and petrographic components, as shown in Table 6.8.

Table 6.8 Coking behavior of petrographic components of coal

Coking behavior	Petrographic constituents
Reactive	Vitrinite, liptinites
Intermediate	Semifusinite
Inert	Fusinite, micrinite and other inertinites
	Highly coalified vitrinites
	Mineral matter

The liptinites are mainly aliphatic, while vitrinites are more aromatic. There is a very large difference in melting point between aromatic and aliphatic compounds having the same number of carbon atoms, as shown in Table 6.9.

Table 6.9 Comparative melting behavior of aromatic and aliphatic compounds of comparable numbers of carbon atoms

Number of C atoms	Aliphatic compound (m.p.)	Aromatic compound (m.p.)
10	Decane (-31°)	Naphthalene (80°)
14	Tetradecane (6°)	Phenanthrene (100°)
16	Hexadecane (20°)	Pyrene (150°)
18	Octadecane (28°)	Chrysene (254°)

The liptinites may therefore be lower melting than the more aromatic, and possibly more crosslinked) vitrinites. In coals the liptinites may melt and act as plasticizers. In polymers there is a characteristic temperature, the *glass transition temperature*, T_g, below which there is no longer enough thermal energy to allow free rotations of the segments of a polymer chain. A plasticizer is a nonvolatile solvent added to a polymer to reduce T_g. Normally T_g decreases as the plasticizer concentration increases. In practical use this behavior can be noticed as a reduction, as plasticizer concentration increases, in the lowest temperature at which a polymer can be flexed, or in a reduction of the applied stress needed to achieve a given elongation. In a sense, a plasticizer acts like a "molecular ball bearing" allowing the other structures to slide past each other.

A model for the possible structural rearrangement in caking coals is the behavior of syn-[2.2](1,4)naphthalenophane

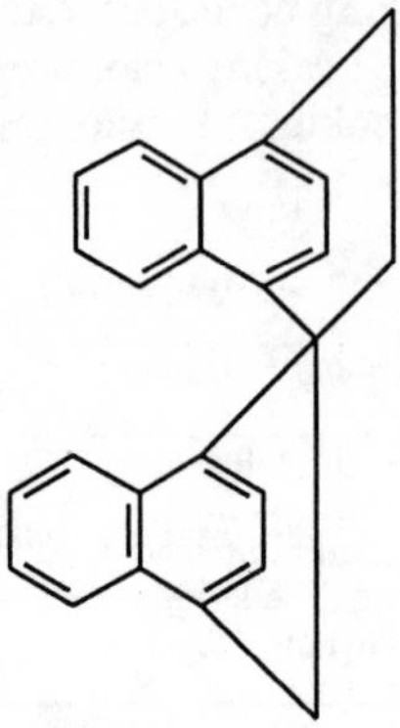

This compound is observed to melt at 240° and to resolidify at ~250° to the anti- isomer

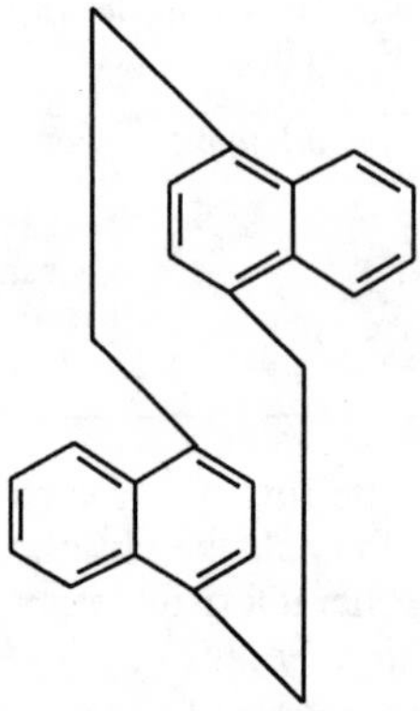

The reaction is believed to proceed via the formation of a diradical

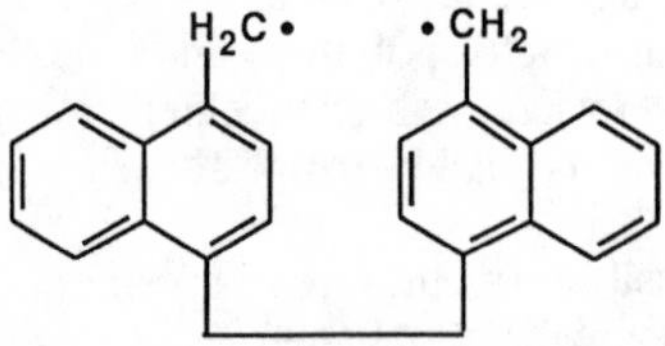

which then rearranges to the anti- isomer. Somehow the two naphthalene ring systems must slip past each other. Notice that these naphthalenophane compounds have $f_a = 0.83$ and $R = 2$, being an approximate model for, say, a high volatile A bituminous coal. In a coal the movement of one aromatic ring system past another, as occurs in the naphthalenophanes, would be facilitated by a plasticizer.

The formation of a graphitic coke is dependent on having a metaplast in which some structural alignment exists. Such ordering in the liquid phase is typical of the class of compounds known as liquid crystals. The liquid structure of bituminous coals presumably contains enough structural ordering of aromatic units that when a partial melting occurs the ordering is preserved into the liquid phase, with the formation of a nematic liquid crystal. As metaplast evaporates or cracks the partially aligned molecules in the nematic liquid crystal can then line up to form the desired graphitic solid. If the bituminous coal were to melt to an isotropic liquid, it is unlikely that further heating of the isotropic liquid would result in the desired, ordered graphitic structure.

The inorganic components of coal

A significant distinction between coal on the one hand and oil and natural gas on the other is that coal contains appreciable amounts of inorganic material. In extreme cases, the ash determined in the proximate analysis exceeds 25% on an as-received basis. No ash residue is produced from the combustion of natural gas. Some fuel oils produce an ash, but generally the amount is less than 0.1% and the ash generally has little effect on the practical uses of the oil. With coal, however, the ash produced while the carbonaceous portion of the coal is being burned must be collected and arrangements must be made for its environmentally acceptable disposal. The magnitude of the problem is illustrated by considering that the burning of one tonne of coal having 25% ash would produce 250 kg ash. In some cases the inorganic components react to form low-melting phases which deposit on the walls or steam tubes of boilers, interfering with the efficient operation of the unit.

Ash is the product of various reactions among the inorganic constituents of coal occurring when the coal is burned. The *inorganic constituents* are the diverse inorganic materials in the coal, including mineral grains, cations associated with ionized functional groups, and cations held in coordination complexes with the heteroatoms in the coal structure. *Mineral matter* refers to the discrete grains of inorganic substances, such as quartz or clay, incorporated in the coal as a separate and distinct phase. We have seen that it is possible to correct analytical data to an ash-free basis by multiplying by 100/(100 - A). However, the weight of ash produced during the burning of coal does not necessarily equal the total weight of inorganic constituents originally present in the coal, because of a variety of reactions which can occur during ashing:

Dehydration: $FeSO_4 \cdot nH_2O \rightarrow FeSO_4 + nH_2O$
Oxidation: $FeS_2 + 3\,O_2 \rightarrow FeSO_4 + SO_2$
Decomposition: $CaCO_3 \rightarrow CaO + CO_2$
Sulfur capture: $CaO + SO_2 + 0.5\,O_2 \rightarrow CaSO_4$

The magnitude of the difference between the weight of ash and the weight of inorganic constituents originally in the coal will depend on the contribution that each of these reactions makes during the formation of the ash. For example, the oxidation of pyrite results in a weight gain, while the decomposition of calcium carbonate results in a weight loss. Two coals having identical ash determinations in the proximate analysis may not necessarily have had the same amounts of inorganic constituents before ashing.

The motivation for removing algebraically the moisture and ash from the analytical data was to be able to compare different coals strictly on the basis of the combustible material, without the effects of the essentially inert water and inorganic constituents. But since identical ash contents of two different coals may not reflect identical amounts of inorganic constituents, a preferable approach for the careful characterization of coals is to correct analytical data for the presence of the inorganic constituents, rather than for ash.

To make such a correction would require the weight of the inorganic constituents. Until recently, however, there was no straightforward way to isolate these materials from coal without significantly altering them in the process. Consequently, methods were developed to calculate the weight of inorganic constituents from the weight of ash. Since these efforts were mainly directed toward bituminous coals, in which essentially all of the inorganic constituents are minerals, the approaches developed were for the calculation of mineral matter content from ash. A variety of equations have been proposed for calculating the mineral matter; one in widespread use is the *Parr formula*:

$$MM = 1.08\,A + 0.55\,S$$

where MM represents mineral matter, A ash, and S the sulfur content of the coal. The conversion of analytical data to a mineral-matter-free basis (or moisture-and-mineral-matter-free) is done in exactly the same fashion as described earlier, except that the mineral matter is used instead of the ash.

The origin of inorganic components in coal

There are three general sources of inorganic components in coal. The first is the original plant material from which the coal was formed. Virtually all plants contain some inorganic material; a good example is provided by the ash left after a wood fire. Some plants actively concentrate certain inorganic species. An example is the modern-day scouring rush, which concentrates silicon dioxide in its stems. The scouring rush is a descendant of the plants which grew extensively in the coal swamps of 300,000,000 years ago. The inorganic accumulations in plants are known as *phytoliths*.

A second source is transportation of minerals into the swamp while plant debris is accumulating and slowly being transformed to coal. Mineral grains can be washed in by water flowing through the swamp. As the flowing water becomes stagnant in the swamp the mineral grains settle into the organic debris on the bottom. The minerals themselves come from erosion or weathering of rocks in the surrounding countryside. Clay minerals and quartz may accumulate in coal by this mechanism. Wind is sometimes a transporting agent. Some coals contain volcanic ash which was evidently transported to the coal swamp wind activity from regions of active vulcanism.

A third source of inorganic constituents in coal is the formation on minerals *in situ* in the coal by reactions occurring in water percolating through cracks and fissures in the coal. Pyrite is a mineral which sometimes accumulates in coal in this way.

The minerals in coal

At least 125 minerals have been identified in coal. Less than 20 are found in sufficient quantity to be important for the practical use of coal. These 20 or so can be grouped into six families.

Clay minerals are usually the predominant minerals in coal, often accounting for more than 50% of the total minerals in a coal sample. The family of clay minerals is very extensive. Clays are hydrous oxides of aluminum and silicon which contain a variety of other cations in the structure. Some of the elements whose cations may be incorporated in clays are iron, sodium, calcium, potassium, and magnesium. Most coal mineralogists believe that the clay in coal washed into the swamp during the accumulation and coalification of the plant matter. Clays contribute aluminum and silicon to the coal (and to the ash) along with lesser quantities of other elements such as potassium.

The principal *carbonates* are salts of iron, magnesium, and calcium, or mixtures of these elements. (An example of a mixed carbonate is dolomite, $CaMg(CO_3)_2$.) Carbonates form by precipitation caused by direct reaction

$$Fe^{+2} + CO_3^{-2} \rightarrow FeCO_3$$

or by pH changes

$$Ca^{+2} + 2\,HCO_3^- \rightarrow CaCO_3 + CO_2 + H_2O$$

Most of the *silica* in coal is in the form of quartz. Quartz can account for up to 20% of the total minerals in coal. Some silica may originate from phytoliths, but most is probably washed into the swamp.

The principal *sulfide* mineral is pyrite. (There are two minerals of the formula FeS_2 but with different crystal structures: pyrite and marcasite. Usually no attempt is made to differentiate between them and all the $FeS2$ is called pyrite.) Most of the pyrite in coal forms by precipitation after the coal has formed. Sulfides generally amount to less than 5% of the total minerals in most coals. The sulfide minerals receive far more attention than would be expected from their concentration, because of the concern for the formation of sulfur oxides during coal combustion. If efforts are not made to capture these oxides, their release to the environment contributes to air pollution, such as the formation of acid rain. In addition, sulfur in coking coals is undesirable because the sulfur can eventually find its way into the coke and thence into the reduced metal, where its presence weakens the metal in subsequent forming operations such as hot rolling. Spontaneous combustion of coal stockpiles is due, in some cases, to heat released by the air oxidation of the pyrite. On the other hand, sulfide minerals in coal are considered to be good catalysts in the hydrogenation of coal to produce synthetic liquid fuels.

Salt is fortunately rare in United States coals, but is found in some coals in the United Kingdom, central Europe, and Australia. Salt can exacerbate corrosion of metals (as best shown by the rusting of car bodies in regions where salt is applied to the roads during winter). In addition, the chlorine in the salt can appear as hydrogen chloride in the products of coal combustion, the hydrogen chloride then dissolving in any available moisture to form corrosive hydrochloric acid.

Sulfates are in low concentration in most coals. The principal sulfates are those of iron and calcium.

The principal inorganic elements contributed to coal by the various minerals are summarized in Table 6.10.

Table 6.10 Contribution of inorganic elements to coal by mineral families

	Si	Al	Fe	Ca	Mg	Na	K	S
Clays	*	*					*	
Carbonates			*	*	*			
Silica	*							
Sulfides			*					*
Salt						*		
Sulfates			*	*				*

The special case of low-rank coals

Coals of lignite and subbituminous rank contain appreciable amounts of oxygen. Some of this oxygen is present as carboxyl groups that can act as ion-exchange sites to accumulate cations from water percolating through the coal, not unlike ion exchange units used for water softening in the home. In particular, sodium, calcium, magnesium, and potassium can be accumulated in this way. In general, 90-100% of the sodium, 75-90% of the magnesium, 50-75% of the calcium, and 35-50% of the potassium present in low-rank coals can be present as cations associated with carboxyl groups in the coal structure. The remainder of the magnesium and calcium is present in carbonate minerals and the remaining potassium is in the clays.

Variations of ash content and composition with rank

It is difficult to generalize about the amount of ash determined by proximate analysis as a function of rank because much depends on the highly variable factors of the environment of deposition of organic matter. One coal swamp might have considerable clay washed in from surrounding regions while another swamp would not. In addition, the factors affecting precipitation of minerals in coal from water percolation through the coal seam can also vary greatly from one coal to another. Whether the water is high in iron or not, for example, might affect the amount of pyrite accumulated in the coal.

For similar reasons it is difficult to generalize about the composition of ash as a function of rank. However, low-rank coals contain appreciable calcium, sodium, magnesium, and some potassium on carboxyl ion exchange sites. Carboxyl groups do not exist in bituminous coals and anthracite. Generally the amount of pyrite tends to increase somewhat with rank, at least through the bituminous coals. Since there are no ion-exchange sites in bituminous coals, the principal inorganic constituents will be silicon, aluminum and iron. The ash of low-rank coals, compared to bituminous, will contain more calcium, magnesium, and sodium and therefore proportionately less silicon, aluminum, and iron.

Further reading

Berkowitz, N. (1979). *An Introduction to Coal Technology.* New York: Academic Press; Chapters 2,3,4.

Bustin, R. M., Cameron, A. R., Grieve, D. A., and Kalkreuth, W. D. (1985). *Coal Petrology: Its Principles, Methods, and Applications.* St. John's, Newfoundland: Geological Association of Canada.

Streitwieser, A. Jr. and Heathcock, C. H. (1985). *Introduction to Organic Chemistry*. New York: Macmillan; Chapter 34.

van Krevelen, D. W. (1961). *Coal: Typology - Chemistry - Physics - Constitution*. Amsterdam: Elsevier; Chapters II, IV, VII, IX.

Chapter 7

Processing and use of natural gas

This chapter begins the third unit of the book. Attention now turns to the uses of fuels, and how the composition and properties of fuels influence the ways we use them. Again we begin with the simplest case, natural gas. Three important concepts are introduced here for natural gas and will then be amplified in later chapters for petroleum and coal. First, most fuels require some processing before they are ready for use. Second, the combustion chemistry is introduced. Third, we must recognize that natural gas, petroleum, and coal are not only fuels useful for their energy content but are also very useful sources of a variety of organic chemicals and materials.

Natural gas processing

Some purification steps are usually applied to natural gas before it is put into the pipeline system for sale to users. These operations are directed to removing the principal components other than methane: hydrogen sulfide, the other hydrocarbons, and moisture. The amount of processing, and the specific processing operations used, depends on the composition of the gas as it comes from the well.

Usually the first step in gas processing is to remove moisture. This is done for two reasons. First, some of the subsequent processing steps operate below 0°, and it is necessary to prevent freezing during these operations. Second, moisture in the gas introduced to the pipeline could cause "icing" of fuel lines, valves, and gauges during cold weather.

Moisture removal is accomplished by passing the gas through dehydrators. These units are tanks or towers in which the gas is contacted with compounds that will absorb moisture. One class of compounds used in this application are the glycols:

$$HOCH_2CH_2OCH_2CH_2OH$$

$$HOCH_2CH_2OCH_2CH_2OCH_2CH_2OH$$

$$HOCH_2CH_2OCH_2CH_2OCH_2CH_2OCH_2CH_2OH$$

which are diethylene glycol, triethylene glycol, and tetraethylene glycol, respectively. These compounds are used for two reasons. First, their abundance of oxygen atoms provides numerous sites for moisture absorption and retention by hydrogen bonding:

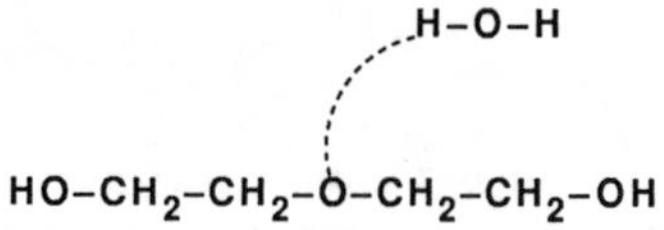

$$H-O-H$$

$$HO-CH_2-CH_2-O-CH_2-CH_2-OH$$

Second, these compounds have very low vapor pressures. Therefore one impurity (the glycol) is not added inadvertently into the gas stream while removing another (the moisture). Moisture absorption is usually carried out in countercurrent flow extractors (Fig. 7.1).

Sweetening of natural gas is done to remove hydrogen sulfide. Sweetening operations take advantage of the fact that hydrogen sulfide is a weak acid, so that it can be absorbed and retained by bases. As with dehydration, there is also concern that sweetening does not simply replace one impurity with another. Reagents to be used for sweetening should be mildly basic compounds with low vapor pressures. An example of a reagent used in sweetening is diethanolamine:

$$(HOCH_2CH_2)_2NH + H_2S \rightarrow (HOCH_2CH_2)_2NH_2{}^+HS^-$$

Sweetening the gas necessarily adds to its cost, because of the investment in the equipment for sweetening and the recurring expenses of operating the sweetening unit. When the hydrogen sulfide content of gas is very high, it can be recovered and sold to sulfuric acid manufacturers. The recovery depends on the oxidation of hydrogen sulfide to sulfur, with the subsequent sale of the elemental sulfur:

$$H_2S + 1.5\,O_2 \rightarrow H_2O + SO_2$$
$$2\,H_2S + SO_2 \rightarrow 2\,H_2O + 3\,S$$
- - - - - - - - - - - - - - - - - - - -
$$\text{net: } 3\,H_2S + 1.5\,O_2 \rightarrow 3\,H_2O + 3S$$

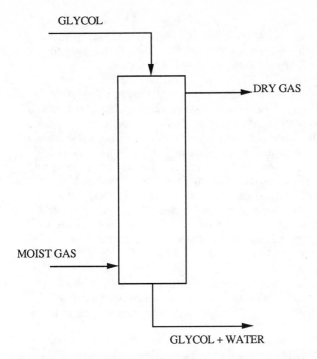

Fig. 7.1 Diagram of a countercurrent flow operation for removing moisture from natural gas. The moist gas flows upward through the column while the glycol flows downward. Glycol absorbs the moisture, allowing the dried gas to exit the top of the unit, while the glycol-water mixture leaves from the bottom.

If the gas is also wet (*i.e.*, containing other light hydrocarbons, not wet in the sense of containing moisture), then the sweetened gas is passed through a light oil absorber. This unit contacts the gas with hexane at -20° and 3.5 MPa. These stringent conditions are necessary because the hydrocarbons in wet gas are neither acidic nor basic, nor can they participate in hydrogen bonding. Thus the strategies employed in dehydration and sweetening are not applicable. Instead, the approach is to force the hydrocarbons to dissolve in a suitable solvent. Hexane is chosen on the basis of the rule of thumb "like dissolves like." The temperature and pressure conditions are chosen because the solubility of gases in liquids increases with decreasing temperature and increasing pressure - hence a low-temperature, high-pressure process.In addition, the low temperatures insure that hexane has a low vapor pressure. Despite the precautions to avoid introducing hexane vapors into the gas stream, some hexane inevitably appears in the gas. This hexane is removed by a heavy oil absorber.

The solution of hydrocarbon gases in hexane is called *rich light oil*. Distillation of the rich lignite oil yields ethane, for sale to the chemical industry, and propane and butane, which can be sold as a fuel (liquefied petroleum gas, or *LPG*) or sold to the chemical industry.

Some natural gases will yield a condensible liquid product containing pentane and higher hydrocarbons. This liquid is called *natural gasoline*. Natural gasoline can be recovered either by compressing the gas stream until the pressure is sufficient to liquefy the relatively non-volatile higher alkanes, or by distillation of the fluid from the heavy oil absorber. At one time natural gasoline was desirable as an additive to gasoline obtained from crude oil distillation, because the natural gasoline increased the vapor pressure of the blended product, thereby improving cold-weather starting characteristics. As we will see in Chapter 8, natural gasoline does not provide good combustion performance in modern high-compression engines. Consequently natural gasoline is now used mainly as a feedstock for petrochemical production. Although the composition of natural gasoline varies somewhat from one source to another, a typical composition might be 20% C_3-C_4, 30% C_5, 25% C_6, 20% C_7, and 5% C_8.

The sequence of gas processing steps is summarized in Figure 7.2. After this sequence of purification steps the gas is sometimes known as *stripped gas*. At this point it is ready to go into the pipeline for sale to users.

Combustion of natural gas

Natural gas is the premium hydrocarbon fuel. It has the highest calorific value of any of the hydrocarbon fuels, 56 MJ/kg. The sweetened gas has no sulfur and leaves no ash on combustion. Gases are easy to meter and handle. For most users, the gas can be taken directly from a distribution system so that no on-site storage facilities are needed.

Combustion of natural gas depends on the ability of free radicals to participate in *chain reactions,* that is, reactions in which a free radical reactant generates a new free radical as a product. The new radical participates in a second step which in turn generates a radical product, and so on through third, fourth, ... and n-th steps, a process of *chain propagation*. The "chain" of reactions is not infinitely long because competing reactions may result in the destruction of radicals, a process known as *chain termination*. A stable reaction is achieved if the rate of formation of radicals is equal to their rate of removal. In some cases one reactant free radical may give rise to two or more new free radicals among the products. The situation producing two product radicals from from one reactant radical is *chain branching*. Now instead of a stable combustion reaction, the potential exists for an explosion to occur. Many free radical chain reactions are exothermic. If the heat of reaction is not dissipated, the increased temperature increases reaction rates, further

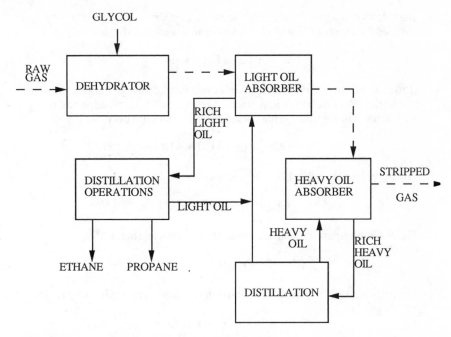

Fig. 7.2 Flow diagram of a natural gas processing operation. The principal by-products are ethane and propane, which are valuable feed materials for the petro-chemical industry. The product sold as fuel is the stripped gas.

exacerbating the situation. In many branching chain reactions, there is a balance between the rapid increase in numbers of radicals from chain branching processes on the one hand and the destruction of radicals by various chain termination steps on the other hand. Whether the net process is a relatively slow, controlled reaction or an explosion depends upon where the balance between chain branching and chain termination lies.

When methane reacts with an excess of air, the overall combustion process proceeds to carbon dioxide and water:

$$CH_4 + 2 O_2 \rightarrow CO_2 + 2 H_2O$$

The mechanism of this reaction is still not fully understood. Some sources in the combustion literature show over 50 individual steps in the conversion of methane to carbon dioxide and water. Since the O=O bond is usually stronger than any of the bonds in the reacting hydrocarbon, combustion reactions are initiated by breaking one of the bonds in the reactant molecule, rather than by dissociating oxygen. At very high temperatures the reaction is probably initiated by cleavage of a C-H bond to form a methyl radical.

$$CH_4 \rightarrow CH_3\cdot + H\cdot$$

However, at lower temperatures this simple bond cleavage may be too slow, and instead the reaction is initiated by a reaction of both methane and oxygen molecules. Thus a hydrogen abstraction process would occur as

$$CH_4 + O_2 \rightarrow CH_3\cdot + HO_2\cdot$$

The hydroperoxy radicals can attack another methane molecule

$$CH_4 + \cdot O_2H \rightarrow CH_3\cdot + H_2O_2$$

They may alternatively participate in other reactions such as

$$CH_3OO\cdot + \cdot O_2H \rightarrow CH_3OOH + O_2$$

The methyl radicals can react with oxygen to form a methylperoxy radical:

$$CH_3\cdot + O_2 \rightarrow CH_3OO\cdot$$

The methyl radical attacks the oxygen molecule at one of the lone pairs of electrons on one of the oxygen atoms. The reaction of methyl radicals with oxygen is a key reaction in the combustion process.

The methylperoxy radicals, $CH_3OO\cdot$, undergo further reactions to two intermediate combustion products, methanol and formaldehyde. An important first step is the generation of the methoxy radical, $CH_3O\cdot$, which can be formed in several reactions. One such process is the reaction of methylperoxy and methyl radicals:

$$CH_3OO\cdot + CH_3\cdot \rightarrow CH_3O\cdot + \cdot OCH_3$$

Hydrogen abstraction *by* the methoxy radical leads to methanol

$$CH_3O\cdot + CH_4 \rightarrow CH_3OH + CH_3\cdot$$

while hydrogen abstraction *from* the methoxy radical leads to formaldehyde

$$CH_3O\cdot + R\cdot \rightarrow H_2C=O + RH$$

In this instance the symbol $R\cdot$ is taken to mean some other radical, not necessarily an alkyl radical, present in the system.

At high temperatures, reactions involving direct attack on the methane molecule by other radical species present in the system are all contributors to the overall combustion process. Examples of such reactions are the attack by hydroxy, hydrogen , or oxygen radicals:

$$CH_4 + \cdot OH \rightarrow CH_3\cdot + H_2O$$
$$CH_4 + H\cdot \rightarrow CH_3\cdot + H_2$$
$$CH_4 + O\cdot \rightarrow CH_3\cdot + \cdot OH$$

Any free radical reaction can participate in three generic types of processes: *chain propagation*, in which one new radical is generated for each one consumed and which keeps the process going; *chain branching*, which offers the possibility of speeding up the process by generating more than one radical for each one consumed; and *chain termination*, which stops the process as radicals combine with each other.

The supply of methyl radicals consumed in the formation of methylperoxy radicals can be regenerated from methane by a hydrogen abstraction process as, for example,

$$CH_4 + \cdot OH \rightarrow CH_3\cdot + H_2O$$

Ideally, to maintain steady combustion, chain propagation is the desired step. The formation of the methylperoxide radical shown above is the key to chain propagation via the reaction

$$CH_3OO\cdot + CH_4 \rightarrow CH_3OOH + CH_3\cdot$$

This again is a hydrogen abstraction reaction. At very high temperatures the peroxy radical decomposes via breaking of the O-O bond; but at lower temperatures, the peroxy radical may be stable enough to attack another molecule of reactant, as shown above. The methyl radical formed in this step can then react with oxygen to regenerate the methylperoxide radical and keep the process going. The formation of methyl hydroperoxide represents a further weakening of the O-O bond (relative to the strong O=O bond in the oxygen molecule) and eventually breakage of the O-O bond can occur. The O-O bond in the hydroperoxide has a bond dissociation energy of about 184 kJ/mol; this value is less than half the bond dissociation energies of various types of C-H bonds (Chapter 3). The methyl hydroperoxide decomposes to intermediates, e.g.,

$$CH_3OOH \rightarrow CH_3O\cdot + \cdot OH$$

$$CH_3OOH \rightarrow H_2C=O + H_2O$$

The •OH radical is one of the most important oxidants in fuel combustion systems containing the elements oxygen and hydrogen. Other oxidants, usually of lesser importance, are HO_2•, H•, and O•. At this point further attack can occur, as for example the reaction of the methoxy radical with another oxygen molecule:

$$CH_3O• + O_2 \rightarrow O=CHOO•$$

At first sight it might seem preferable to encourage chain branching in order to accelerate the combustion process. For example, consider the sequence of reactions

$$CH_3• + O_2 \rightarrow CH_3OO•$$
$$CH_3OO• + CH_4 \rightarrow CH_3OOH + CH_3•$$
$$CH_3OOH \rightarrow CH_3O• + •OH$$
$$CH_3O• + CH_4 \rightarrow CH_3OH + CH_3•$$
$$•OH + CH_4 \rightarrow CH_3• + H_2O$$

- -

net: $CH_3• + 3\ CH_4 + O_2 \rightarrow 3\ CH_3• + CH_3OH + H_2O$

The single methyl radical on the left hand side of the net equation has produced three methyl radicals in the products. Each of these three methyl radicals could in principle initiate the same sequence of reactions just shown above, thus leading to the production of 9, then 27, then 81 .... radicals. Assuming that appropriate conditions of temperature, pressure, and reactant concentration were available, after twenty such cycles the number of radicals produced would be 3.5×10^9 from each methyl radical initiating the process. The reaction accelerates so rapidly we observe it as an explosion.

If the amount of oxygen is limited, incomplete combustion to carbon monoxide occurs:

$$2\ CH_4 + 3\ O_2 \rightarrow 2\ CO + 4\ H_2O$$

(Notice that O_2/CH_4 ratio is 1.5, compared with 2.0 in the case of complete combustion.) The incomplete combustion is not a desirable reaction, from the perspective of uses of natural gas as a fuel. First, much less heat is evolved, -519 kJ/mol vs. -804 kJ/mol. Second, the carbon monoxide is an extremely toxic gas which represents a potentially fatal hazard. There is no reason why one would deliberately carry out the incomplete combustion of natural gas in a combustion system. However, the partial combustion of methane represents an alternative route to the very useful mixture of carbon monoxide and hydrogen, discussed below in the section on non-fuel uses of natural gas.

When methane is burned in a very limited amount of oxygen, one of the reaction products is solid carbon:

$$CH_4 + O_2 \rightarrow C + 2 H_2O$$

If this reaction were to occur in a combustion system it would be very undesirable, first because of the limited amount of heat evolved and second because the carbon would deposit on surfaces as soot. However, the formation of carbon from fuel-rich flames is a very useful one in industry because the solid products, carbon blacks, have a variety of valuable applications, to be discussed in the next section.

Soot is essentially completely aromatic, but does not quite attain the structure of graphite. Some hydrogen remains in soot, and C_8H is a reasonable empirical formula of soot. With the hydrogen atoms present, the planes of aromatic carbon do not attain the perfect vertical stacking of graphite, but rather the stacking is disordered.

Carbon blacks

Carbon blacks are colloidal carbon materials which usually form as spherical particles (or aggregates of spherical particles). They are produced as an amorphous powder. As produced, carbon blacks are very light and fluffy, with bulk densities as low as 0.06 g/cm^3. A distinguishing feature of carbon blacks is their small particle size, which is in the range of 10 - 1000 nm. Carbon blacks have a wide range of uses, including as a pigment in ink, a pigment in rubber (most tires are about 25% by weight carbon black), and as a filler in plastics to increase resistance to wear and abrasion. Carbon blacks have a variety of trivial names depending on their specific method of manufacture. Carbon blacks can be manufactured by the incomplete combustion of various hydrocarbons, but natural gas has used widely because it is relatively inexpensive and usually readily available. Manufacture of carbon blacks from petroleum products is also an important source of this material. The yield of a carbon black varies with the manufacturing process, but is typically in the 2-6% range.

Channel blacks are produced by burning methane in a limited supply of air. The carbon deposits on cold iron channels, which move slowly with a reciprocating motion past scrapers, so that the deposited carbon is scraped off the channels and is collected in hoppers. The process for manufacturing channel blacks was developed in 1872, and remained in use essentially unchanged at least until 1945. Originally a gas made by the partial thermal decomposition of coal was used, but this fuel was supplanted by natural gas. The yield of carbon was about 10-15% at best. The channel black particles are 20 - 30 nm in diameter. They are used as pigments in high quality printing inks, and as additives for rubber used in large truck tires. *Furnace blacks* are produced by burning methane or oil in air inside a furnace. The fine carbon particles, produced in yields of 20-40%, are recovered by electrostatic precipitation. Furnace blacks are typically 30 - 80 nm in diameter and are used as polymer additives to improve the tensile strength, tear resistance, and abrasion resistance of the polymers. About 75% of all carbon blacks manufactured are furnace blacks. (The rest of the production is about evenly divided between the other two types.) *Thermal blacks* are produced by the thermal decomposition of methane on hot refractory surfaces, with temperatures of 1400 - 1650°. A 40-50% yield of thermal blacks can be obtained. The particle sizes of thermal blacks are greater than furnace blacks, and can be in the range 200-500 nm. These large particles are not as suitable for use as polymer or rubber reinforcement. Thermal blacks are used in the manufacture of wire insulation, dry cells, and carbon paper.

Carbon blacks have a totally aromatic structure ($f_a = 1.00$). However, instead of the perfect alignment of aromatic planes found in the graphite structure, there is no long-range structural order in carbon blacks. The microstructure of carbon blacks consists of small packets of turbostratic crystallites, which are not aligned with each other. Each turbostratic stack consists of a few planes of atoms. Continuing the convention of Chapter 6, representing the edge-on view of an aromatic unit as a straight line, the structure of a carbon black could be represented as

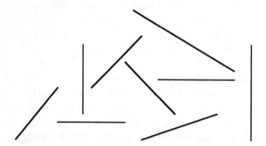

Fig. 7.3 A schematic view of the aromatic planes in carbon black. Although no structural ordering is shown in this diagram, the carbon layers near the surface of the spherical particles tend to align parallel to the surface.

The structural order in carbon black depends on the method of preparation. Generally the atomic planes near the surface tend to align parallel to the surface. In graphite the π electron systems in each aromatic plane interact to provide the weak bonding between planes. Because the aromatic planes are not aligned in carbon blacks, the weak π interactions can be fulfilled by adsorption of gases such as oxygen or hydrogen. Channel blacks may contain

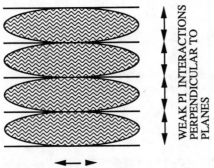

WEAK PI INTERACTIONS PERPENDICULAR TO PLANES

STRONG COVALENT INTERACTIONS WITHIN PLANES

Fig. 7.4 Graphite structure viewed edge-on showing covalent interactions within planes of atoms and weak π interactions (wavy lines) between planes.

12-18% of material which volatilizes on heating the carbon black to 1000°. The strong interaction of gases with carbon blacks is not a simple, reversible physical adsorption; rather, it is an example of *chemisorption*, an adsorption process which involves chemical bonding interactions. Thus when oxygen is chemisorbed onto carbon blacks, it is desorbed as carbon monoxide or carbon dioxide rather than as oxygen. Chemisorption is partly responsible for the ability of carbon blacks to reinforce polymers. To reinforce a rubber, the carbon black particles must be small, 20-50 nm, and able to be "wetted" by the rubber, implying that there must be string surface interactions between the components of the rubber and the surface of the carbon black particles. Carbon black additives can improve the tensile strength of natural rubber by 40%. They can improve the tensile strength of styrene-butadiene rubber, a major component of tires, by a factor of 10. Indeed, the use of carbon blacks to improve the physical properties of tires is the reason why tires are black.

Other non-fuel uses of natural gas

Synthesis gas and methanol

The most important non-fuel use of natural gas is in *steam reforming*. At temperatures of about 900°, methane reacts with steam to form hydrogen and carbon monoxide:

$$CH_4 + H_2O \rightarrow 3\,H_2 + CO$$

The effect of temperature on the steam reforming reaction is illustrated by the data shown in Table 7.1.

Table 7.1 Effect of temperature on composition of product in steam reforming (starting material an equimolar mixture of steam and methane)

Temperature	CH_4	H_2O	H_2	CO
710°	8.8%	8.8%	61.8%	20.6%
940°	0.8%	0.8%	73.8%	24.6%

In industrial scale gas reactions it is often desirable to use increased pressure, because operation at increased pressure decreases the size of the reactor needed, and hence potentially decreases the capital investment in the reactor. The potential savings in using a smaller, but higher pressure reactor, may be offset by the reactor itself needing to have thicker walls and higher mechanical integrity to contain the increased pressure, design features which increase the cost of the reactor. Thus a balance must be struck between the cost savings from the smaller size and the cost increase from the higher integrity. However, in steam reforming two moles of reactants form four moles of products. From LeChatelier's principle we would expect an increase in pressure to shift the equilibrium to the left. Nevertheless in industrial practice the reaction is carried out at 850-900° and 3 MPa. A catalyst of nickel supported on magnesium or aluminum oxide is used. Unconverted methane is recycled to the reactor.

The major application of steam reforming is the manufacture of hydrogen. The carbon monoxide can be destroyed by further reaction with steam in the *water gas shift reaction*::

$$CO + H_2O \rightarrow CO_2 + H_2$$

The thermodynamics of the water gas shift reaction indicate that low temperatures would drive the equilibrium to the right side of the equation.

Unfortunately, reaction rates at low temperatures are too low for practical use of the reaction, so that in practice the water gas shift is run at high temperatures in the presence of a catalyst.

Like hydrogen sulfide, carbon dioxide is mildly acidic and is readily absorbed by weak bases, leaving essentially pure hydrogen as the gaseous product. Carbon dioxide can also be removed by scrubbing the gas with water at high pressures. Hydrogen is very valuable as a starting material in the production of ammonia for use in fertilizers. Hydrogen reacts with nitrogen at 400-600° and 20 MPa:

$$3\,H_2 + N_2 \rightarrow 2\,NH_3$$

Hydrogen is also in demand as a rocket fuel; one of the largest consumers of hydrogen in the United States is the National Aeronautics and Space Administration.

The carbon monoxide - hydrogen mixture produced in the steam reforming reaction is known as *synthesis gas*. Synthesis gas is an extremely versatile material used to produce a wide variety of fuels and useful chemicals. In Chapter 11 we will discuss in detail the chemistry of synthesis gas and how its composition can be adjusted by the water gas shift reaction. One example of the commercial application of synthesis gas is in the production of methanol

$$CO + 2\,H_2 \rightarrow CH_3OH$$

The reaction proceeds at 300-400° and pressures of 26-36 MPa. Methanol is a good solvent having numerous uses in the chemical industry. In the future, methanol may prove useful as a liquid fuel for vehicles to replace petroleum-derived fuels. In the past most methanol was made by heating wood in the absence of air (which was the source of its common name, wood alcohol). Today about 99% of all methanol is produced from synthesis gas.

An alternative route to hydrogen / carbon monoxide mixtures is the partial combustion of methane with oxygen in the presence of supported nickel catalysts at 850-900°. With new copper/zinc supported catalysts the reaction proceeds rapidly at temperatures of 210-240°. Again, by-product carbon dioxide is removed by scrubbed with monoethanolamine. The net reaction is

$$2\,CH_4 + O_2 \rightarrow 4\,H_2 + 2\,CO$$

Because of the increase in moles of products relative to reactants, an increase in pressure would drive the equilibrium to the left. However, since the overall equilibrium is so favorable, this reaction can be carried out at 4 MPa with acceptable conversions to the desired products. Operation at increased pressure allows a reduction in the size of the reactor needed to achieve a desired throughput. When the eventual aim is to produce ammonia, the oxidation process can be run with a mixture of air and oxygen, so that the ratio of nitrogen (contributed by the air) and hydrogen (from the partial oxidation followed by water gas shift) is the desired value for the ammonia synthesis reaction.

Acetylene

To introduce the chemistry of acetylene production and use, it is helpful first to digress to discuss the thermal stabilities of hydrocarbons, and specifically the free energies of formation. The reaction

$$C \text{ (graphite)} + 2 H_2 \text{ (g)} \rightarrow CH_4 \text{ (g)}$$

has a $\Delta G°_f$ of -50 kJ/mol at 298 K and 0.1 MPa. The negative free energy of formation indicates that methane is stable with respect to the elements at 298 K. However, ΔG becomes less negative at higher temperatures, and becomes zero ~900 K, or about 600°. Above 600° methane tends to decompose back to the elements. Increased pressure will shift the equilibrium to the right. As one passes along a homologous series of hydrocarbons, ΔG becomes less negative with increased molecular size. This change of ΔG with increased size is observed for homologous series of alkanes, alkenes, alkynes, and aromatics. However, the change of ΔG with respect to temperature is different for different compound classes, as illustrated in Figure 7.5. Thus there is a temperature range in which alkenes are more stable than alkanes, and, at higher temperatures, the order of thermal stability is aromatic > alkene > alkane. High temperature pyrolysis (or partial combustion) of alkanes tends to give alkenes and aromatics. Acetylene is one of the very few organic compounds for which ΔG_f becomes increasingly negative (or, strictly speaking, less positive) with increasing temperature. At 298 K, ΔG is +209 kJ/mol, and in fact acetylene can decompose explosively, especially if compressed. The ΔG_f finally becomes negative when the temperature reaches about 4000°.

At temperatures in the range of 1500° methane cracks to acetylene

$$2 CH_4 \rightarrow C_2H_2 + 3 H_2$$

The reaction becomes thermodynamically feasible about 1200°. However, it is critical that the reaction products be quenched immediately to prevent further decomposition of the acetylene to carbon

$$C_2H_2 \rightarrow H_2 + 2 C$$

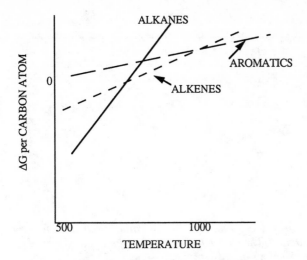

Fig 7.5 Variation of free energy of formation per carbon atom as a function of temperature for alkanes (solid line), alkenes (short dashes), and aromatics (long dashes). As temperature increases aromatics eventually become the most stable of these compounds.

Acetylene is used as a fuel gas in oxyacetylene welding torches. The flame is extremely hot because a great deal of heat is liberated when a highly unstable hydrocarbon like acetylene is burned. Its main use is in the production of vinyl chloride:

$$C_2H_2 + HCl \rightarrow CH_2CHCl$$

Vinyl chloride is the monomer from which the ubiquitous plastic poly(vinyl chloride), generally known as PVC, is produced. Larger alkanes can also be used for acetylene production. Regardless of the feedstock, the energy required for the cracking reaction is obtained from burning a portion of the feed. Both carbon monoxide and hydrogen are formed as by-products in acetylene production. It is critical that the reaction products be quenched rapidly to avoid destruction of the acetylene in subsequent reactions.

Further reading

Barnard, J. A. and Bradley, J. N. (1985). *Flame and Combustion*. London: Chapman and Hall; Chapters 1,5,6,7

Donnet, J. P. and Voet, A. (1976). *Carbon black*. New York: Marcel Dekker Inc.

Kohl, A. and Riesenfeld, F. (1985). *Gas Purification*, 4th edn. Houston: Gulf; Chapters 2,11,14.

Chapter 8

Petroleum Products

This chapter discusses the principal method for separating crude oils into useful products - distillation. The characteristics and major applications of each of the distillation fractions are also treated.

There are essentially no applications in which crude oil is used exactly as it comes from the ground. Presumably, crude oil could in principle be used "as-is", but for the most efficient use of fuel it is preferable to have as narrow a range of fuel properties as possible and to design the combustion equipment to make use of a fuel having those specifications, rather than to try to use fuel of widely varying properties. Thus crude oil is separated into a range of products, each having specific use. We have seen previously that crude oil contains hundreds to thousands of individual compounds. In principle, one could separate each of these individual compounds if one wished to go to the trouble. In practice the separations of crude oil represent a compromise between, on the one hand, the desire to achieve a limited range of fuel specifications and, on the other hand, being able to offer a product at an economically realistic price. For example, a high purity grade of hexane, a compound which is one of many components of gasoline, costs about 50 times as much as a comparable volume of gasoline.

Crude oil is passed through several operations designed to separate it into fractions, to purify the products, and, in some cases, to increase the yield of the more valuable products relative to others for which there is not so great a demand. Collectively these processes represent the technology of oil refining. A refinery will be based on a number of individual unit operations. Modern refineries are designed with flexibility for which operations are run, depending upon the quality of the oil being refined and on the demands of the market for specific products. In some extreme cases, refinery operations may change on a weekly basis.

Desalting

Many of the geologic traps that collect petroleum as it migrates through the earth contain formations of salt. During drilling and pumping operations used to extract oil from the earth, some salt inevitably gets mixed with the oil. Salt is undesirable for the same reasons it was an undesirable component of coal: during subsequent processing the chloride ion may become converted to hydrochloric acid, which can be highly corrosive to the processing equipment.

The first operation in most refineries is *desalting*. Water, in amount of 5-15% of the amount of oil, is mixed with the oil at 90-150°. The pressure is maintained at whatever value is required to prevent the water from vaporizing at the operating temperature. A water-oil emulsion forms as a result, with the salt dissolving into the water. After the emulsion settles, the water layer, now containing the salt, is drawn off.

Distillation

Distillation is the key operation in petroleum refining. It is used to separate the crude oil into *cuts*, products such as gasoline, kerosene, and fuel oil. Distillation can also be used to achieve finer separations of some of these product streams.

The boiling points of the various compound classes in crude oil increase with increasing molecular weight. This relationship is shown in Figure 8.1 for the n-alkanes, and in Figure 8.2 for some alkylbenzenes.

Examining Figure 8.1 for the compounds through C_{10} shows that the curve increases steeply in the early part of the series, and then less steeply among the larger molecules. The reason is that each additional $-CH_2-$ unit represents a large increase in molecular weight when the parent compounds are small, but with the larger alkanes the proportional increase in molecular weight by addition of an "extra" $-CH_2-$ is much less. For example, the molecular weight of ethane is 88% higher than that of methane, and the boiling point is 73° higher; but when comparing nonane and decane the molecular weight of the latter is 11% higher, and the boiling point is increased only by 25°. A rough rule of thumb for relating boiling points of the n-alkanes is that the compound of molecular weight 100 (heptane) has a boiling point of *roughly* 100° (the actual boiling point of heptane is 98°) and that, among the liquid alkanes, there is a difference of *roughly* 20° for each -CH_2- additional (or deleted) relative to heptane.

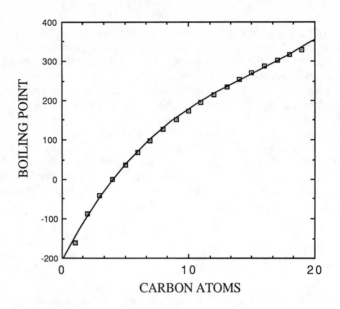

Fig. 8.1 Variation of boiling point with number of carbon atoms among the n-alkanes.

The increase in boiling point with molecular weight is in part a reflection of the increased London force interactions between the larger molecules. We have seen previously that the greater the apparent surface area of a molecule, the greater will be the attractive London force (Chapter 4). In the alkanes, for example, each added $-CH_2-$ unit in the structure increases the area of contact, and therefore increases the total London force interactions between two molecules. The increase obtained for each added $-CH_2-$ unit is about 4-6 kJ/mol. The branched chain alkanes are more compact than their straight chain isomers; consequently, the London force interactions between molecules are weaker and the branched isomers have lower boiling points and melting points than their straight chain isomers. Many distillation systems are in use in refineries; however, most crude oil distillation can be classified into one of three generic types: single stage, two stage, or two stage with vacuum tower distillation. The specific practice used varies from one refinery to another, depending on the composition and boiling range of the crude oil to be processed and on the mixture of salable products desired. Despite the diversity of refinery practice, it is generally

174

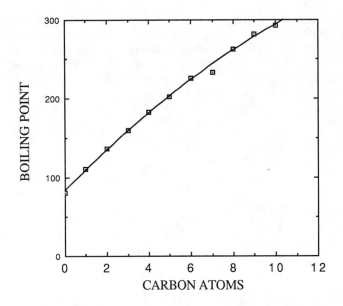

Fig. 8.2 Variation of boiling point for the n-alkylbenzenes
as a function of the number of carbon atoms in the alkyl group.

true that fractional distillation is the most versatile and, usually, the cheapest means of separating volatile mixtures.

Principles of distillation

The fundamental physical basis of distillation is that when a liquid containing two (or more) components is heated to its boiling point, the composition of the vapor at equilibrium will be different from that of the liquid. The vapor will be enriched in the more volatile component(s). If the vapor is withdrawn from the system and condensed separately, the liquid formed by condensation of the vapor will have a higher proportion of the more volatile component than did the original liquid. Thus a mixture of liquids can be separated on the basis of their relative volatilities. Although we have discussed volatility relationships within classes of compounds, it is important to remember that the volatility relationships among different classes of compounds does not necessarily correlate with molecular weight. For example, the naphthene decalin, $C_{10}H_{18}$, with a molecular weight of 138, has a boiling point of 195°, while the paraffin decane, $C_{10}H_{22}$, has a molecular weight of 142 but boils at 174°.

Consider a liquid containing two components. It should be borne in mind that the principles are equally applicable to liquids having a greater number of components, although the analysis of the problem becomes more complex. The relationships among vapor composition, liquid composition, and temperature for a liquid containing two components, arbitrarily called A and B, is shown in Figure 8.3.

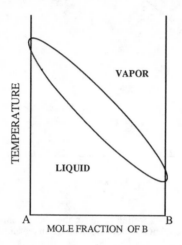

Fig. 8.3 Vapor-liquid equilibrium for a two-component liquid, In this example the boiling point of component A is much higher than

The key feature of Fig. 8.3 is that, at any given temperature, the equilibrium compositions of the vapor and liquid will be different. In this instance, component B was arbitrarily selected to be the more volatile component (its boiling point is lower than that of component A).

Let us consider an A-B mixture having a composition represented by x_1 on the diagram in Figure 8.4. When this mixture is heated to its boiling point, T_1, the vapor at equilibrium will be richer in component B, which is the more volatile of the two components. The liquid will therefore be richer in component A. If the vapor, which has the composition represented by y_1, is then removed from the system and condensed, the liquid will have composition x_2, and will be richer in component B than was the original liquid mixture. If the new liquid, represented by composition x_2, is boiled (at its boiling point T_2), a vapor of composition y_2 and even further enriched in

176

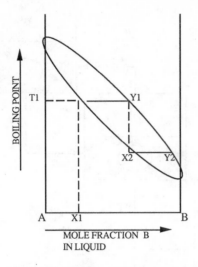

Fig. 8.4 Heating of a two-component liquid
results in the formation of a liquid relatively
rich in the less volatile component, and a
vapor enriched in the more volatile component.

component B will be produced. A succession of steps condensing and
reboiling the vapors can in principle produce a liquid (by condensation)
which contains only B. At the same time, the liquid phase which did not
vaporize in each of these steps will be continually enriched in component A,
and eventually would consist only of A.

The notion of *fractional distillation* can be developed by considering
each individual vaporization, condensation, and reboiling to be carried out in
separate process vessels, as shown schematically in Figure 8.5.

In the sequence of operations shown in Fig. 8.5, heat is supplied to
vessel 1 so that the liquid to be fractionally distilled, of composition x_1, boils.
The vapor, which has composition y_1, is passed into vessel 2. Since $T_1 > T_2$, at
equilibrium vessel 2 will contain a liquid having composition x_2 and a vapor
having composition y_2. (The heat required for the fractional distillation
process is supplied only to stage 1; heat for the subsequent stages is obtained
from the vapors from previous steps.) The same argument applies to a third
stage, where, by extension, the temperature T_3 would be such that $T_2 > T_3$
and the vapor and liquid would have equilibrium compositions of y_3 and x_3,
respectively. The effect of this sequence of operations on compositions is
shown in Figure 8.6.

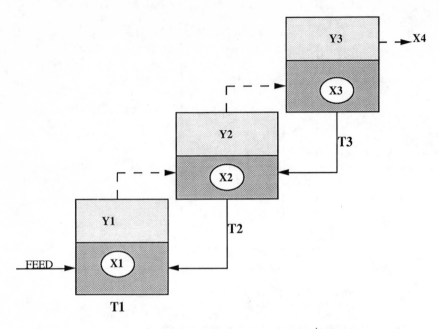

Fig. 8.5 Schematic diagram of a fractional distillation process. Liquids are designated by the symbol **X**, darkly shaded blocks, and solid lines; vapors are designated by the symbol **Y**, lightly shaded blocks, and dashed lines.

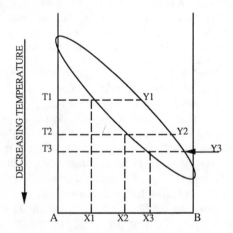

Fig. 8.6 Successive changes in the composition of liquid and vapor in the fractional distillation scheme illustrated in Figure 8.5.

178

Successive enrichment of the vapors in component B proceeds from stage 1 through 2, 3 ... etc. Enrichment of the liquid in component A proceeds in reverse from the highest stage through 3, 2, and 1. To obtain the maximum efficiency of separation some of the product from a given stage is recycled to the previous stage, *e.g.*, liquid of composition x_4 is recycled to stage 3. The efficiency can be expressed in terms of the *reflux ratio*, which is defined as the moles of material recycled per mole removed. Maximum efficiency is obtained when the reflux ratio is infinite, which corresponds to a situation in which the moles of material removed from the process is zero. Obviously a reflux ratio of infinity is useless in practice!

In industrial distillation practice, the entire operation is carried out in a single vessel, rather than the separate units introduced in Fig. 8.5 as a conceptual approach. A distillation column or tower is a vertical, cylindrical steel vessel. The individual distillation stages are carried out on plates inside the column. Various designs of plates are employed; Figure 8.7 shows an example of the bubble cap plate.

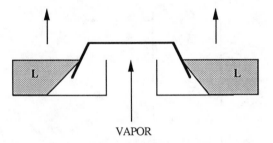

VAPOR

Fig. 8.7 A bubble cap plate used in a distillation column. The vapor ascending from the plate immediately below is caught by the cap and forced to bubble through the liquid (shaded areas marked L) before passing upward to the next plate.

Each plate provides a site for contacting liquid and vapor. A distillation column affords much greater simplicity of construction and operation than a sequence of individual vessels (Fig. 8.5), but has the attendant disadvantage that an equilibrium is never truly attained on each plate. Usually the separation attained on each plate is 80 - 90% of the theoretical equilibrium separation.

Single stage distillation

This operation begins by heating the crude oil in a *pipe still*. The feed may be preheated to temperatures in the range 315-370°. The hot oil is then fed into the *distillation column*, which is often heated by steam. A simple column may be 4 m in diameter and 25 m tall, and may contain 15 to 30 plates. The volatile components of the crude oil vaporize. As the vapors pass up the column, they equilibrate with liquids which condensed higher in the column and are falling back down. This process can be depicted as shown in Figure 8.8.

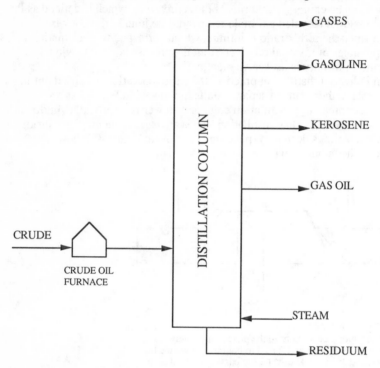

Fig. 8.8 A simple distillation column, showing fractionation of a crude oil into the major product streams.

The most volatile material (i.e., the material with the lowest boiling range) leaves the top of the column as vapor and is then condensed. (The vapor stream exiting the top of a distillation column is called the *overhead*.) This product is *gasoline*. Inside the distillation column the vapors and condensed liquids establish equilibria. The more volatile the fraction, or the lower its boiling range, the higher in the column it will equilibrate with liquid. By providing a way to draw streams from various heights in the column, it is possible to obtain a sequence of products all separated - fractionated - on the basis of boiling points.

In some cases it desirable to adjust carefully the boiling range of one or more of the particular products. The various streams drawn from the side of the column can be fractionated further in a *side stream stripper*. Again steam is used as the heat source.

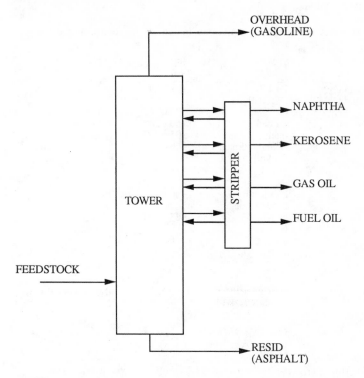

Fig. 8.9 A distillation unit with a side stream stripper, to adjust the boiling ranges of selected product steams.

The side stream liquids contain lower boiling components in solution. These low boiling materials are removed by fractional distillation and carried with the steam into the main distillation tower.

A limiting factor in a simple distillation process is that thermal cracking of the oil begins at temperatures around 360°. The thermal cracking can lead to the deposition of carbon in process equipment, which can in turn lead to the formation of "hot spots" in the equipment, eventually causing failure. Thus the distillation is concluded at temperatures below 360°. The residual material is known as *reduced crude*.

An example of the petroleum fractions which could be produced in a simple distillation is given in Table 8.1. It should be understood that the petroleum industry has spawned a variety of informal terms for the fractions, and that the temperature limits of the boiling ranges of the fractions may also vary somewhat with different refinery practices. Thus the information in Table 8.1 is not rigidly codified in the way that, for example, the rank classification of coal is.

Table 8.1 Examples of petroleum fractions produced by distillation of crude oil

Fraction	Typical uses	Boiling range, °C
Light gas, C_1 and C_2	Fuel	-160 - 30
C_3 and C_4	Fuel (LPG)	
Light naphtha	Gasoline	30 -150
Heavy naphtha	Jet fuel	150 - 205
Kerosene	Fuel	205 - 260
Light gas oil	Heating furnaces	260 - 315
Heavy gas oil	Fuel oil, feed to cracking units*	315 - 425
Reduced crude		
Lube oils	Lubrication	~400
Vacuum gas oil	Feed to cracking units	425 - 600
Resid	Heavy fuel oil, asphalt	>600

*The cracking process is discussed in Chapter 9.

Generally the boiling range of gasolines is 60 - 200°, and that of diesel fuels, 200 - 380°.

Two-stage distillation

Often there is a need to produce and market a wider range of products than can be obtained from single-stage distillation even with a side stream stripper. Two-stage distillation actually uses three distillation columns: a *primary tower*, which operates at 0.3 MPa, a *secondary* (or atmospheric) *tower*, which operates at atmospheric pressure, and a *stabilizer*. The vapors from the primary tower are fed to the stabilizer, where they are separated into *light gasoline* and *light ends* (principally propane and butane). The side streams from the primary tower are *light kerosene* and *naphtha*. The material which does not vaporize in the primary tower, the *bottoms*, is fed to the atmospheric tower. The side streams from the atmospheric tower are *kerosene*, *light diesel*, and *heavy diesel*. The material which does not distill in the atmospheric tower is the residuum, often referred to as *resid*.

A block diagram of the two-stage distillation operation is shown in Figure 8.10:

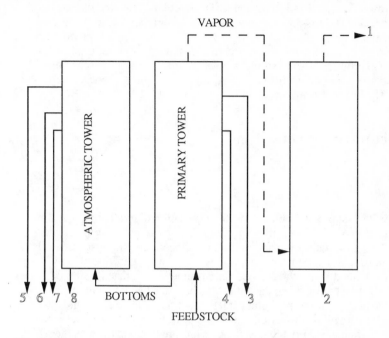

Fig. 8.10 A two-stage distillation operation. The principal products are 1, light gases; 2, light gasoline; 3, naphtha; 4, light kerosene; 5, kerosene; 6, light diesel oil; 7, heavy diesel oil; and 8, residuum.

Two-stage distillation with vacuum tower

In some refinery operations there is a market for materials having very high boiling ranges, an example being lubricating oils. As we have seen, the operation of conventional distillation columns at high temperatures can result in deposition of carbonaceous solids from thermal cracking. Formation of coke inside a distillation column would not only waste some of the oil but could interfere with the operation of the column, requiring that it be shut down for cleaning or repair.

   To distill materials of high boiling points without extremely high temperatures requires that the pressure be reduced. Thus a vacuum distillation operation (a *vacuum tower*) can be added to the two-stage distillation system. A vacuum tower is required to produce lubricating oil (*lube oil*) fractions. In this case the resid from the atmospheric tower would be pumped into the vacuum tower. The pressure in the vacuum tower is 7-17 kPa. Because of the very low pressure, the volume of vapor is quite large, and vacuum towers can be 15 m in diameter. The overhead from the vacuum tower is *gas oil*. The gas oil is obtained at ~150° from the vacuum tower; its boiling range at atmospheric pressure is 315-425°. The lube oil fractions are taken off the vacuum tower as side streams. Usually three or four lube oil fractions are produced. An example is *cylinder stock*, a highly viscous oil used to lubricate the pistons in slow speed engines such as stationary steam engines. The bottoms from the vacuum tower is *asphalt*.

Petroleum products

Gasoline

Normally the first product obtained in the distillation of crude oil is gasoline. The production of gasoline is the aspect of refining that calls for the greatest effort and has the greatest impact on the energy economy. In most crude oils the content of hydrocarbons having boiling points in the gasoline range is 10-30%. In the United States the demand for gasoline is such that 40-50% of the crude oil must be converted to gasoline, which means that much of the gasoline is manufactured from other fractions of the crude oil, using processes which will be discussed in Chapter 9. In Europe, where fuel is more expensive than in the United States, the cars are smaller, and the average family owns fewer cars, the requirement for gasoline is lower but the need for heating oil is higher. Therefore, different refining strategies are used in Europe and the United States.

The specifications of gasoline vary somewhat with the specific crude being refined and with the type of distillation operation used, but generally the boiling range of gasoline is 25-150°. (This boiling range is equivalent to the C_5 - C_9 n-alkanes.) Gasolines are sometimes classified as *motor gasoline*, for use in automobiles and light trucks, and *aviation gasoline*, for use in piston-engine aircraft.

We have seen previously that natural gasoline is the liquid condensed from wet natural gas. The gasoline produced from the distillation of crude oil, without any other chemical treatments, is *straight run gasoline*. There are two major problems associated with both natural and straight run gasoline. First, the yield of these materials is not adequate to meet market demands for gasoline. Second, their combustion performance in modern automobile engines is not satisfactory.

The first high performance, eight cylinder vee-block engine (V-8) was introduced in 1932 by Ford. Throughout the 1930's there were continuous efforts to improve performance, especially horsepower rating, of automobile engines. Three factors determine the horsepower of an engine: the effective average pressure in the cylinders during the operating cycle, the total swept volume (*i.e.,* the volume displaced by all the pistons moving through the full length of the stroke), and the average piston speed. The average pressure in the cylinders is of particular concern for fuel chemists.

Increasing the pressure increases the thermal efficiency; that is, more of the energy released by fuel combustion is available to act on the pistons and therefore more work is done per unit of fuel consumed. One design approach to increasing the average cylinder pressure is to increase the compression ratio. The compression ratio is a measure of the volume change occurring as the piston moves from the bottom of its stroke to the top (Fig. 8.11)

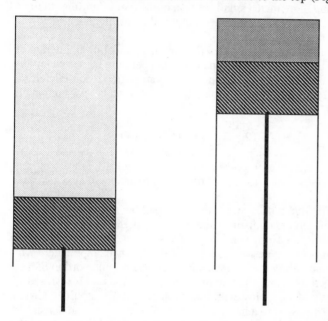

Fig. 8.11 When the piston is at the bottom of its stroke (as in the left-hand sketch) the gas pressure in the cylinder is relatively low, as indicated by the light shading. With the piston at the top of its stroke, the gas pressure is increased, as indicated by the darkly shaded area in the right-hand sketch. The compression ratio is the ratio of the cylinder volume when the piston is at the bottom of the stroke to the volume when the piston is at the top.

A "high compression" automobile engine may have a compression ratio of 10:1. To develop the relationship of engine design to fuel composition and structure it is necessary to investigate the events occurring when the fuel burns.

185

Engine knock

After the air-fuel mixture has been drawn (or injected) into the cylinder and compressed, it is ignited by a spark plug. (This method of igniting the fuel is why such engines are sometimes called *spark ignition,* or SI, engines.) Initially the spark plug ignites only a tiny portion of the air-fuel mixture, in the immediate vicinity of the spark plug gap. This ignition produces a flame of roughly hemispherical shape which ignites further molecules in the immediate vicinity of the flame; that burning layer ignites another, and so on until all of the fuel has been consumed. The entire combustion process requires less than one second, but even so, for a properly tuned engine there is no instant when any more than a small fraction of the fuel is burning. In this way the energy of the fuel is released smoothly and uniformly and not violently.

When the air-fuel mixture in the burns in the cylinder of an internal combustion engine, several things happen. The temperature increases rapidly, thus increasing the pressure of the mixture. The combustion reaction yields a greater number of moles of products than there were of reactants. For octane, for example,

$$C_8H_{18} + 12.5\ O_2 \rightarrow 8\ CO_2 + 9\ H_2O$$

In this example, 17 moles of products are formed for each 13.5 moles of reactants. The increase in the number of moles of products will also increase the pressure.

As the flame front moves away from the tip of the spark plug, the remaining mass of unburned air-fuel mixture will experience increasingly high pressures as a result of the combined effects of increasing temperature and increased number of moles of products. In a well-tuned engine there will be no instant of time during the combustion stroke when some portion of the air-fuel mixture is not burning, so that the energy from the fuel is released uniformly and not violently. However, it is possible that the unburned air-fuel mixture can be compressed to a point at which the temperature and density are high enough to cause all of the remaining unburned mixture to ignite at once. The resulting explosion is severe enough that it is audible to the driver, and is the phenomenon we know as *engine knock*. Engine knock is undesirable because it wastes fuel and puts mechanical stresses on engine components which lead to premature wear or breakage.

An increase in the compression ratio of the engine will increase the pressure in the cylinder at the instant of ignition. Since the pressure is already higher than in the cylinder of an engine with lower compression ratio, the chances that the pressure will build to a point of knocking are also greater. That is, a high compression ratio engine has a greater tendency for knocking (assuming for the moment that identical fuels are burned in each engine). As automobile performance increased over the years with the development of higher compression engines, there became a need for gasolines with less tendency to knock.

We examined in Chapter 7 some of the combustion chemistry of methane. In some respects the combustion of higher hydrocarbons, such as the components of gasoline, is similar to methane combustion. It is likely that the combustion reaction is initiated in the same way:

$$RH + O_2 \rightarrow R\bullet + \bullet O_2H$$

with a hydrogen abstraction reaction generating an alkyl radical. However, the larger and more complex structure of the alkyl radical relative to the methyl radical offers the opportunity for reactions which cannot occur with methyl radicals. These new reactions include β-bond scission:

$$R'CH_2CH_2R\bullet \rightarrow R'\bullet + CH_2=CHR$$

to form an alkene and a new alkyl radical; a further hydrogen abstraction to form an alkene:

$$RCH_2CH_2\bullet + O_2 \rightarrow RCH=CH_2 + \bullet O_2H$$

and the alkyl radical can react directly with oxygen to form an alkylperoxy radical:

$$R\bullet + O_2 \rightarrow ROO\bullet$$

The alkyl radicals that are consumed in this process can be regenerated by hydrogen abstraction reactions such as the reaction with the hydroxyl radical:

$$RH + \bullet OH \rightarrow R\bullet + H_2O$$

The formation of the alkylperoxy radicals is an major step in the combustion process because these radicals then participate in a variety of reactions, the most important of which is the conversion of the alkylperoxy radical to a hydroperoxyalkyl radical, $\bullet R'OOH$. This conversion can take place via internal hydrogen transfer, as for example

$$RCH_2CH_2CH_2CH_2OO\bullet \rightarrow RCH_2CH\bullet CH_2CH_2OOH$$

a reaction which proceeds via a six-membered cyclic intermediate.

The hydroperoxyalkyl radical decomposes to various intermediate, partially oxidized products of combustion, including heterocyclic oxygen-containing compounds, aldehydes, and ketones. The O-O bond in peroxides is a relatively weak bond, and is able to undergo cleavage to generate an alkoxy radical:

$$ROOH \rightarrow RO\cdot + \cdot OH$$

The alkoxy radicals can decompose to aldehydes, as in the formation of formaldehyde:

$$RCH_2O\cdot \rightarrow R\cdot + H_2C=O$$

They can also lose a hydrogen via hydrogen abstraction to form aldehydes (or ketones) as, for example

$$RCH_2O\cdot + O_2 \rightarrow RCH=O + \cdot O_2H$$

Alkoxy radicals also participate in hydrogen abstraction reactions, resulting in the formation of alcohols:

$$RO\cdot + R'CH_2R'' \rightarrow ROH + R'CH\cdot R''$$

Here the R' and R'' symbols are used to indicate that the alkyl portions of the molecule do not necessarily have to be identical.

Aldehydes formed in these reactions may lose the hydrogen from the carbonyl carbon atom in a hydrogen abstraction reaction:

$$RCH=O + R'\cdot \rightarrow RC\cdot=O + R'H$$

At high temperatures the new radical generated in this process decomposes to carbon monoxide:

$$RC\cdot=O \rightarrow R\cdot + CO$$

At lower temperatures, a two-step process results in the formation of peroxy carboxylic acids:

$$RC\cdot=O + O_2 \rightarrow RCOOO\cdot$$
$$RCOOO\cdot + R'H \rightarrow RCOOOH + R'\cdot$$

Like other species containing the peroxide bond, the peroxy carboxylic acid can undergo cleavage at the O-O bond

$$RCOOOH \rightarrow RCOO\cdot + \cdot OH$$

and the carboxy radical undergoes a further decomposition to the final reaction product, carbon dioxide:

$$RCOO\bullet \rightarrow R\bullet + CO_2$$

The new alkyl radical generated in this reaction can of course itself participate in any of the sequence of reactions just discussed.

Combustion studies of typical gasoline components showed that branched chain alkanes are less likely to knock than straight chain alkanes. Recall the combustion reaction for methane:

$$CH_4 + O_2 \rightarrow CH_3\bullet + HOO\bullet$$

For this reaction to occur, the oxygen molecule has to approach the methane molecule to interact in a bimolecular reaction. The approach of the oxygen and methane molecules involves the interaction of London forces between the temprary dipoles induced in these two molecules. Recall that a branched chain alkane has a lower surface area than the corresponding straight chain alkane. Thus the branched chain compound will have reduced London force interactions with the oxygen molecule relative to the straight chain compound, and consequently a reduced likelihood of the bimolecular reaction with oxygen initiating combustion.

The second step of the methane combustion process was the reaction of the methyl radical with oxygen

$$CH_3\bullet + O_2 \rightarrow CH_3OO\bullet$$

The methyl radical is the least stable of the possible types of free radicals. In contrast, the combustion of a branched chain alkane can form, in the analogous reaction, relatively stable 3° radicals. The two factors, reduced London force interactions and greater stability of 3° radicals, reduce the likelihood of the air-fuel mixture burning so rapidly that an explosion (knocking) occurs.

We have seen in Chapter 7 that radical species important in the combustion process include $\bullet OH$, $O\bullet$, and $HO_2\bullet$. For all of these radicals the activation energy for reaction with a C-H bond decreases in the order 1° > 2° > 3°. For example, reactions with the $\bullet OH$ radical have activation energies of about 28 kJ/mol for a 1° C-H bond, 23 kJ/mol for a 2° C-H, and only about 20 kJ/mol for 3°.

The standard compound by which fuel performance is judged is 2,2,4-trimethylpentane, which is commonly but erroneously referred to as "octane" or "isooctane." The 2,2,4-trimethylpentane is assigned an *octane number* of 100; heptane is assigned an octane number of 0. The octane number of a fuel is equal to the percentage of 2,2,4-trimethylpentane in a blend of 2,2,4-

trimethylpentane and heptane which has the same combustion performance (knocking behavior) as the test fuel. The higher the octane number the less tendency of a fuel to knock, when measured in standardized test conditions. Thus the higher the compression ratio of an engine, the higher would be the octane number of the gasoline required for good performance. In turn, the higher the octane number of the gasoline, the higher the proportion of branched chain alkanes required in the fuel.

Unfortunately, branched chain alkanes are relatively rare in nature. The principal source is the isoprenoid compounds such as terpenes. Some branched chain compounds may arise during cracking of kerogen via the recombination of two 2° radicals, such as

$$CH_3-CH_2-\overset{\bullet}{C}H + H\overset{\bullet}{C}-CH_3 \longrightarrow CH_3-CH_2-CH-CH-CH_3$$
$$\underset{CH_3}{|} \quad \underset{CH_3}{|} \quad \quad \quad \underset{CH_3}{|} \underset{CH_3}{|}$$

Although some small amount of branched chain compounds occur in crude oil, the octane number of straight run gasoline is in the range 30 - 70. This gasoline was adequate when automobiles first became widespread, in the 1920's, because engine compression ratios in those days were about 4:1. With modern engines, having compression ratios as high as 10:1, significantly improved gasolines are required. Many of today's automobile engines require gasolines with octane numbers in the mid-80's, and high performance engines will need octane numbers of 90-100. Refining processes are needed to increase the proportion of branched chain compounds in the gasoline.

The octane number of alkanes decreases as the carbon chain is lengthened but increases as chain branching increases. Alkenes have higher octane numbers than the corresponding alkanes. The octane number of an alkene in which the double bond is in the interior of the chain will be higher than that of the isomer in which the double bond is at one end of the chain. Cycloalkanes have higher octane numbers than alkanes, and aromatic hydrocarbons tend to have still higher octane numbers. Thus improvements in gasoline combustion characteristics are obtained by the conversion of long chain to shorter chain alkanes, by increasing chain branching, and by the formation of cycloalkanes or aromatics. The chemical nature of these processes is discussed in Chapter 9.

In commercial practice, many factors are taken into account in the production of gasoline, including having an appropriate concentration of molecules with high vapor pressures, to insure good engine starting during cold weather; having a low concentration of high boiling compounds which might be liable to extensive thermal cracking and deposition of carbonaceous solids in the engine, and the elimination of impurities which might cause corrosion or produce gummy or tarry deposits. Nevertheless, the knocking tendency of the fuel is the issue of paramount importance.

Leaded gasoline

In the early 1920's it was discovered that the antiknock behavior of gasoline could be improved by adding tetraethyllead, $(CH_3CH_2)_4Pb$. Tetraethyllead is an example of an *organometallic compound*, a compound in which there are direct covalent bonds between a metal atom and a carbon atom. The first synthesis of tetraethyllead made use of the Grignard reagent ethylmagnesium bromide:

$$4\ C_2H_5MgBr + 2\ PbCl_2 \rightarrow (C_2H_5)_4Pb + 4\ MgClBr + Pb$$

However, this process was superseded by the reaction of chloroethane with an alloy of lead and sodium:

$$4\ C_2H_5Cl + 4\ NaPb \rightarrow (C_2H_5)_4Pb + 4\ NaCl + 3\ Pb$$

Tetraethyllead decomposed by homolytic bond cleavage at ~140°. The radicals which are produced promote even, steady combustion and facilitate the combustion reactions at lower temperatures. For example, the reaction of chlorine with methane to form chloromethane and hydrogen chloride normally occurs at 250° via a free radical mechanism. If 0.02% tetraethyllead is added to the methane, the identical reaction occurs at 140°. The role of the tetraethyllead in this example is to generate chlorine radicals by the reactions

$$(CH_3CH_2)_4Pb \rightarrow Pb + 4\ CH_3CH_2\bullet$$
$$CH_3CH_2\bullet + Cl \rightarrow CH_3CH_2Cl + Cl\bullet$$

after which the chlorine radicals can initiate the chain reaction process

$$CH_4 + Cl\bullet \rightarrow CH_3\bullet + HCl$$
$$CH_3\bullet + Cl \rightarrow CH_3Cl + Cl\bullet$$

In an analogous manner, the combustion reactions can be facilitated because the ethyl radicals released from the decomposition of the tetraethyllead initiate the chain reactions. The effect of tetraethyllead additions on changing the octane number of pure alkanes and cycloalkanes is shown in Table 8.2.

Gasoline for which the octane number was increased above 80 by the addition of tetraethyllead was known as *ethyl gasoline*. Since tetraethyllead boosts the octane number, it could be added to 2,2,4-trimethylpentane to increase the octane number above 100. Indeed this is one approach to defining octane numbers >100 for fuels. The effect of tetraethyllead additions to 2,2,4-trimethylpentane is shown in Table 8.3.

Table 8.2 The effect of addition of tetraethyllead (3 mL/gallon) on octane numbers of pure compounds

Compound	Octane number	Octane number with lead addition
pentane	68.0	84.0
2-methylbutane	92.0	101.5
cyclopentane	101.5	108.0
hexane	25.0	65.5
2-methylpentane	73.5	93.0
3-methylpentane	74.5	93.5
2,2-dimethylbutane	92.0	106.0
2,3-dimethylbutane	103.5	113.0
methylcyclopentane	91.5	105.5
cyclohexane	83.0	97.5

Table 8.3 Increase in octane number achieved by addition of tetraethyllead

Addition, mL/gallon	Octane number
0	100
1	109
2	113
3	116
4	118
5	119
6	120

Use of tetraethyllead can result in deposition of lead or lead oxide (from the reaction of lead released by decomposition of the tetraethyllead with oxygen during combustion) on cylinder walls, valves, and spark plugs. In addition, the mildly abrasive lead oxide can contribute to erosion of spark plugs and exhaust valves. To counteract this problem, 1,2-dibromoethane or 1,2-dichloroethane was added to the gasoline. The reaction of these compounds with lead or lead oxide under combustion conditions led to the formation of lead bromide or lead chloride, which are volatile at the temperatures of combustion and are swept out of the cylinder with the other combustion products. A mixture of tetraethyllead, 1,2-dibromoethane, 1,2-dichloroethane, and a dye became known as *ethyl fluid*. A typical concoction contained 63% tetraethyllead, 26% 1,2-dibromoethane, 9% 1,2-dichloroethane, and 2% of the desired dye. This mixture could be added to gasoline to increase the octane number. The legal limit for tetraethyllead addition to automobile gasoline was 3.0 mL/gallon. Higher levels of lead were used in leaded aviation gasolines.

Because of the environmental concern associated with lead emissions from vehicles and the poisoning of catalysts in automobile exhaust systems by lead, there is much interest in finding compounds to improve the anti-knock characteristics of gasoline without the attendant problems of the tetraethyllead. Two compounds which have been investigated in this connection are 2-methyl-2-propanol (t-butyl alcohol) and 2-methoxy-2-methylpropane (methyl t-butyl ether). 2-Methyl-2-propanol has, for example, an octane number of 108, compared with some straight-run gasolines which have octane numbers of ~70. However, it is questionable whether an adequate supply of t-butyl alcohol would be available to treat the total production of gasoline.

We will see in Table 9.1 that many compounds have octane numbers in excess of 100. Some gasolines, particularly those used for piston-engine aircraft, are routinely used with octane numbers >100. Since the standard of performance, 2,2,4-trimethylpentane, is defined to have an octane number of 100, methods are needed to assign octane numbers higher than 100. One approach is that just discussed: to compare the performance of the test fuel with 2,2,4-trimethylpentane containing a known amount of tetraethyllead. A second approach depends on measuring engine performance. In this test the load on the engine is increased until a load is reached at which the engine knocks. With smooth operation, a plot of pressure in the cylinder as a function of time would be like Figure 8.12.

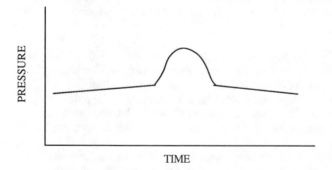

Fig. 8.12 A schematic illustration of the change in cylinder pressure with time during the smooth combustion of a fuel; pressure rises smoothly to a maximum and then decreases.

When knocking occurs, the same graph would have the appearance of Figure 8.13 (next page).

The cylinder pressure is measured at the point of incipient detonation, that is, just at the instant before knocking occurs. This pressure is the knock-limited *indicated mean effective pressure*, or knock-limited IMEP. This measurement allows the calculation of a *performance number* by the equation

$$PN = 100 \cdot IMEP / IMEP_o$$

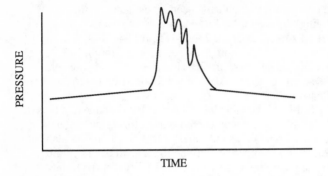

Fig. 8.13 When knocking occurs during combustion, cylinder pressure does not rise and decline smoothly as a function of time, but rather becomes erratic after ignition.

where PN is the performance number and $IMEP_0$ is the indicated mean effective pressure for 2,2,4-trimethylpentane. This method provides for the determination of performance numbers (which are essentially equivalent to octane numbers) for very high octane fuels such as aviation gasolines.

Volatility

In addition to combustion performance, a gasoline should provide good cold weather starting. That is, some of the components of gasoline should have sufficiently high vapor pressures so that even at low temperatures encountered in winters some of the material will vaporize in the cylinder and be ignited by the spark plug. This requirement indicates that the gasoline should have a high proportion of low-molecular weight, or short chain, compounds.

On the other hand, a high concentration of low-boiling components could, during hot weather, vaporize in the fuel line, filling the line with vapor and preventing the delivery of liquid fuel to the engine. This condition is sometimes known as *vapor lock*.

In essence, a gasoline has to be a compromise between providing good cold weather starting while at the same time not being susceptible to vapor lock, and do both while delivering acceptable octane ratings. Fuel refiners may change the blends of the gasoline with changes in the season, or deliver gasolines having slightly different vaporization characteristics to regions having different climates. For winter driving, or for use in cold climates, a gasoline should have a high proportion of components with good volatility. Alternatively, for summers or hot climates, a higher proportion of lower volatility components unlikely to cause vapor lock is desired.

194

Naphtha

Naphtha is a distillation cut obtained at 95-150°. Naphtha is seldom used as a fuel, but can be blended with gasolines which consist of low molecular weight (and therefore high vapor pressure) components to reduce the overall volatility and reduce the likelihood of vapor lock.

Naphthas are widely used as solvents in the chemical industry. Naphthas are also cracked to produce lighter compounds. Propane and butane which are obtained from naphtha cracking can be used as LPG. However, the most important use of naphtha is the production of ethylene, which is the organic compound used in greatest quantity by the chemical industry. Some of the many uses of ethylene will be discussed later in the book. Thermal cracking of naphtha causes unzipping of the alkane molecules by repeated β-bond scission reactions, with the formation of large quantities of ethylene.

Kerosene

Kerosene was once used very widely as an illuminant, as the fuel in kerosene lamps, before electricity became widely available, especially in rural areas. At one time, kerosene was the most valuable product from crude oil because of this application in illumination. With the near-simultaneous rise of rural electrification and the ready availability of automobiles during the early decades of the twentieth century, gasoline became far more important than kerosene. First, the market for kerosene as an illuminant declined as electric lighting increased in use. Second, the burgeoning popularity of automobiles brought with it a steadily increasing demand for gasoline. In fact, the substantially greater desirability of gasoline relative to kerosene led to the development of processes for cracking kerosene to gasoline. Today kerosene is used as a fuel for jet aircraft, and consequently is again an important product of crude oil.

While many gasolines are predominantly composed of alkanes, some kerosenes tend to be rich in cycloalkanes. Recall that the cycloalkanes generally have higher boiling points than the corresponding alkanes of the same number of carbon atoms; indeed kerosene is a higher boiling cut than gasoline. One the basis of the ternary composition classification of crude oils, we would expect paraffinic-naphthenic or aromatic-naphthenic crude to give higher kerosene yields than paraffinic or aromatic-intermediate crudes.

High concentrations of aromatic compounds are undesirable in kerosene. Aromatic compounds are more likely to produce soot on combustion. Soot formation is a detriment for either of the major uses of kerosene - illumination or as a jet engine fuel. A high concentration of aromatic compounds is undesirable in jet fuels because the aromatics burn with a smoky or sooty flame. Although alkanes do not have this problem, some of the larger alkanes in the jet fuel boiling range have relatively high melting points (*e.g.,* tridecane, -6°) and can solidify at low temperatures, as for example in high altitude flight.

Fuels used for gas turbine engines (such as jet engines) are mainly required to remain liquid under any anticipated operating condition. In the case of jet engines used on modern aircraft, the operating conditions place severe constraints on the fuel properties. The fuel must remain liquid at pressures in the range 10-100 kPa and temperatures from -75 to 50°. Thus low boiling materials would vaporize at the higher temperatures and lower pressures, while very high boiling materials would likely freeze at the lower temperatures.

Kerosene used for domestic heating (sometimes called household kerosene) must have a carefully controlled boiling range. If the boiling range is too low, vaporization of material inside the combustion device can lead to explosion of the air-vapor mixture upon ignition. This problem has occurred when, for example, the fuel rank of a kerosene heater was inadvertently filled with gasoline. In extreme cases the ensuing explosion and fire has caused the destruction of homes and loss of life. On the other hand, too high a boiling range leads to poor vaporization and inefficient combustion. The safety of household kerosene can be expresses in terms of the *flash point*, the temperature at which a liquid will ignite when an open flame is applied to the vapor over the liquid surface. Household kerosene should have a flash point of about 50°.

Since the combustion products from kerosene-fired heaters used for domestic heating are often vented into a room, it is essential that smoke, soot, and carbon monoxide formation not occur. As a rule, the higher the hydrogen content of the fuel, the "cleaner" (*i.e.,* the less soot and carbon monoxide) is the flame. Thus kerosenes which are high in alkanes, rather than aromatics, are preferred.

Soot formation is a potential problem in the combustion of kerosene and jet fuels. In all cases, soot formation represents the loss of some of the energy which might have been liberated in the complete combustion of the fuel.When kerosene is used for home heating or lighting, soot formation is an aesthetic problem. The emission of soot to the atmosphere from combustion sources, including jet aircraft, represents an air pollution problem. As a rule, aromatics have the highest sooting tendency, napthenes are intermediate, and paraffins are least likely to produce soot. Alkenes tend to form soot more readily than the paraffins. Therefore low concentrations of aromatics and naphthenes are desired for "clean" burning. Some of the intermediates in soot formation are very likely polyalkenes and polycyclic aromatic compounds. In this sense, the formation of soot is similar to the formation of carbon deposits on the surface of catalysts, a problem discussed in Chapter 9. It is also similar to the formation of aromatic compounds from polyalkenes discussed in

Chapter 3 in connection with kerogen catagenesis. Hydrogen redistribution is one of the processes at work, since the formation of the polyalkenes or polycyclic aromatic compounds results from a loss of hydrogen from the original fuel molecules.

Diesel fuel

The diesel engine was designed to run on cheaper and poorer quality fuels than a gasoline-powered engine. In fact, much of the early developmental work on diesel engines focused on the use of pulverized coal or coal tars as possible fuels. Modern diesel engines operate on a "middle distillate" (so called because the boiling range is intermediate between the low-boiling gasoline and naphtha and the higher-boiling fuel oils) *diesel oil* or *gas oil*. The gas oil name derives from an alternate use of this material, cracking to light gaseous products A typical yield of middle distillate, boiling at 230-340°, from crude oil is 6%.

Diesel engines belong to the category of internal combustion engines known as *compression ignition* (CI) engines. The name derives from the fact that CI engines depend on compression of the air-fuel mixture to ignite the charge. In a sense, we require a CI to knock, whereas of course knocking is very undesirable in an SI engine. We have seen that knocking tendency increases with increasing compression ratio of an engine. Since ignition on compression is desired in a diesel engine, the compression ratios are much higher than typical SI engines. Many diesel engines have compression ratios in the range of 14:1 to 20:1. The heat needed to ignite the charge is obtained from the heat of compression of the mixture. An analogy is the experience of having the metal casing of a manually operated air pump become palpably hot while, for example, inflating a bicycle tire.

The higher boiling range of diesel fuel relative to gasoline indicates that larger molecules will be present in the diesel fuel. The diesel fuel may contain alkanes in the C_{15}-C_{17} range and aromatics as large as alkylated naphthalenes. We should therefore also expect that diesel fuel will be of higher density (lower API gravity) and higher viscosity relative to gasoline. For example, octane has an API gravity of 70° and a viscosity of 0.54 mPa•s at room temperature, whereas hexadecane (also called cetane) has a gravity of 51° and viscosity of 3.34 mPa•s. Many diesel fuels have API gravities in the range 33-44°, even lower than would be anticipated from the values for the alkane compounds in the diesel fuel boiling range. The lower value of the API gravity relative to the alkanes reflects the presence of substantial quantities of aromatics. As a rule of thumb, alkanes have much higher API gravities than aromatics having the same number of carbon atoms, as summarized in Table 8.4.

Table 8.4 Comparative API gravities of alkanes and aromatics

Carbon atoms	Alkane	API gravity	Aromatic	API gravity
10	Decane	62°	Naphthalene	-8°
11	Undecane	59°	1-Methylnaphthalene	7°
12	Dodecane	57°	2-Ethylnaphthalene	9°

Unfortunately, the operation of a diesel engine is not simply a matter of injecting the fuel and air together, because in such a case the air-fuel mixture would burn in an uncontrolled fashion analogous - in a strict sense - to knocking in an SI engine. In such circumstances the best that could be hoped for is to have the engine run smoothly at one load and one speed. (In fact, some model airplane engines operate in just this way because they do have an almost constant load and run at the same speed all the time.) For engines in most practical applications, particularly automobiles and light trucks, this problem is overcome by using a fuel injector. Air is admitted to the cylinder and heated by compression. The fuel is then injected into the hot air. The air-fuel mixture becomes a very heterogeneous mix of fine, intermediate, and large-sized droplets of fuel (Fig. 8.14).

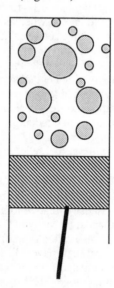

Fig. 8.14 A schematic diagram of the cylinder in a diesel engine, showing a distribution of sizes of fuel droplets.

At various spots within this heterogeneous mixture the air/fuel ratio will be locally correct to insure ignition. The liquid fuel injected into the hot air ignites on contact with the air. Fuel injection continues for most of the power stroke in the cylinder. Over wide ranges of air/fuel ratios there will be at least some locations in the cylinder in which the correct ratio needed to obtain ignition will prevail. Consequently, reliable ignition and smooth performance can be obtained over a wide range of engine loads.

The evaporation of the fuel, once injected into the hot air in the cylinder, is crucial to maintain good mixing of the air and fuel. Furthermore, vaporization precedes the actual combustion reactions. The heat required to vaporize the fuel is a portion of the heat released by the combustion reactions. Thus *volatility* is an important property of the fuel, but because the evaporation is effected by the heated air in the cylinder the diesel fuel need not be as inherently volatile as gasoline. For this reason diesel fuel can be taken as the distillation cut between kerosene and the even higher boiling fuel oils. Evaporation of the diesel fuel is promoted by "atomizing" the fuel to achieve a high ratio of surface to volume in the fuel droplets. The volatility of the fuel must be controlled because fuels of high volatility can cause vapor lock on the low pressure side of the fuel pump, and fuels of low volatility can cause problems when trying to start the engine in cold weather. In addition, fuels of low volatility are also undesirable because, as a rule, low-volatility fuels are those with high concentrations of polycyclic aromatic compounds. These compounds are responsible for smoke and soot emissions during combustion.

A high specific gravity of the fuel provides a greater vehicle mileage per volume of fuel (*e.g.*, miles per gallon). Consider a gasoline having a calorific value of 48 MJ/kg. The density of this fuel will be about 0.74 kg/L. In comparison, a diesel fuel may have a lower calorific value on a mass basis, about 45 MJ/kg, but a higher density, 0.88 kg/L. When the calorific values of the two fuels are compared on a volumetric basis, the diesel fuel is the superior fuel, 39 MJ/L vs. 35 MJ/L for gasoline. The volumetric calorific value for the diesel fuel contributes to the superior fuel economy often observed for diesel engine vehicles.

The ignition quality of a diesel fuel is measured by *delay*, the time lapse between the beginning of ignition and the onset of a significant pressure rise in the cylinder. The shorter the delay, the better the ignition quality. A good ignition quality is essentially the opposite of good knock resistance. The ignition quality is often expressed in terms of the *cetane number*, which is determined by comparing the combustion performance of a test fuel to that of a mixture of hexadecane (cetane) and 1-methyl-naphthalene. The cetane number of the fuel is the percentage of hexadecane in the test mixture giving the same combustion performance as the fuel. Typical diesel fuels for automobile engines and other applications requiring high speed have cetane numbers of about 50.

The desired molecular structures for a good fuel for a CI engine are essentially the opposite of those desired for an SI engine fuel. The fuel for the CI engine should have abundant straight chain alkanes, whereas these are undesirable in the SI fuel. We have also seen that aromatics in gasoline have very high octane numbers. This reversal of desirable fuel properties when comparing diesel fuel and gasoline suggests an inverse relationship between cetane and octane numbers, as illustrated in Figure 8.15.

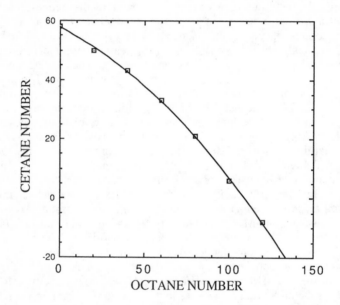

Fig. 8.15 The inverse relationship between cetane and octane numbers.

The cetane numbers of a variety of compounds are shown in Table 8.5. As a rule, for compounds having the same number of carbon atoms, the cetane number drops in the order n-alkane > alkene > cycloalkane > alkyl aromatic. For example, compare dodecane, 7-dodecene, dicyclohexane, and 1-phenylhexane. Among the alkanes, cetane number drops with decreasing chain length; compare octadecane, tetradecane, and decane, for example.

We have seen how tetraethyllead is used as an antiknock additive to gasoline. Since there is a relationship, at least in concept, between knocking in gasolines and ignition quality in diesel fuels, this relationship suggests that

Table 8.5 Cetane numbers for pure organic compounds

Compound	Cetane number
Octadecane	103
Hexadecane	100
Tetradecane	96
1-Octadecene	90
Dodecane	88
1-Hexadecene	84
1-Tetradecene	83
Decane	77
7-Dodecene	71
Octane	64
1-Decene	60
Heptane	56
1-Phenylnonane	50
Dicyclohexyl	47
Decalin	42
1-Octene	40
1-Phenylhexane	26
Methylcyclohexane	20
1-Phenylpentane	8
Diisopropylbenzene	-12

there could be, by analogy, materials added to diesel fuels to enhance the ignition qualities. These compounds are called *ignition accelerators*; the ones most frequently used are pentyl nitrate (also called amyl nitrate), ethyl nitrate, and ethyl nitrite. An indication of the performance improvements obtained is shown in Table 8.6.

Table 8.6 Improvements in Pennsylvania gas oil derived from addition of amyl nitrate

Volume percent added	Cetane number
0	34
1	50
2	57

A comparison of Tables 8.3 and 8.6 shows that a much greater amount of the ignition accelerator is required compared to the addition of tetreethyllead to gasoline, but that in both cases a law of diminishing returns applies, such that the incremental improvement in octane or cetane number obtained from increased concentration of the fuel additive becomes smaller and smaller as more and more additive is used. Ignition accelerators are also sometimes called *pro-knocks*.

Residual diesel oils (*i.e.*, those of high boiling range) are used in the remarkable diesel engines on ships known as *marine diesels*. These engines have a very high thermal efficiency, up to 42% with a turbocharger. They may be as larger as a five-story building with each piston being about 1 m in diameter. These engines operate at low speeds, about 100 rpm, and are coupled directly to the propellors.

Fuel oils

Fuel oil, burned in furnaces for home heating, process heating in industry, or raising steam in electric power plant boilers, is often a residue of distillation rather than a distillation cut. Fuel oils are graded on the basis of viscosity. To achieve a desired viscosity of oil, a distillation residue can be blended with some proportion of one of the distillation cuts. The classification of fuels oils uses a numbering system as well as a grade. Unfortunately the terminology is imprecise, and sometimes gas oil or diesel oil are also called fuel oils. In the numerical classification system, the increasing numbers imply increasing viscosity. This system is illustrated in Table 8.7.

Table 8.7 Classification of fuel oils

	No.1	No. 2	No.4	No. 5	No. 6
API gravity	40°	32°	21°	17°	12°
Pour point	<-18°	<-18°	-12°	-1°	18°
Sulfur	0.1%	0.4 - 0.7%	0.4 - 1.5%	≤2.0%	≤2.8%
Ash	Trace	Trace	0.02%	0.05%	0.08%
Calorific val., MJ/L	38.2	39.3	40.7	41.2	41.8
Viscosity, mPa•s	1.6	2.7	15	50	360
Color	Light	Amber	Black	Black	Black

No. 1 fuel oil is comparable to kerosene. No. 2 oil is also a distillate product. The other oils are all resids. The No. 5 and No. 6 oils are sometimes called *bunker oils*. They are of such high viscosity that they usually have to be heated to be pumped through the nozzles of the burners in combustion equipment. We have seen that diesel oil may contain alkanes through at least C_{16}; consequently, bunker oils may have compounds with >20 carbon atoms. Such large molecules are not too different in composition from the waxes, and in fact some fuel oils derived from crudes having high wax contents can actually solidify during storage.

Oils that are highly viscous or even solid during storage are characterized by the *pour point*, which is the lowest temperature at which the oil will still flow (as, for example, when poured from a beaker). Some bunker oils have pour points in the range of 20°.

Lubricating oils

Lubricating oils ("lube oils") have very high boiling ranges, and are usually produced by a vacuum distillation step following two-stage distillation. In a typical operation, the resid from a two-stage distillation unit is fed to a vacuum tower at ~425° and pressures of 7-18 kPa. The overhead from the vacuum tower is *gas oil*. Lube oils are taken off as side streams. The vacuum tower bottoms are asphalt. A lubricating oil might consist of 20-25% alkanes of straight and branched chains; 45-50% cycloalkanes of one, two, or three rings, possibly with alkyl side chains; 25% of alkylated two, three, or four ring hydroaromatic compounds (*i.e.,* compounds in which one or more of the rings is aromatic but the others are aliphatic); and 10% high molecular weight aromatic compounds.

Lube oils may amount only to about 2% of the total crude, but they are very desirable products because of their high selling price. An enormous array of products is sold for lubrication; at one time in the United States there were 1,156 different lube oil blends and 271 greases on the market. Lube oil is one of the few commodities for which sales have decreased as the performance has improved. From the end of World War II through the mid-60's the number of miles driven per barrel of lube oil doubled. During this period the demand for gasoline increased by 40%, but the increase in lube oil demand was only 20%.

There are four desirable properties for lube oils. *High temperature stability* is required so that the oil is not thermally decomposed by the heat generated from friction on the parts that the oil is lubricating. *Low temperature fluidity* enables the oil to flow between moving parts even at very low temperatures (as when starting a car during the winter). The dependence of viscosity on temperature should be minimal, so that the fluidity of the oil will be nearly constant from a cold start through the high temperatures generated from operations at high speeds. *Adhesiveness* is important for insuring that the oil will cling to metal surfaces even under very high shear. Fortunately, long-chain alkanes of high boiling points have most of these properties.

In many cases the distillation streams from the vacuum tower will receive further treatment to improve the properties of the oils. *Solvent extraction* is used to remove compounds which are not alkanes. A solvent frequently used in this operation is furfural,

$$\text{(furan ring)}\!-\!\text{CHO}$$

The extraction is carried out in a countercurrent column, which is similar in concept to the countercurrent dehydration of natural gas shown in Fig. 7.1.

Aromatics, cycloalkanes, and compounds containing nitrogen or sulfur will dissolve in the solvent. None of these compounds contribute to the lubricating qualities of the oil, so their removal improves the quality of the oil. The solvent is recovered for reuse by *steam stripping*, passing steam through the solvent - extract mixture to cause the furfural to distill with the steam.

Waxes contain long alkane chains and in principle could be good lubricants. However, waxes tend to solidify and precipitate from the oil, interfering with the operation of whatever equipment the oil is supposed to be lubricating. To avoid this, lube oils are also treated by *dewaxing*. Two processes are used for lube oil dewaxing. In *solvent dewaxing* oil is treated with 2-butanone. The oil dissolves in this solvent. Chilling the solution causes the wax to solidify and precipitate. The wax is removed by filtration. Steam stripping the filtrate removes the 2-butanone, which can be recycled, and leaves the dewaxed oil. This process can reduce the pour point to -25°.

The process of *urea complexation* involves treatment of the oil with urea at about 40°. The urea reacts with the ester functional group in the wax molecules to form a complex insoluble in the oil. Filtration removes the urea-wax complex. The complex can be decomposed to recover the urea by reacting with water at 75°. Urea complexation can produce oils with pour points of -55°. Alkanes having seven or more carbon atoms can combine with urea to form crystalline complexes. The urea molecules crystallize to produce a cylindrical channel in which the hydrocarbon molecule can fit. The channel is stabilized by hydrogen bonding between the individual urea molecules. The hydrocarbon molecule is not chemically bonded to the urea, but rather is physically trapped in the channel, and held in place by van der Waals forces. Because the urea must form a channel for the hydrocarbon, the amount of urea in the complex increases with increasing chain length of the alkane. For example, heptane complexes with six moles of urea, while octacosane, $C_{28}H_{58}$, complexes with 21 moles. The shape of the channel is such that n-alkane molecules can be accommodated in the complex but branched-chain compounds cannot. Consequently, urea complexation can also be used for reducing the amount of straight-chain alkanes in gasoline, to improve the combustion performance, and from jet fuel to reduce the freezing point.

If the oil contains significant amounts of oxygen, nitrogen, or sulfur compounds, their removal can be accomplished by *clay treatment*. The oil is allowed to percolate slowly through a bed of clay, typically fuller's earth, a very porous aluminum silicate. The clay absorbs the NSO's.

In some lubricating applications it is useful to have a material having a higher viscosity than a lube oil. Semi-solid lubricating materials are called *greases*. A grease is made by blending lube oil with a "metallic soap", which is a salt of long chain carboxylic acids with aluminum, calcium, cobalt, zinc, or other multivalent cation.

Waxes

The major use of waxes is impregnation of paper and cardboard, especially for food preservation. (Originally the major use of wax was the manufacture of candles.) A material related to waxes, but usually having lower melting point, is *petrolatum*. Petrolatums have melting ranges of 40-80° and are semisolids at room temperature. This material is best known to most people as the petrolatum product Vaseline (and related brands). The major use of petrolatums is in the pharmaceutical and cosmetic industries. Waxes have melting points much more sharply defined (50-60°) than petrolatums and are hard, brittle solids at room temperature. The alkane molecules in waxes have 18-56 carbon atoms.

In addition to their use for preparing food storage and handling materials, waxes can also be useful as feedstocks for chemical processing. The thermal cracking of waxes produces 1-alkenes in the C_6-C_{20} range. The waxes are cracked at 540-565° and 0.2-0.4 MPa, with residence times of about 5-15 sec. The products are alkenes of all possible chain lengths from ethene up. The process of wax cracking is basically one of β-bond scission, which begins with a hydrogen abstraction

$$\text{------CH}_2\text{CH}_2\text{CH}_2\text{-----} \rightarrow \text{------CH}\bullet\text{CH}_2\text{CH}_2\text{------} + \text{H}\bullet$$

(here the dashed line is used to imply a longer segment of the alkane molecule not shown in the structure; recall that waxes may have ~30 carbon atoms) followed by the bond scission

$$\text{------CH}\bullet\text{CH}_2\text{CH}_2\text{------} \rightarrow \text{------CH=CH}_2 + \bullet\text{CH}_2\text{------}$$

The radical produced in the bond scission process is capped by hydrogen abstraction

$$\text{------CH}_2\bullet + \text{------CH}_2\text{CH}_2\text{CH}_2\text{------} \rightarrow \text{------CH}_3 + \text{------CH}\bullet\text{CH}_2\text{CH}_2\text{------}$$

or by recombination

$$\text{------CH}_2\bullet + \bullet\text{CH}_2\text{------} \rightarrow \text{------CH}_2\text{CH}_2\text{------}$$

We have previously discussed unzipping, during which successive β-bond scission reactions produce very high yields of ethene. Careful selection and control of reaction conditions prevents unzipping from occurring during wax cracking. Cracking of a naphtha for ethylene production is done at much higher temperatures (750-900°); since bond breaking is an endothermic process, these higher temperatures favor an increased amount of bond breaking. Wax cracking is done at higher pressures. Typically naphtha cracking is done at 0.1 MPa. The higher pressures provide a greater number of molecules per unit volume, and therefore an increased number of molecular collisions to provide a greater chance of hydrogen abstraction or recombination reactions. Thus the lower temperatures of wax cracking make bond breaking less favorable while at the same time the higher pressures favor reactions which would end the unzipping process.

The 1-alkenes produced in wax cracking have several uses in the chemical industry. The alkenes react with a mixture of carbon monoxide and hydrogen in a process known as *hydroformylation* to produce aldehydes. For example,

$$CH_3CH=CH_2 + H_2 + CO \rightarrow CH_3CH_2CH_2CHO$$

The reaction is run at 110-180° and 20-30 MPa in the presence of a cobalt catalyst. The aldehydes are subsequently reacted with hydrogen in the presence of cobalt or nickel catalysts to give the corresponding alcohols:

$$CH_3CH_2CH_2CHO + H_2 \rightarrow CH_3CH_2CH_2CH_2OH$$

The long chain alcohols are valuable as plasticizers, while the shorter chain compounds are useful solvents.

The 1-alkenes can react with benzene in the presence of hydrogen fluoride at 5-70° to produce alkylated benzenes in 90% yield:

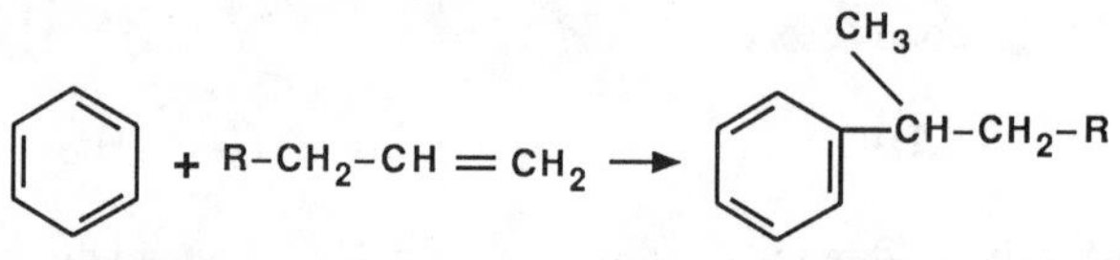

The preferred alkenes for this process are the C_{10}-C_{14} compounds. The use of HF as a catalyst causes the reaction to proceed via carbocations; the cation first formed rearranges by a hydride shift to move the positive charge further inside the chain. The products of this reaction are known as *detergent alkylate*. In a subsequent reaction with fuming sulfuric acid they are converted to alkylbenzenesulfonic acids. The sodium salts - the sodium alkylbenzenesulfonates - are the predominant surface-active detergents for use in domestic laundry products.

Asphalt

Asphalts are residues from the distillation of crude oils. Typically asphalts are mixtures of aromatics, long-chain alkanes, and high molecular weight NSO's. Asphalts may have pour points exceeding 95°.

The vacuum tower bottoms are a good source of asphalt. The resid is treated with liquid propane at 70° and 3.5 MPa. All of the components of the resid except the asphalt dissolve in the propane. Once the asphalt has been separated, the release of pressure allows the propane to evaporate, yielding a product called *residual lube oil*. The residual lube oil may then be treated by solvent extraction, dewaxing, or clay treatment.

The vacuum tower bottoms are a good source of asphalt. The bottoms are treated with liquid propane at 3.5 MPa and 70°. All of the compounds in the bottoms dissolve in the propane except the asphalt. Once the asphalt has been separated from the propane solution, a reduction of pressure allows the propane to evaporate, leaving a product called *residual lube oil*. This product can then be treated further by the processes just discusses: solvent extraction, dewaxing, or clay treatment. A block flow diagram for a propane deasphalting process is given in Figure 8.16.

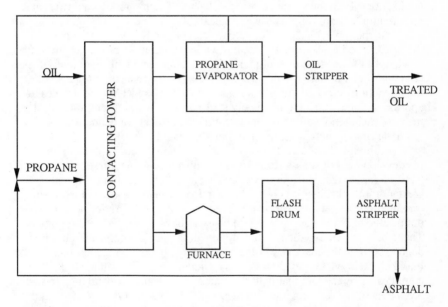

Fig. 8.16 Schematic flow diagram of a process for treating oils to recover asphalt by propane extraction.

In this operation the contacting tower usually is operated at 55 - 80° and 3 - 4 MPa.

Petroleum cokes

The final product of petroleum processing is a family of highly carbonaceous solids known as petroleum cokes. Coke is not a direct product of distillation of crude oil, but rather is formed either inadvertently during catalytic processing of petroleum fractions or deliberately in an effort to convert heavy products into lighter, more salable distillate liquids. These processes are discussed in Chapter 9. Coke is therefore a by-product of a refinery, a product which is usually salable but is not a premium product. Coke, is however, often a large volume product from a refinery. Roughly 40% of petroleum coke production is used as a solid fuel, another 40% is used to manufacture carbon anodes for the aluminum smelting industry, 10% is used to manufacture graphite electrodes for use in electric arc furnaces in the steel industry, and the remaining 10% finds a variety of small uses. In each of these applications, both the structure of the coke and the amount of impurities are important in determining its use and saleability.

As the molecules in a liquid hydrocarbon feedstock undergo cracking, the lighter products escape the reactor while the remaining, heavier molecules condense to form structures of increasing molecular weight. Eventually these structures become so large that a solid, coke, is formed as the product. As these reactions proceed, there is a tendency for the condensing molecules to approximate the thermodynamically stable form of carbon, graphite. That is, the condensing molecules attempt to form planar sheets of condensed aromatic rings which develop a parallel alignment.

Although there is a tendency to try to establish the graphite structure, the development of such a structure is not always possible. First, not all possible molecular configurations are flat, and therefore either cannot condense into planar sheets or condense into non-planar sheets which then cannot align. Second, in some cases even if planar sheets are formed, they may not be able to rearrange in the liquid phase to develop the parallel alignment. That ability to achieve the parallel alignment of planar sheets in the liquid will depend both on the reactivity of the species reacting and on the viscosity of the liquid phase in which the alignment is trying to occur. The situation is complicated by the fact that both the reactivity and structure and the physical properties of the reacting medium (especially the viscosity) play a role in the development of coke structure, and these two factors can vary independently of each other.

In the worst case essentially no alignment of molecular sheets occurs. In such a situation the properties of the solid coke will be the same in all directions, the product thus being an isotropic coke. Usually, however, there will be at least some alignment of molecular sheets. The alignment confers directionality to the physical properties. The regions of molecular alignment in the liquid are termed *mesophase*, the term implying an intermediate phase between the relatively small organic molecules of the precursors and the fully solidified coke. Regions of mesophase in the liquid can grow by coalescence to form ultimately very extensive regions. Upon solidification, the produce is an anisotrpoic semicoke. Since the extent of mesophase development depends on structure, reactivity, and liquid viscosity, all of which can vary independently, it is possible to produce cokes having varying degrees of molecular alignment. The type of coke produced in delayed coking operations in a refinery (Chapter 9) has some mesophase development, but not a great deal. The solid is a highly porous material often referred to as *sponge coke*. As a result of the highly porous structure sponge coke burns readily and thus is a useful fuel, provided that the sulfur content is not excessively high. Sponge cokes are mainly produced from resids which are paraffinic or naphthenic.

In addition to its use as a fuel, sponge coke is also used to manufacture anodes for the electometallurgical smelting of aluminum. Anodes used in this service must be good conductors of electricity and have good mechanical strength at high temperatures. The anodes are slowly consumed by reaction with the oxygen liberated from the electrolytic decomposition of aluminum oxide; for economic reasons it is not feasible to use anodes made of graphite. Manufacture of the anodes involves grinding the sponge coke, mixing with a binder of pitch (the mixture containing 60-70% coke), and heating to 1100-1200°.

Resids which are highly asphaltic form a more isotropic, low porosity coke. This material assumes a nearly spherical form somewhat reminiscent of the lead pellets used in shotguns, and is known as *shot coke*. The spherical particles of shot coke sometimes aggregate into large clusters. Shot coke is generally an undesirable product.

In general, the higher the degree of mesophase development, the greater is the anisotropy of the resulting coke.

Further reading

Anderson, R. O. (1984). *Fundamentals of the Petroleum Industry.* Norman, OK: University of Oklahoma; Chapter 21.

Barnard, J. A. and Bradley, J. N. (1985). *Flame and Combustion,* 2d edn. London: Chapman and Hall; Chapters 5,6,7,11.

Bland, W. F. and Davidson, R. L. (eds.) (1967). *Petroleum Processing Handbook.* New York: McGraw-Hill; Section 3.

Castellan, G. W. (1964). *Physical Chemistry*. Reading, MA: Addison-Wesley; Chapter 14.

Fieser, L. F. and Fieser, M. (1961). *Advanced Organic Chemistry*. New York: Reinhold Chapters 2,4,7.

Heywood, J. B. (1988). *Internal Combustion Engine Fundamentals*. New York: McGraw-Hill; Chapters 1,9,10

Kent, J. A. (ed.) (1974). *Riegel's Handbook of Industrial Chemistry*. New York: Van Nostrand Reinhold; Chapter 14.

Meyers, R. A. (ed.) (1986). *Handbook of Petroleum Refining Processes*. New York: McGraw-Hill ; Part 7.

Royal Dutch Shell (1983). *The Petroleum Handbook*, 6th edn. Amsterdam: Elsevier; Chapters 4,5.

Singer, J. G. (1981). *Combustion*, 3rd edn. Windsor, CT: Combustion Engineering; Chapter 2.

Taylor, C. F. (1985). *The Internal Combustion Engine in Theory and Practice*, 2d edn., Cambridge MA: M.I.T. Press; Volume II, Chapters 1,2,3,4.

Vollhardt, K. P. C. (1987). *Organic Chemistry*. San Francisco: Freeman; Chapter 2.

Chapter 9

Processing of petroleum fractions

Distillation allows the separation of crude oil into a variety of useful products, ranging from LPG to asphalt. Often subsequent processing steps are necessary to increase the yield of products which have strong market demand, or to improve the performance of some of the products. This chapter discusses processes for addressing these needs. Thermal cracking provides a further extension of free radical chemistry introduced previously. Catalytic processing introduces the chemistry of carbocations and reaction processes at catalyst surfaces.

Thermal cracking

One of the problems associated with natural gasoline and straight run gasoline is that the total yield of these products from most crude oils is not adequate to meet the demand for gasoline. This situation has existed since before the First World War. Consequently, methods are needed to increase the yield of products in the gasoline range at the expense of other, less desirable petroleum fractions. The n-alkanes in the C_5 to C_9 range are important components of gasoline; to examine the strategies for increasing the yield of gasoline we can consider the ways of increasing the amount of alkanes in this range. Two broad options are available: cracking molecules having greater than nine carbon atoms, or building up molecules from starting materials having fewer than five carbon atoms. In practice both options are used. We will begin with a discussion of cracking processes.

Industrial cracking reactions are of two general types: thermal cracking and catalytic cracking. Thermal cracking processes were first developed around 1913. Thermal cracking enabled refiners to meet the increasing demand for gasoline in the 1920's and 30's. Thermal cracking reactions proceed very much in manner of the free radical cracking processes we have discussed in the catagenesis of kerogen. A significant difference between thermal cracking and kerogen catagenesis is that industrial reactions are run at significantly higher temperatures to increase the rate. Thermal cracking reactions in industry involve the same types of free radical reactions we have discussed previously: hydrogen abstraction, disproportionation, β-bond scission, recombination, and hydrogen capping.

The initial reaction is the homolytic cleavage of a C-C bond to form two free radicals. One of the radicals produced in the initiating reaction attacks a second molecule of alkane, abstracting a hydrogen. Because a 2° radical is more stable than a 1° radical, the hydrogen is removed from one of the interior carbon atoms:

$$RCH_2\cdot + CH_3CH_2CH_2CH_2CH_2CH_2CH_2CH_3 \rightarrow RCH_3 + CH_3CH_2CH_2CH_2CH_2CH\cdot CH_2CH_3$$

The products are a new, smaller alkane and a 2° radical. The 2° radical then undergoes a β-bond scission reaction producing the expected 1-alkene and a 1° radical.

$$CH_3CH_2CH_2CH_2CH_2CH\cdot CH_2CH_3 \rightarrow CH_3CH_2CH_2CH\cdot + CH_2=CHCH_2CH_3$$

The small 1° radical can continue to undergo β-bond scission until ethylene is produced. Thus ethylene, sometimes in large amount, is produced as a by-product.

Free radicals do not ordinarily undergo isomerization. However, long chain 1° radicals can convert to a more stable 2° radical if the chain coils into such a configuration that the 1° radical can abstract a hydrogen from one of the carbon atoms in the interior of the chain.

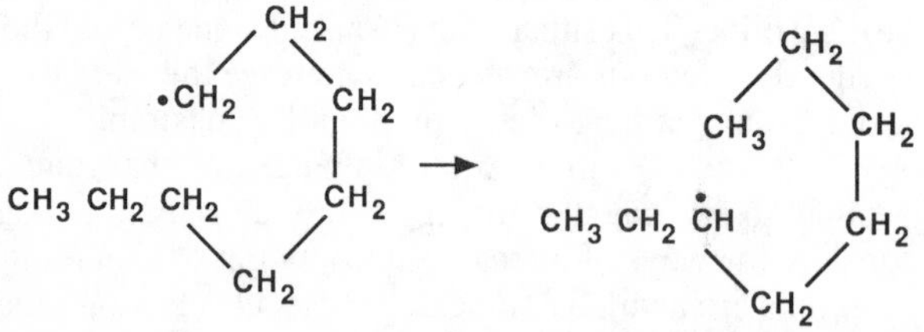

The carbon atom from which the hydrogen is abstracted is usually five or six atoms backward along the chain.

Free radical recombination is very rapid, and in principle could bring cracking to a stop fairly quickly. However, in the reaction conditions used in industrial cracking processes the concentration of non-radical hydrocarbon molecules is much greater than the concentration of radicals. Consequently the collision of a radical with a hydrocarbon is much more likely than the collision of a radical with another radical.

Mixed-phase cracking

A block flow diagram for a mixed-phase cracking operation is shown in Figure 9.1.

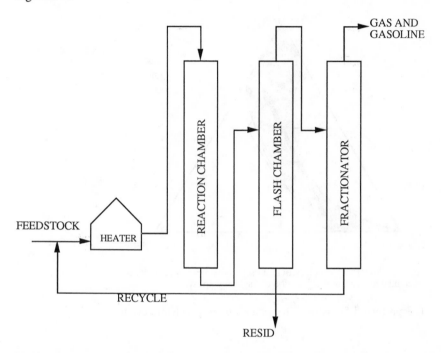

Fig. 9.1 Schematic flow diagram for mixed-phase cracking to gas, gasoline, and resid.

The feedstock used for this process is kerosene, gas oil (which has a boiling range between kerosene and lube oil), resid, or sometimes whole crude oil. The feedstock is heated rapidly, after which it flows to the reaction chamber. The reaction chamber is maintained at 400-480° with pressures of 2.5 MPa. The choice of temperature is dictated by the nature of the charge. The pressure is chosen to be high enough to maintain a virtually homogeneous phase with a high liquid/vapor ratio. Increasing the temperature in the reaction chamber therefore necessitates an increase in pressure to prevent excessive vaporization of feedstock. The overhead from the flash chamber is sent to a fractionating tower to obtain the gasoline fraction. The fractionating

213

tower bottoms are recycled. The flash chamber bottoms are withdrawn to use as a heavy fuel oil.

The tendency for compounds to undergo cracking is in the order alkanes > alkenes > alkadienes > cycloalkanes > aromatics. Using the ternary classification diagram introduced in Chapter 5, we can now see how the tendency of a crude oil to undergo cracking varies with its composition (Fig. 9.2)

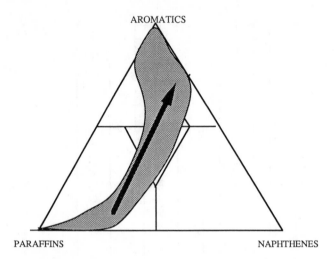

AROMATICS

PARAFFINS   NAPHTHENES

Fig. 9.2 The ternary oil classification scheme; tendency to undergo cracking decreases in the direction of the heavy arrow.

The preferred feedstocks for cracking are paraffinic crudes.

Visbreaking

Viscosity breaking (generally known as *visbreaking*) is a mild thermal cracking process used for the reduction of viscosity, the lowering of the boiling range, and the reduction of the pour point of heavy resids. The ability to meet all three of these objectives in a single process derives from the relationships among molecular size, boiling and melting behavior, and viscosity introduced in Chapter 5.

A block flow diagram of a visbreaking process is shown in Figure 9.3. With a resid as feedstock, the feed will be heated to 500-525° and 0.35-0.70 MPa. The temperature at the exit side of the reaction chamber is 440-455°.

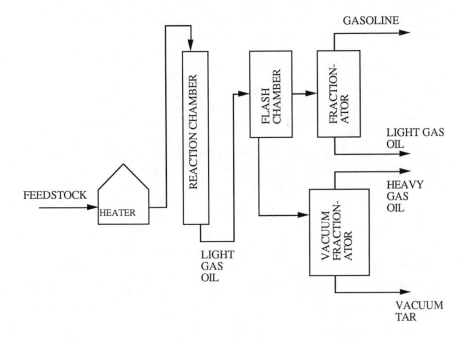

Fig. 9.3 Schematic flow diagram of a visbreaking process, with production of gasoline gas oils, and tar.

Cracked products from the reaction chamber are fed to a flash distillation chamber. The overhead from the flash distillation chamber is separated into gasoline and gas oil (the bottoms) in a fractionator. Liquids from the flash chamber are fed to a vacuum fractionator, which separates them into a heavy gas oil and a residual tar.

Visbreaking is primarily used for feed preparation of lower-boiling gas oils for catalytic cracking operations. About 8-10% of the charge is converted to gasoline. The gasoline yield from visbreaking is incidental; the principal production of gasoline would come from the subsequent catalytic cracking operation.

Alkene formation in thermal cracking

The thermal cracking process, like thermal cracking in nature, begins with C-C bond cleavage more or less randomly along the carbon chain. Thus it is entirely possible for some large alkanes to cleave near the end of the chain, producing appreciable quantities of products having less than five carbon atoms (that is, too light to be in the gasoline range). In addition, alkenes can be formed via disproportionation reactions or β-bond scission reactions:

$$CH_3CH \cdot CH_3 + CH_3CH \cdot CH_3 \rightarrow CH_3CH_2CH_3 + CH_3CH=CH_2$$

or

$$CH_3CH_2CH_2CH_2CH \cdot CH_2CH_3 \rightarrow CH_3CH_2CH \cdot + CH_2=CHCH_2CH_3$$

The formation of alkenes is not necessarily an undesirable event. Depending on the marketing situation, they can be converted to a wide variety of useful materials, as illustrated for propylene in Figure 9.4.

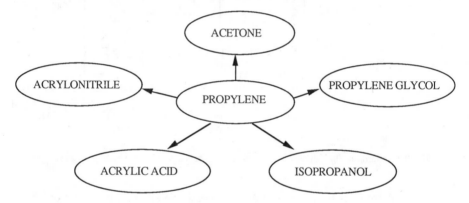

Fig. 9.4 A schematic diagram illustrating some of the uses to which a product of thermal cracking - propylene - can be put. The alkenes are valuable feedstocks for the petrochemical industry for the production of solvents and polymers.

The higher alkenes have boiling points so close to the parent alkane that large-scale separation is totally impractical. For example, the boiling points of the C_7 compounds are heptane, 98°; 1-heptene, 94°; 2-heptene, 98°; and 3-heptene, 96°. These by-product alkenes remain in the fuel stream. Thermal cracking of alkanes in the C_{14} to C_{34} range will produce, under mild conditions, alkenes in the C_6 to C_{20} range.

Alkenes have higher octane numbers than the corresponding alkanes. No only does thermal cracking approximately double the yield of gasoline, as compared with simple distillation, but the quality of gasoline produced from thermal cracking is superior to straight-run gasoline. The presence of the alkenes in the gasoline from thermal cracking helps to increase the octane number relative to straight-run gasoline.

Ethylene is unquestionably the most important alkene, and indeed the most important organic compound, in industrial chemistry. Ethylene is produced by thermal cracking reactions at 750-900° in which the hydrocarbon feedstock "unzips" by successive β-bond scission processes to ethylene. In essence, β-bond scission is the reverse reaction of free radical polymerization. Thus in principle one might expect polymerization and unzipping to be competetive processes in thermal cracking. That polymerization is not important in commercial processing is a result of the careful selection of temperatures. Polymerization is an exothermic process at relatively low temperatures, but the variation of ΔH with reaction temperature is such that the reaction becomes endothermic above 720°. Since polymerization is endothermic above 720°, its reverse reaction is exothermic. Thus the high reaction temperatures of ethylene production favor cracking instead of polymerization.

Delayed coking

Delayed coking is a process to obtain additional distillate products by the thermal cracking of resids. In essence, a hydrogen redistribution process allows disproportionation of the feedstock to the carbon-rich coke and the relatively hydrogen rich distillate liquids and gases. Delayed coking operations are run with very long residence times to insure completion of the cracking and hydrogen redistribution processes. Roughly 75% of the feed is converted to useful lighter products and the remaining 25% forms coke.

The charge for delayed coking is pumped into a furnace where it is heated rapidly to temperatures over 480°. The preheated feed is then fed into the bottom of a large, cylindrical reactor which might be 4 - 6 m in diameter and 18 - 27 m tall. The cracked products normally leave the reactor quickly; the coke builds up in the reactor until the reactor has been filled with coke to some pre-determined level. Normally a delayed coking operation will have two or four reactors, so that at least one reactor will be on line while the other is being "de-coked". When the reactor has filled with coke, steam is passed through the reactor to vaporize any feedstock that had not been converted. The coke is then cooled by injection of water and is removed by using water lances with high pressure (>20 MPa) water streams to cut the coke mass apart and away from the reactor walls.

Delayed coking is sometimes facetiously referred to as the "refiner's garbage can." Many types of feedstocks, including resids and any unsalable products with boiling points >340°, can be treated in delayed coking.

Polymerization and alkylation

An alternative to using the alkenes of five or fewer carbon atoms as chemical feedstocks is to recombine them to form new products in the desired C - C gasoline range. Industrially the process is important because it increases the yield of gasoline. The discussion is important for fuel chemistry because it represents the introduction of a new reacting species, *carbocations*.

Polymerization

The combination of two alkene molecules is known industrially as *polymerization*. In reality the formation of gasoline components represents a dimerization reaction, although with appropriate reaction conditions more extensive polymerization can occur. We can begin by considering 2-methylpropene (isobutylene). In the presence of an acid, the double bond can be protonated to form a carbocation

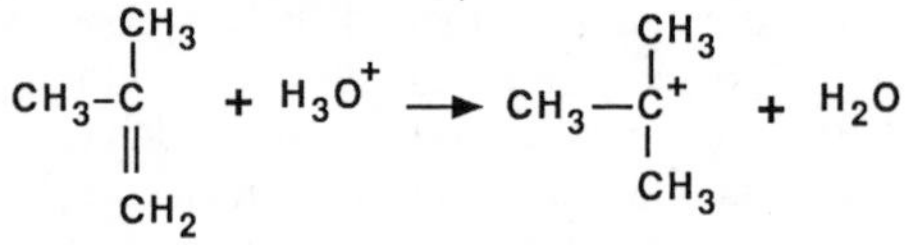

$$
\underset{\underset{CH_2}{\overset{\|}{\underset{|}{C}}}}{\overset{CH_3}{\underset{|}{CH_3-C}}} + H_3O^+ \longrightarrow \underset{\underset{CH_3}{\overset{}{\underset{|}{C}}}}{\overset{CH_3}{\underset{|}{CH_3-C^+}}} + H_2O
$$

Notice that there are two options for protonation of the double bond. That is, in principle, one might expect the formation of

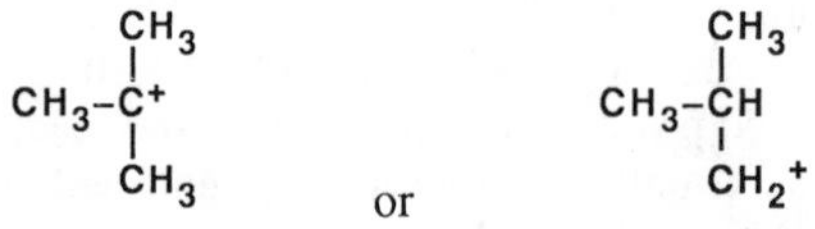

$$
\underset{\overset{|}{CH_3}}{\overset{CH_3}{\underset{|}{CH_3-C^+}}} \qquad or \qquad \underset{\overset{|}{CH_2^+}}{\overset{CH_3}{\underset{|}{CH_3-CH}}}
$$

The protonation reaction will proceed to form the most stable carbocation. The stability of carbocations is in the same order as free radicals, i.e., 3° > 2° > 1°. Therefore the protonation of the double bond in 2-methylpropene forms the 3° carbocation exclusively. The increased stability of the 3° carbocation relative to the 1° ion derives from the presence of three methyl groups, each of which can release electron density toward the carbon bearing the positive charge and thus partially stabilize the charge.

Because carbocations are electron-deficient and are positively charged (recall that free radicals are electron-deficient but are electrically neutral) they will seek out and react with materials containing an apparent concentration of electron density, such as a double bond. Thus the carbocation can attack a second molecule of 2-methylpropene

218

$$\underset{\substack{\text{CH}_3}}{\overset{\substack{\text{CH}_3}}{\text{CH}_3\text{-}\underset{|}{\overset{|}{\text{C}}}{}^+}} + \text{H}_2\text{C}\!=\!\overset{\substack{\text{CH}_3}}{\underset{|}{\text{C}}}\text{-CH}_3 \longrightarrow \text{CH}_3\text{-}\overset{\substack{\text{CH}_3}}{\underset{\substack{\text{CH}_3}}{\text{C}}}\text{-CH}_2\text{-}\overset{\substack{\text{CH}_3}}{\underset{+}{\text{C}}}\text{-CH}_3$$

The product of this reaction is itself a carbocation. In the new dimeric carbocation the charge is also stabilized by three alkyl groups. In principle, it could attack yet another molecule having a double bond. Indeed this continual attack does occur if the acid catalyst is anhydrous hydrogen fluoride or concentrated sulfuric acid. The product of such reaction is poly(isobutylene), a very tacky material used in formulating adhesives for sealing tapes.

If the polymerization of 2-methylpropene is run in 60-65% sulfuric acid, at 70°, there is sufficient water present in the reaction mixture so that the dimeric carbocation loses a proton to a water molecule:

$$\underset{\substack{\text{CH}_3}}{\overset{\substack{\text{CH}_3}}{\text{CH}_3\text{-}\underset{|}{\text{C}}}}\text{ - }\underset{\substack{\text{H}}}{\overset{}{\text{CH-}\underset{|}{\text{C}}}}\overset{+}{\underset{\text{CH}_3}{}} \longrightarrow \text{CH}_3\text{-}\overset{\substack{\text{CH}_3}}{\underset{\substack{\text{CH}_3}}{\text{C}}}\text{-CH}\!=\!\overset{\substack{\text{CH}_3}}{\text{C}}\text{-CH}_3$$

and

$$\text{CH}_3\text{-}\overset{\substack{\text{CH}_3}}{\underset{\substack{\text{CH}_3}}{\text{C}}}\text{-CH}_2\text{-}\overset{\substack{\text{CH}_3}}{\underset{\substack{+ \; \text{H}}}{\text{C}}}\text{-CH}_2 \longrightarrow \text{CH}_3\text{-}\overset{\substack{\text{CH}_3}}{\underset{\substack{\text{CH}_3}}{\text{C}}}\text{-CH}_2\text{-}\overset{\substack{\text{CH}_3}}{\text{C}}\!=\!\text{CH}_2$$

Thus the reaction stops with dimerization. About 80% of the product is 2,4,4-trimethyl-1-pentene; the remaining 20% being the isomer with the double bond in the 2-position. The two dimers, 2,4,4-trimethyl-2-pentene and 2,4,4-trimethyl-1-pentene (also known as diisobutylenes), can be hydrogenated to the same product, 2,2,4-trimethylpentane. We have seen that this compound is the "octane" or "isooctane" used as the standard of combustion performance in gasoline, with an octane number defined to be 100. The production of branched chain alkanes such as 2,2,4-trimethylpentane by dimerization of by-product alkenes can be a valuable route not only to increasing the yield of gasoline but also to increasing its octane number.

The 2-methylpropene is obtained as part of the C_4 cut of cracked gasolines. This fraction will also include other C_4 compounds: 1-butene, 2-butene, butane, and 2-methylpropane (isobutylene). Isobutylene is much more soluble in 60-65% sulfuric acid than the other C_4 compounds, so that it selectively absorbed in the acid. Heating a solution of isobutylene in sulfuric acid to 100° results in complete polymerization in about 1 minute reaction time. Polymerization is also carried out with propylene as a feedstock, in which case the principal product is 2,3-dimethyl-2-butene.

A block flow diagram of an industrial polymerization process is illustrated in Figure 9.5.

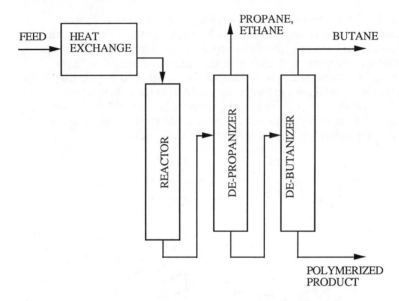

Fig. 9.5 Schematic diagram of a polymerization process, for formation of diisobutylenes.

The temperature range is 150-220° and pressures are 1-8.5 MPa. The product has octane numbers of 93-99.

Alkylation

Alkylation is a process related to polymerization; in this case, however, the reactants are an alkene and an alkane. The key to alkylation reactions is that alkanes with a 3° hydrogen can lose that hydrogen atom to a carbocation. The reaction of propene with 2-methylpropane provides an example.

Propylene can be protonated in a manner analogous to the 2-methylpropene reaction introduced previously.

$$CH_3-CH=CH_2 + H_3O^+ \longrightarrow CH_3-\overset{+}{C}H-CH_3 + H_2O$$

The resulting 2° carbocation readily reacts with the 3° hydrogen atom of 2-methylpropane.

$$CH_3-\overset{+}{C}H-CH_3 \;+\; H-\underset{\underset{CH_3}{|}}{\overset{\overset{CH_3}{|}}{C}}-CH_3 \;\longrightarrow\; CH_3-CH_2-CH_3 \;+\; \underset{\underset{CH_3}{|}}{\overset{\overset{CH_3}{|}}{\overset{+}{C}}}-CH_3$$

(Notice that the reaction is driven by the opportunity to form the more stable 3° carbocation at the expense of the 2° ion.) The reactive product formed is the 3° carbocation; propane is a by-product of the reaction. The 3° carbocation will then react with the double bond in the propylene

$$CH_3-\underset{\underset{CH_3}{|}}{\overset{\overset{CH_3}{|}}{C}}{}^+ \;+\; H_2C=CH-CH_3 \;\longrightarrow\; CH_3-\underset{\underset{CH_3}{|}}{\overset{\overset{CH_3}{|}}{C}}-CH_2-\overset{+}{C}H-CH_3$$

and the new 2° carbocation formed as a result will react with yet another 3° hydrogen from another molecule of 2-methylpropane (isobutane).

$$CH_3-\underset{\underset{CH_3}{|}}{\overset{\overset{CH_3}{|}}{C}}-CH_2-\overset{+}{C}H-CH_3 \;+\; H-\underset{\underset{CH_3}{|}}{\overset{\overset{CH_3}{|}}{C}}-CH_3 \;\longrightarrow\; CH_3-\underset{\underset{CH_3}{|}}{\overset{\overset{CH_3}{|}}{C}}-CH_2-CH_2-CH_3 \;+\; \underset{\underset{CH_3}{|}}{\overset{\overset{CH_3}{|}}{\overset{+}{C}}}-CH_3$$

The carbocation generated in this process allows the reaction to continue. The 2,2-dimethylpentane (neoheptane) is a valuable gasoline constituent with high octane number.

Many industrial alkylation processes are based on anhydrous hydrogen fluoride as the acid catalyst. Typically the alkenes fed to the process are in the C3 - C5 range. The reaction temperature is 0-40° and the pressure is 0.1-8.5 MPa. The pressure is kept high enough to maintain the reactants in the liquid phase. When isobutane is reacted with propylene, the product is a mixture of 60-80% 2,2-dimethylpentane, 10-30% 2-methylhexane, and ~10% 2,2,3-trimethylbutane (which has the trivial name triptane). The product, *motor fuel alkylates*, has octane numbers of 90-115. Specific examples are shown in Table 9.1.

Catalytic cracking

Thermal cracking provides a way of increasing the yield of gasoline. Recombination reactions involving 2° radicals also produce an increased proportion of branched chain compounds, so that the octane number of

Table 9.1 Products of alkylation of alkenes with isobutane

Alkene	Products	Octane number
Propene	2,3-Dimethylpentane	90
	2,4-Dimethylpentane	84
2-Methylpropene	2,2,4-Trimethylpentane	100
1-Butene	2,3-Dimethylhexane	75
	2,4-Dimethylhexane	68
2-Butene	2,3,4-Trimethylpentane	100
	2,2,4-Trimethylpentane	100
	2,2,3-Trimethylpentane	103

gasolines produced directly from thermal cracking may be about 75. While this is a substantial increase in octane number compared to the value of about 30 for some straight run gasolines, 75 is still too low for satisfactory performance in most modern SI engines. Thus processes are needed to increase still further the proportion of branched chain structures in the gasoline.

Thermal cracking processes were developed shortly before the First World War, and were used to meet increased gasoline demand in the following decades. As with many aspects of the fuel and chemical industries, the situation was drastically changed with the coming of the Second World War. The outbreak of the war enormously increased the demand for gasoline for vehicles and aircraft. Furthermore, high performance piston aircraft engines needed gasoline with octane numbers of 100 or higher. (At the same time, of course, the demand for diesel oil for tanks and trucks, fuel oil for ships, and all kinds of lube oils increased.) The oil industry responded with the development of large scale catalytic cracking, along with the fluidized bed reactor system. The development of fluidized bed catalytic cracking units represents one of the outstanding achievements of chemical engineering.

Carbocation rearrangements

We have seen that in some ways carbocation reactions are similar to reactions of free radicals. For example, the order of stability, $3° > 2° > 1°$, is the same for both radicals and carbocations. The alkylation process represents an example of hydrogen abstraction by a carbocation, analogous to hydrogen abstractions by radicals. Later we will see examples of β-bond scission reactions of carbocations. However, there is one very great difference between the reactions of radicals and carbocations - structural rearrangement. In most reactions of free radicals, the radical initially formed will retain its basic carbon framework throughout the subsequent reactions. For example, if a hydrogen abstraction reaction leads to the formation of the 2° radical $CH_3CH_2CH \cdot CH_2CH_2CH_3$, this species will remain a 2° radical until it undergoes some reaction at the radical site. In contrast, a carbocation will undergo a rearrangement of the carbon skeleton to form a more stable carbocation. For example, a 2° carbocation may rearrange to form a 3° carbocation.

Consider the carbocation $CH_3CH^+CH_2CH_2CH_2CH_3$. Once this ion has formed, the alkyl group with its pair of electrons can in essence exchange places with the positive charge:

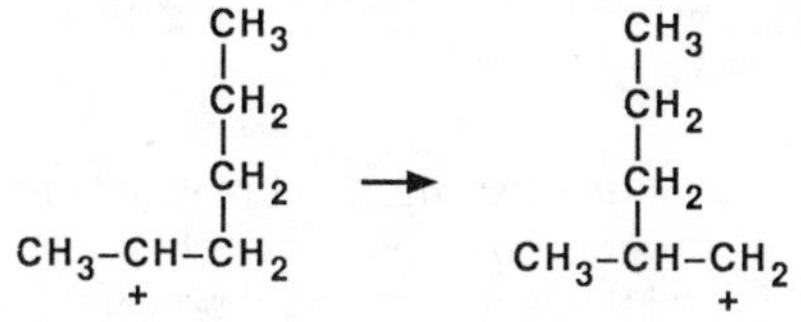

```
        CH3                      CH3
         |                        |
        CH2                      CH2
         |                        |
        CH2          →           CH2
         |                        |
 CH3-CH-CH2              CH3-CH-CH2
      +                          +
```

The new carbonium ion formed in this process can be stabilized quickly by a shift of a hydride

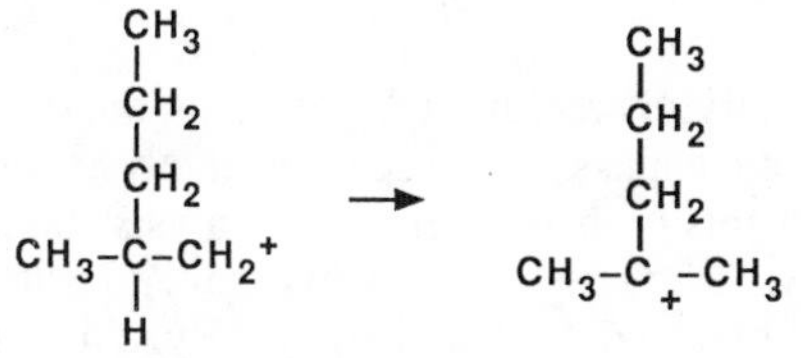

```
        CH3                      CH3
         |                        |
        CH2                      CH2
         |                        |
        CH2          →           CH2
         |                        |
 CH3-C-CH2+             CH3-C -CH3
      |                      +
      H
```

Capping the 3° carbocation with a hydride yields 2-methylpentane as a product

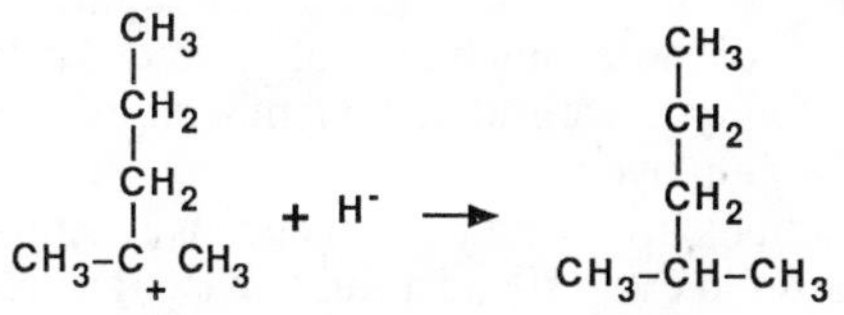

```
        CH3                          CH3
         |                            |
        CH2                          CH2
         |                            |
        CH2      + H-   →            CH2
         |                            |
 CH3-C CH3               CH3-CH-CH3
      +
```

Thus the formation of a 2° carbocation at the 2-position in the straight chain compound hexane leads, via rearrangement, to the branched chain isomer 2-methylpentane.

An alternative view of the rearrangement shown above is that it may proceed via a *nonclassical carbocation*. In either case, however, the net effect of the rearrangement is to create a branched chain compound from one with a straight chain. Doing so for compounds in the gasoline range would increase the octane number. Put the other way around, if we wish to increase the octane number of gasoline, we an do so by finding processes which will increase the proportion of branched chain hydrocarbons. Since carbocation

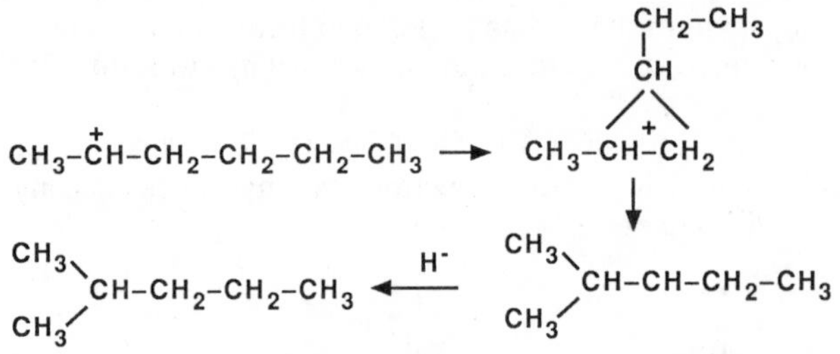

$$CH_3-\overset{+}{C}H-CH_2-CH_2-CH_2-CH_3 \longrightarrow$$

Top right structure:
$$\begin{array}{c} CH_2-CH_3 \\ | \\ CH \\ \diagup \quad \diagdown \\ CH_3-CH-\overset{+}{C}H_2 \end{array}$$

$$\downarrow$$

$$\begin{array}{c} CH_3 \diagdown \\ CH-\overset{+}{C}H-CH_2-CH_3 \\ CH_3 \diagup \end{array}$$

$$\xleftarrow{\quad H^-\quad}$$

$$\begin{array}{c} CH_3 \diagdown \\ CH-CH_2-CH_2-CH_3 \\ CH_3 \diagup \end{array}$$

rearrangement is a route to conversion of straight to branched chains, in essence the need to increase the proportion of branched chain hydrocarbons in the gasoline means that the processing reactions should proceed via carbocations rather than by free radicals.

Cracking catalysts

The reactions of interest in catalytic cracking take place on the surfaces of heterogeneous catalysts. Before discussing the specific catalysts used in cracking, some general comments on heterogeneous catalysts should be made. For any heterogeneous catalyst to be used in industry, several criteria must be fulfilled. First, of course, the material must actually catalyze the reaction, which is to say that the material must provide a rate of reaction which is high enough to achieve the rates of conversion needed for a commercial-scale reactor. Second, the catalyst must show good *selectivity*, which means that it must induce the formation of desired products and suppress, or at least not enhance, the formation of those which are not desired. Third, it must have sufficient physical stability to withstand the temperatures and pressures at which it is likely to be employed.

Almost all reactions involving heterogeneous reactions occur at the interface between the solid catalyst and a fluid phase. The reactions catalyzed in five stages. The reactant must first diffuse to the surface of the catalyst. It must then be adsorbed on the catalyst. The adsorbed reactant species then undergoes a reaction in which it is converted to a product molecule. The product must desorb from the catalyst and finally diffuse away from the catalyst surface into the bulk of the process stream. Any one of these five steps can potentially be the rate-determining step for the overall process. For many reactions the third step, the conversion of adsorbed reactant into product on the catalyst surface, is the rate-determining step. In such instances, the reaction is said to proceed via the *Langmuir-Hinshelwood mechanism*.

There are two general ways of generating a carbocation relevant to petroleum processing. The first is by protonation of a double bond:

$$CH_3CH{=}CH_2 + H^+ \rightarrow CH_3CH^+CH_3$$

and the second is by heterolytic bond cleavage:

$$CH_3CH_2CH_3 \rightarrow CH_3CH^+CH_3 + H^-$$

The protonation of the double bond requires a good proton donor, that is, a strong Bronsted acid. Abstraction of a hydride ion in heterolytic bond cleavage requires a material having a powerful affinity for an electron pair (which would come with the hydride ion), in other words, a strong Lewis acid. Bronsted acids generate carbocations from alkenes, while Lewis acids generate carbocations from alkanes. To be sure of handling any material in a feed, it would be desirable to have a catalyst that could function as a Bronsted or as a Lewis acid.

To illustrate the structural relationships in carbocation-generating cracking catalysts (i.e., acidic catalysts) we can begin with the structure of silicon dioxide.

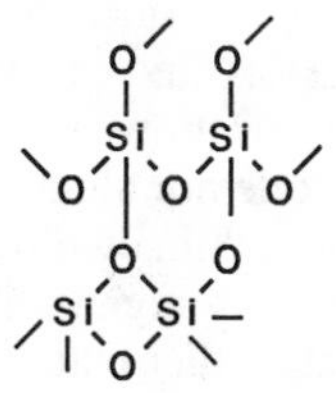

The structure of solid silicon dioxide is essentially an infinite network of SiO_4 tetrahedra, in which each silicon atom is joined to four others via Si-O-Si bonds. At the surface it is not possible for oxygen atoms to fulfill both valencies by bonding to other Si atoms. The surface oxygen atoms could however bear hydrogen atoms

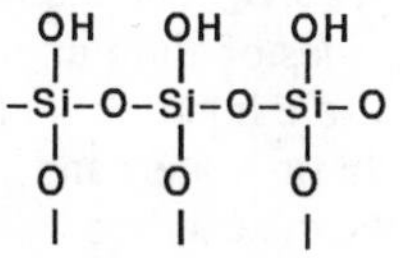

The hydrogen atoms are fairly strongly bonded to the surface oxygen atoms, and the structure shown above is a rather weak Bronsted acid.

Let us now consider what happens if some of the silicon atoms are replaced by aluminum atoms.

```
     OH        OH
      |         |
   -Si-O-Al-O-Si-
      |    |    |
      O    O    O
      |    |    |
```

Here the aluminum atoms, with only three pairs of electrons, provide sites having strong Lewis acidity. Furthermore, this structure has the potential of forming also a strong Bronsted acid. Interaction of the aluminum atom with an oxygen atom bonded to a neighboring silicon can weaken the O-H bond.

```
     OH        OH                        OH       H+
      |         |                         |       O
   -Si-O-Al-O-Si-       →             -Si-O-Al-O-Si-
      |    |    |                         |    |    |
      O    O    O                         O    O    O
      |    |    |                         |    |    |
```

The weaker the O-H bond, the more easily is the hydrogen removed and hence the greater is the Bronsted acidity. Thus catalysts whose structures are based on aluminosilicates, mixed aluminum and silicon oxides, offer the desired property of being good Bronsted and good Lewis acids, and consequently potentially useful catalysts for generating carbocations.

It is significant that neither silicon dioxide nor aluminum oxide by themselves make satisfactory cracking catalysts. Silicon dioxide does not catalyze cracking. Aluminum oxide provides very high initial rates of cracking but deactivates very quickly, so that further reactions do not occur. A simple mechanical mixture of the two compounds does not make a suitable cracking catalyst. However, when SiO_2 and Al_2O_3 are crystallized together, even with only a small proportion of Al_2O_3 in the SiO_2, an active catalyst results. The first catalyst used in catalytic cracking was a clay mineral, montmorillonite. At one time the most widely used cracking catalyst was a synthetic blend of 87% SiO_2 - 13% Al_2O_3. These catalysts were made by mixing solutions of sodium metasilicate, Na_2SiO_3, and aluminum sulfate to co-precipitate a mixture of the hydrated oxides of aluminum and silicon. Stirring for less than a minute causes the whole liquid to set to a gel; on drying at 100° the gel loses about 90% of its weight (as water) and yields a hard solid. To minimize abrasion in fluidized bed units, the catalyst was fabricated in spherical particles. The catalysts are highly porous, with surface areas of 100-300 m^2/g and pore diameters of 3 - 10 nm. Most such catalysts have now been supplanted by the zeolites.

A family of aluminosilicate materials widely used as catalysts are the *zeolites*. The zeolites are naturally occurring aluminosilicates. The basic structural units in zeolites are AlO_4 and SiO_4 tetrahedra:

These tetrahedra are linked in three dimensions to form the sodalite structure:

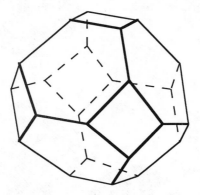

Fig. 9.6 The sodalite structure, formed from aluminum and silicon oxide tetrahedra. The darker lines project toward the reader.

The zeolite structure consists of sodalite "cages" linked to form a "supercage" (next page). The truncated octahedra comprising the sodalite structure illustrated in Figure 9.6 are joined together to make the zeolite structure (Fig. 9.8). Molecules encountering the catalyst have access to the supercage via "windows." The size of the windows permits only molecules of certain sizes and shapes to enter the catalyst. For example, in a cracking catalyst straight chain alkanes can enter the windows whereas branched chain alkanes, which we would wish to preserve against further cracking, cannot. Very large alkanes may be unable to enter the windows but can crack on the surface of the catalyst. The products from the first cracking step of the large alkanes may then be small enough to enter the catalyst, where they can undergo further reactions. The volume of the supercage is sufficient enough to retain the reactant molecules long enough for them to react while they are in the cage (Fig. 9.9).

Zeolites are naturally occurring aluminosilicates, of which the most common forms are the minerals faujasite and mordenite. In all, there are about 30 zeolites found in nature. Most zeolites contain some water of crystallization; the anhydrous form has a large, regular pore structure consisting of interconnecting channels or tunnels. A crystalline lattice based only on AlO_2 structures would have a net negative charge; thus zeolites also contain cations of the alkali and alkaline earth elements, particularly sodium,

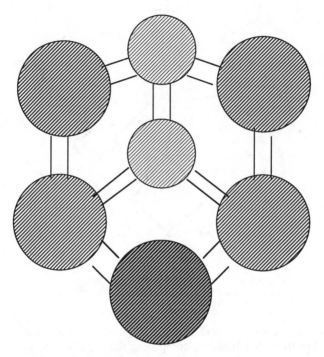

Fig. 9.7 The supercage is made of connected sodalite structures. In this schematic drawing the sodalite structures of Fig. 9.6 are respresented as shaded circles; the lighter the shading, the farther into the page is the sodalite unit from the reader. For clarity, a sodalite unit which should be present in the top center of the drawing has been omitted.

potassium, and calcium. These cations have considerable freedom of movement within the structure, and readily undergo ion exchange reactions. (Zeolites have been used since the 1930's for water softening, removing Ca^{+2} or Mg^{+2} from the water in exchange for Na^+.) The application of zeolites as catalysts has led to the synthesis of numerous zeolites which are not found in nature. The properties of the zeolite, including the crucial diameter of the pores, can be modified by changing the associated cations. For example, a synthetic zeolite known as Type A containing Na^+ ions has a pore diameter of 0.4 nm, but if the Na is replaced with the larger K^+, the openings of the pores are reduced to 0.3 nm. The Mobil Corporation has been a leader in the synthesis and development of zeolite catalysts. Over 95% of all catalytic cracking units in the United States now use Mobil zeolite catalysts. Syntheticzeolite catalysts must meet certain requirements: Obviously they

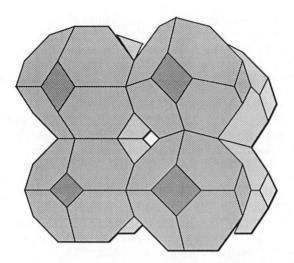

Fig. 9.8 The truncated octahedra of the sodalite structure
(Fig. 9.6) are joined to make the zeolite structure. The intensity
of shading decreases as the structure recedes from the viewer.
The diamond-shaped aperture in the center of the structure is a
window into the supercage.

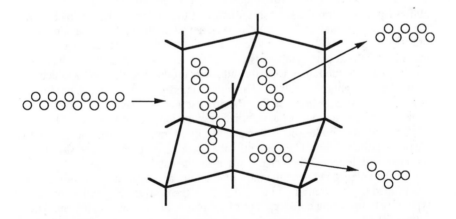

HYDROCARBON   THEN CRACKS INSIDE CAGE   PRODUCTS
DIFFUSES INTO CAGE   DIFFUSE OUT

Fig. 9.9 A hydrocarbon molecule diffuses through a window into the supercage, where
it undergoes cracking; the products then diffuse out.

must be catalytically active - not all synthetic zeolites function as catalysts. They must be stable at the temperatures required for operation in a commercial cracking reactor (roughly 300-550°) - the crystal structure of some zeolites is altered at high temperature. For some applications, the zeolites must also be stable to steam - high temperature steam can also disrupt the crystal structure/

Zeolites are the principal catalysts for catalytic cracking processes. Indeed, the development of zeolite catalysts for industrial cracking processes is probably the single most important development in cracking technology in the last several decades. The catalyst used for industrial processes consists of 3-25% of zeolite crystals, generally about 1 μm diameter, embedded in a matrix of silicon and aluminum oxides. The catalyst particles themselves, when used in fluidized bed cracking units are 20-60 μm in diameter. The use of the SiO_2-Al_2O_3 matrix for supporting the zeolite catalysts is due to two facts: first, zeolites are too expensive to allow the use of pure zeolite particles on an industrial scale; and, second, the zeolites, when used alone, are so active as catalysts that it would be difficult to design practical industrial-scale reactors which could accommodate the heat transfer requirements for smooth operation.

Zeolites are considerably more active than an SiO_2-Al_2O_3 catalyst. An important reason for the superior activity of zeolites is the much greater concentration of catalytically active sites on the zeolite surface compared with an SiO_2-Al_2O_3 catalyst. Furthermore, the porous structure of the zeolite allows for a greater effective concentration of the molecules being cracked in the vicinity of a catalyst site than can be achieved on the surface of a particle of SiO_2-Al_2O_3 catalyst. Even more important than the greater activity provided by the zeolite catalysts is the fact that zeolites give more products in the desired gasoline range, C_5-C_{10}, and fewer products in the C_3-C_4 which would be too light for gasolines. The molecular size of the cracking products depends on a balance between the cracking itself (*i.e.,* cleavage of carbon chains) and hydride transfer to stabilize carbocations. Compared to SiO_2-Al_2O_3 catalysts, zeolites have a greater ability to transfer hydrogen relative to cleaving carbon chains. Consequently, the cracking process tends to stop at products with longer carbon chains when zeolite catalysts are used.

The balance between the rates of cracking and hydrogen transfer is determined by the strength of the acid sites on the catalyst. Regardless of the type of reaction being considered or of the nature of the catalyst, the catalytic activity of any surface requires the formation of bonds between the reactants and the surface. The activities of a series of catalysts used to catalyze a particular reaction therefore correlates with the the strength of the bond formed between the reactant and the active site on the catalyst surface. A SiO_2-Al_2O_3 catalyst contains very strong acid sites. The molecule undergoing cracking is, consequently, very strongly bound to the catalytically active site and the cracking reaction is rapid. Cracking in this case can be much faster than hydrogen transfer. On the other hand, the acid sites on a zeolite surface are weaker than on the SiO_2-Al_2O_3 site. As a result, the binding of molecules to the acid sites on the surface is weaker, and cracking is slow relative to hydrogen transfer.

Another distinction between zeolite and SiO_2-Al_2O_3 cracking catalysts is that fewer alkenes are produced when a zeolite catalyst is used. Not only are zeolites superior at hydride transfer reactions to stabilize carbocations, but they are also superior for transferring hydrogen to alkenes. Most of the alkenes formed during the cracking step become hydrogenated more quickly than they desorb, and thus the alkene concentration in the product stream is low. Both the hydride transfer to stabilize carbocations and the hydrogen transfer to alkenes proceed at high rates on zeolite catalysts because of the greater effective concentration of molecules in the vicinity of an active catalyst site attained by having the reactions proceed in the small pores of the zeolite. We have seen previously that the further loss of hydrogen from an alkene to form alkadienes or alkatrienes is a major step toward coke formation, since the highly unsaturated compounds are precursors for the formation of the polycyclic aromatic structures in coke. Since zeolites are particularly effective at transferring hydrogen to alkenes, zeolites therefore tend to produce less coke.

A comparison of the performance of a zeolite catalyst and an amorphous SiO_2-Al_2O_3 catalyst is shown in Table 9.2

Table 9.2 Comparative product yields (volume percent) for fluidized catalytic cracking of gas oil for amorphous SiO_2-Al_2O_3 and zeolite catalysts

Product	SiO_2-Al_2O_3 Catalyst	Zeolite Catalyst
C_1 and C_2 gases	3.8%	2.1%
C_3 gases	17.6%	13.1%
C_4 gases	20.8%	15.4%
Gasoline boiling to 200°	55.5%	62.0%
Light fuel oil	4.2%	6.1%
Heavy fuel oil	15.8%	13.9%
Coke	5.6%	4.1%

The use of a zeolite catalyst affords a higher yield of gasoline with less coke formation.

In summary, zeolite catalysts have four important features. The crystalline structure of these catalysts results in a uniform pore size, rather than a distribution of pore sizes. The unique structure provides catalytic materials of high surface area and high pore volume. As with all catalysts, the activity is controlled by the composition of the surface, but in the case of zeolites the surface composition can be changed by slight modifications to the synthetic procedure. The selectivity of zeolite catalysts derives from the regularity of their pore structure; selectivity can be changed by substituting a zeolite of different pore structure.

Catalytic cracking mechanisms

Catalytic cracking of alkanes begins with hydride abstraction on a Lewis acid catalyst to form a 2° carbocation:

$$CH_3-CH_2-CH_2-CH_2-CH_2-CH_2-CH_2-CH_2-CH_2-CH_3 \longrightarrow$$

$$CH_3-CH_2-CH_2-CH_2-\overset{+}{C}H-CH_2-CH_2-CH_2-CH_2-CH_3 \ + \ H^-$$

The carbocation then undergoes β-bond scission to form an alkene and a new carbocation. Usually β-bond scission occurs such that the smaller of the fragments contains at least three carbon atoms.

$$CH_3-CH_2-CH_2-CH_2-\overset{+}{C}H-CH_2-CH_2-CH_2-CH_2-CH_3 \longrightarrow$$

$$CH_3-CH_2-CH_2-CH_2-CH = {}^+CH_2-CH_2-CH_3$$

The alkene can be protonated on a Bronsted acid site

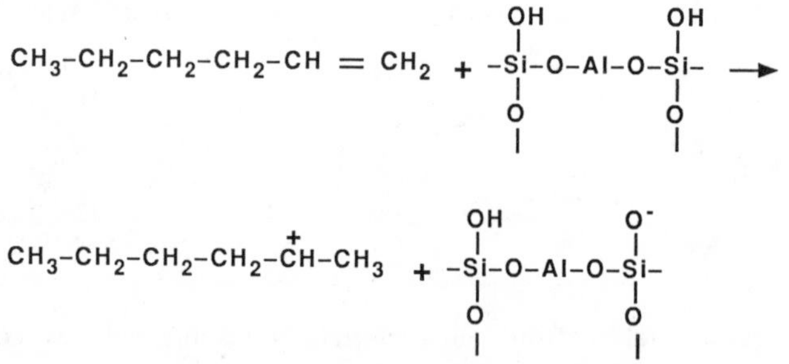

$$CH_3-CH_2-CH_2-CH_2-CH = CH_2 \ + \ \overset{OH}{\underset{O}{-Si}}-O-Al-O-\overset{OH}{\underset{O}{Si-}} \longrightarrow$$

$$CH_3-CH_2-CH_2-CH_2-\overset{+}{C}H-CH_3 \ + \ \overset{OH}{\underset{O}{-Si}}-O-Al-O-\overset{O^-}{\underset{O}{Si-}}$$

and the 1° carbocation formed in the β-bond scission reaction rearranges to the more stable 2° ion

$$CH_3-CH_2-CH_2-CH_2{}^+ \longrightarrow CH_3-CH_2-\overset{+}{C}H-CH_3$$

Both of these carbocations are able to undergo further β-bond scission. The products can in turn react again by protonation and rearrangement. This sequence of protonations, β-bond scissions, and rearrangements can continue until fragments of about six carbons atoms are formed. These fragments abstract hydride from the catalyst surface and become stabilized, neutral molecules. Unlike thermal cracking processes, catalytic cracking does not ordinarily proceed all the way to ethylene. The rapid rearrangement of the 1° carbocation prohibits further bond scission.

Branched chain alkanes and cycloalkanes crack more rapidly than do straight chain alkanes. As a rule of thumb, 3° carbon atom sites are about ten times as reactive as 2° carbon atoms, and about twenty times as reactive as 1° carbon sites. For straight chain alkanes, the longer the carbon chain (and hence the greater number of 2° carbons), the faster is the rate of cracking. The longer the chain length of the compound, the greater will be the coverage of the catalyst surface; the greater the coverage of the surface of the catalyst, the higher the rate of carbocation formation. For example, octadecane cracks about twenty times faster on a zeolite catalyst than does octane.

The cracking process is in essence the reverse of the polymerization and alkylation processes described earlier. It is important to realize how the choice of reaction conditions can be used to favor cracking relative to polymerization (or *vice versa*). Because energy is required to break a chemical bond, the enthalpy change for cracking will be positive. Because one large molecule generates two smaller ones, the entropy change is also positive. Thus both ΔH and ΔS are positive, and for the cracking reaction to occur, that is, for an equilibrium to favor the products of cracking, the temperature must be high. Since

$$\Delta G = \Delta H - T\Delta S$$

for any reaction and ΔG must be negative, that condition can be fulfilled for positive ΔH and positive ΔS only if T is large enough for the negative $T\Delta S$ term to overcome the positive ΔH. On the other hand, in a polymerization process, ΔH is negative because bonds are being formed, and ΔS is also negative because two smaller molecules are combining to form one larger one. In this instance the negative ΔH term will dominate the ΔG provided that temperatures are kept low enough. Thus low temperatures favor polymerization.

This sequence of cracking reactions is accompanied by three important side reactions. The first is *isomerization*. We have already discussed this process in the conversion of hexane to 2-methylpentane via the 2° carbocation. The choice of appropriate reaction conditions can lead to extensive chain branching, as in the conversion of hexane to 2,3-dimethylbutane. *Hydrogenation* can occur in two ways. Hydrogen can be transferred from cycloalkanes to alkenes, as for example

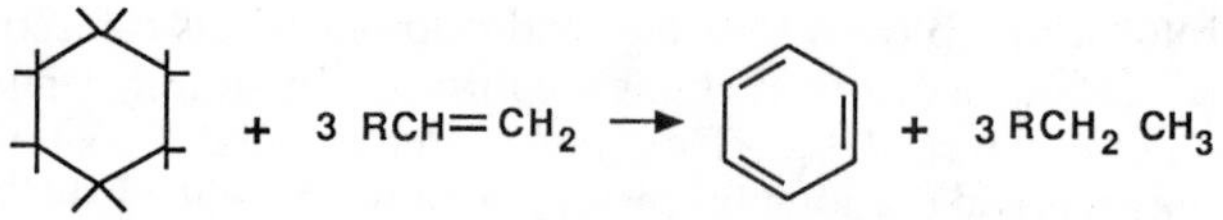

$$\text{cyclohexane} + 3\ RCH{=}CH_2 \longrightarrow \text{benzene} + 3\ RCH_2\ CH_3$$

Hydrogen can also be transferred from one alkene to another:

$$\underset{\displaystyle CH_3}{CH_2{=}CH{-}\overset{|}{C}H{-}CH_3} + \underset{\displaystyle CH_3}{CH_2{=}CH{-}\overset{|}{C}H{-}CH_3} \longrightarrow$$

$$\underset{\displaystyle CH_3}{CH_2{=}CH{-}\overset{|}{C}{=}CH_2} + \underset{\displaystyle CH_3}{CH_3{-}CH_2{-}\overset{|}{C}H{-}CH_2}$$

Hydrogenation is important first because the aromatic compounds formed from the cycloalkanes have high octane numbers, and second because the dienes formed by hydrogen transfer between alkenes are the precursors to coke formation.

Coking is the formation of high molecular weight aromatic or highly unsaturated carbonaceous solids on the surface of the catalyst. Coke formation is a poorly understood process, but since most cokes contain polycyclic aromatic structures, the mechanism outlined below is representative of the process.

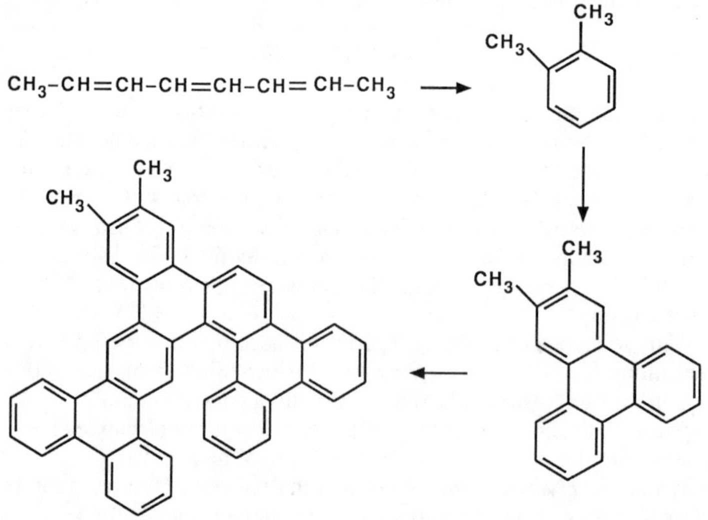

$$CH_3{-}CH{=}CH{-}CH{=}CH{-}CH{=}CH{-}CH_3 \longrightarrow$$

Notice that the coking on a catalyst surface is analogous to the hydrogen redistribution processes discussed during kerogen maturation. The products are a carbon-rich, highly aromatic solid and hydrogen-rich, low molecular weight alkanes. Aromatic carbocations are very stable, because the charge can be delocalized over the aromatic system, rather than being localized on one carbon atom as would be the case for most aliphatic carbocations. The aromatic carbocations can remain on the surface of the catalyst and grow into very large (on a molecular scale) structures. Coking is undesirable because the coke can plug the pores or occlude the surface of the catalyst, reducing or destroying the activity of the catalyst and requiring that the catalyst surface be regenerated by carefully burning off the coke. On the other hand, the coking reactions are exothermic and provide a source of heat to help drive the endothermic cracking reactions. The hydrogen derived from the coking reaction helps to convert alkenes to alkanes. Generally the rate of formation of coke from various types of structures decreases in the order naphthalene derivatives > benzene derivatives > alkenes > cycloalkanes > alkanes. Process streams rich in aromatic compounds produce cokes that are highly aromatic and nearly graphitic in structure. Alkenes and alkanes tend to produce cokes which are of low aromaticity and are amorphous.

Practical aspects of catalytic cracking

Catalytic cracking is commonly carried out in fluidized bed units. Hot vapors of the *cracking stock*, which could be a crude oil or a distillation cut, are passed through a bed of catalyst particles. At a particular flow velocity of the vapors the vapors essentially lift the catalyst particles, the mixture of solid plus vapor behaving like a single phase fluid. This fluidization process insures intimate contact of the vapor with the catalyst particles. Typical process conditions are temperatures of 465-540° and pressures of 0.2-0.4 MPa. The weight ratio of catalyst to vapor in in the range of 5-20. About 80% of all gasoline is produced by cracking, and 85-90% of cracking processes use catalytic cracking.

The major emphasis of catalytic cracking is the reduction of molecular weight of high-boiling materials to make products in the gasoline range. In principle a variety of materials can be used as feedstocks for catalytic cracking, including resids from vacuum distillation. However, not all potential feedstocks are equally conducive to the optimum operation of a catalytic cracking process. Some feedstocks have a very high tendency to form cokes, and are undesirable. For example, alkenes have a higher coking tendency than the comparable alkanes; thus high concentrations of alkenes would be undesirable. Other feedstocks may contain sufficient amounts of trace metals, such as nickel or vanadium, that cause poisoning of the catalyst. Feedstocks with high nitrogen contents are also undesirable, since most organic nitrogen compounds likely to be present in the feed are basic and would adsorb tightly onto the acidic sites of the catalyst surface, again poisoning the catalyst.

In the discussion of hydrogen redistribution processes we have seen that if a system is allowed to proceed to equilibrium the products would be graphite and a hydrogen-rich material such as methane. Clearly cracking processes on an industrial scale are not limited by equilibrium. In most industrial cracking processes isomerization and dealkylation of aromatic compounds could occur only to a moderate extent; and alkylation of alkenes by alkanes or polymerization of alkenes could occur only rarely.

Because of the importance of β-bond scission in cracking processes, we might expect one alkene molecule to be produced each time an alkane is cracked. In practice, however, catalytic cracking over zeolite catalysts results in products which are almost free of alkenes. The near-absence of alkenes is a result of the catalyst promoting not only cracking, but also very rapid hydrogen transfer from cycloalkanes and cycloalkenes to the alkenes (producing aromatic compounds from the original cyclic compounds); from one alkene to another, forming an alkane and a cycloalkene; and from loss of hydrogen from aromatics or polyalkenes during coke formation. The hydrogen transfer appears to occur directly between the compound donating hydrogen and the olefin. For example, if decahydronaphthalene (decalin) were to be passed over a catalyst, no hydrogen gas is formed. Further, a cracking catalyst is not usually a good catalyst for hydrogenation, and the passing of a mixture of hydrogen gas and an alkene over a zeolite results in no hydrogenation of the alkene. However, if decalin and an alkene were to be passed together over the zeolite, the alkene would be hydrogenated to an alkane.

Summary

Straight run gasoline has octane numbers as low as ~30. For many crudes the yield of straight run gasoline is 20% (i.e., 100 barrels of crude will provide 20 barrels of gasoline). By including a thermal cracking process in a refinery, the yield of gasoline can be increased to 40%, and the octane number raised to ~75. In contrast, if catalytic cracking is employed, in the most favorable cases a 50% yield of ~100 octane gasoline is obtained.

The extent of the cracking, hydrogen transfer, and isomerization reactions is governed by three factors. The chemical structure of the catalyst determines the strength of the Lewis and Bronsted acid sites on the catalyst surface; in turn, the strength of these acid sites determines the ability of the catalyst to generate carbocations by protonation or hydride removal. The access of the cracking stock to the surface of the catalyst (including internal surface in pores) is determined by the physical structure of the catalyst. Since the catalytic reactions are heterogeneous, access to the catalyst surface is critical. Finally, the course of reaction will also be affected by the manner of operation of the cracking reactor, by the choices of temperature, pressure, and residence time.

The continuing need for increasing octane number of gasolines is illustrated by the very significant improvements in engine design and performance. In the 1920's. a typical automobile engine had a compression ratio of 4.4:1. It would run well on 55 octane gasoline, and generate 14 horsepower per liter of cylinder volume. By the 1960's, the era of high-performance "muscle cars", compression ratios had increased to 9.5, producing about 48 horsepower per liter. For these engines, gasolines of ~95 octane were needed. In recent years there has been a concern to be able to use gasolines of lower octane number, particularly to eliminate the use of tetraethyllead as an additive. Use of lower octane gasoline has been accommodated by lowering the compression ratio of engines, to allow regular use of gasoline with octane numbers of ~87.

Catalytic reforming

We have seen that a goal of refining operations is to increase the octane number of gasoline relative to that which might be obtained from straight-run gasoline. The octane numbers of some pure compounds are summarized in Table 9.3.

Table 9.3 Octane numbers of some pure compounds

Compound class	Compound	Octane number
Alkanes	Butane	113
	Pentane	62
	Hexane	19
	Heptane	0
	2-Methylhexane	41
	2,2-Dimethylpentane	89
	2,2,3-Trimethylbutane	113
	Octane	-19
Cycloalkanes	Methylcyclopentane	107
	Cyclohexane	110
	1,1-Dimethylcyclopentane	96
	Methylcyclohexane	104
	Ethylcyclohexane	43
Aromatics	Benzene	99
	Toluene	124
	1,3-Dimethylbenzene	145
	1,3,5-Trimethylbenzene	171
	Isopropylbenzene	132

The data in Table 9.3 provides some indication of changes in molecular structure which would be beneficial for increasing the octane number. There are four categories of such changes. *Isomerization* of straight chain to branched chain alkanes can increase octane number substantially; compare heptane with 2,2,3-trimethylbutane. Conversion of alkanes to cycloalkanes or aromatics involves both the formation of a ring structure and the loss of some hydrogen, a process of *dehydrocyclization*. A comparison of cyclohexane or benzene with hexane shows the increase in octane number possible. *Ring enlargement* of alkylated cyclopentanes to cyclohexanes provides a slight gain in octane number; compare methylcyclopentane and cyclohexane. *Dehydrogenation* accompanying the conversion of cycloalkanes to aromatics can in some cases also increase octane number.

In these examples, the total number of carbon atoms in the molecule has not changed. These reactions are not cracking reactions, since the conversion of large molecules into smaller ones is not occurring. The molecules instead are converted into new forms or configurations, as in the formation of branched chain alkanes from straight chain alkanes. Because new molecular forms are being produced, these reactions are known collectively as *reforming*. Since C-C bonds are fairly strong, about 350 kJ/mol, and since isomers are usually stable at laboratory temperatures, the reforming of molecules requires elevated temperatures and is facilitated by catalysts.

Reforming catalysts

Bifunctional catalysts

A typical reforming catalyst combines a metal with an acidic oxide. An example is platinum dispersed onto aluminum oxide. The formulation of a reforming catalyst to contain both a metal and an acidic oxide derives from the fact that some steps in the mechanism of reforming occur on the metal surface and other steps occur on the oxide surface. The combination of catalysis by a metal with catalysis by an oxide leads to the term *bifunctional catalyst*. Bifunctional catalysts are also sometimes known as dual function catalysts. Zeolites are generally not used as reforming catalysts because they are such powerful cracking catalysts. The goal of reforming is not to reduce molecular size but only to alter molecular configuration.

The metallic portion of the bifunctional catalyst is a transition metal. Transition metals generally adsorb both hydrogen and the hydrocarbon molecules well. Among the best catalysts are palladium, platinum, and nickel. These elements have the desirable characteristics of being able to coordinate with double bonds and to adsorb and dissociate hydrogen. Other metals of the platinum group, such as iridium, rhodium, and ruthenium, are also excellent catalysts but are seldom used in commercial practice. All of these metals are poisoned by organosulfur compounds, which form strong bonds with the metal atoms on the surface and essentially render the surface unavailable to the hydrocarbons or hydrogen. When platinum is used as the metallic component of the bifunctional catalyst, the reforming process is sometimes known as *platforming*. Current technology uses a mixture of rhenium with platinum, because the rhenium tends to lower the rate of coke deposition on the catalyst surface.

The role of the metal is to adsorb and dissociate hydrogen, a process of *dissociative chemisorption*.. Although the addition of hydrogen to a double bond may seem at first sight a straightforward process, in fact mixtures of ethene and hydrogen may be kept for years with no appreciable formation of ethane. Indeed, for practical synthetic or preparative purposes it can be said that the hydrogenation of an alkene does not occur in the absence of an appropriate catalyst. The direct reaction of hydrogen with a molecule such as ethene requires a four-member transition state, which is unstable (in fact, anti-aromatic) compared with the six-member transition state which forms during, for example, ester decarboxylation. Even though the overall hydrogenation of ethylene is highly exothermic, the very high activation energy required to reach the four-membered transition state

$$
\begin{array}{c} H \\ | \\ H \end{array} + \begin{array}{c} CH_2 \\ \| \\ CH_2 \end{array} \quad \longrightarrow \quad \left[\begin{array}{c} H \cdots\cdots CH_2 \\ \vdots \qquad \vdots \\ H \cdots\cdots CH_2 \end{array} \right] \quad \longrightarrow \quad \begin{array}{c} CH_3 \\ | \\ CH_3 \end{array}
$$

effectively results in an extremely slow reaction rate. However, in the presence of a catalyst

$$
\underset{\displaystyle \text{⁄⁄⁄⁄⁄}}{H-H} \longrightarrow \text{⁄⁄Ⓗ⁄⁄⁄Ⓗ⁄⁄⁄} \longrightarrow \underset{\displaystyle \text{⁄⁄Ⓗ⁄⁄Ⓗ⁄⁄⁄}}{H_2C = CH_2} \longrightarrow \underset{\displaystyle \text{⁄⁄⁄⁄⁄}}{CH_3-CH_3}
$$

The process shown above represents the *Bonhoeffer-Farkas mechanism*. The hydrogen is adsorbed onto two surface sites and the H-H bond is broken, forming two dissociated hydrogen atoms in the catalyst. After adsorption of the alkene, a process which may involve the formation of metal-carbon bonds, the individual hydrogen atoms can react with the double bond, leading first to the formation of an *adsorbed alkyl* and then to the desired alkane, followed by its subsequent desorption from the catalyst. An alternative view of this reaction is the *Eley-Rideal mechanism,* in which the alkene is adsorbed on the catalyst surface and then reacts with undissociated hydrogen in the gas phase. Reactions that occur via the Eley-Rideal mechanism take place when the reactant species that has not been adsorbed (*i.e.,* is still in the gas phase) reacts by colliding with the species adsorbed on the catalyst surface or reacts after first undergoing an extremely weak physical adsorption on the catalyst surface. Regardless of mechanism, the reaction would proceed best when the π electrons of the double bond have easy access to the catalyst surface. The rate of hydrogenation of an alkene in the presence of a catalyst is about 10^{20} times greater than in the absence of a catalyst.

The traditional definition of a catalyst is that it increases the rate of a chemical reaction without itself being consumed in the reaction. It is important to recognize that when a reaction system is in equilibrium, the catalyst will affect the rates of both the forward and reverse reactions. In other words, if a substance is a catalyst for the reaction $A \rightarrow B$, that same substance will also catalyze the reverse reaction $B \rightarrow A$. The practical impact of this behavior is that a metal hydrogenation catalyst will also catalyze dehydrogenation of alkanes to alkenes. The ability of the metal to remove or add hydrogen to compounds provides a route from an alkane to an alkene and thence to another alkane.

Lewis acid catalysts
Since isomerization can be initiated by abstraction of a hydride ion from the alkane, reforming can also be catalyzed by strong Lewis acids. For example, anhydrous aluminum chloride has been used to isomerize pentane at reaction temperatures of 100°. Abstraction of a hydride ion from pentane forms a 2° carbocation, which can then rearrange as

$$CH_3-CH_2-CH_2-CH_2-CH_3 \longrightarrow CH_3-CH_2-CH_2-\overset{+}{C}H-CH_3 \longrightarrow CH_3-CH_2-\overset{+}{\underset{\underset{CH_3}{|}}{C}}-CH_3$$

The final product, 2-methylbutane, is formed by recapture of a hydride ion from the catalyst by the 2° carbocation.

The mechanism of catalytic reforming

At least six different reactions occur in the overall process of catalytic reforming:

•Dehydrogenation of cycloalkanes

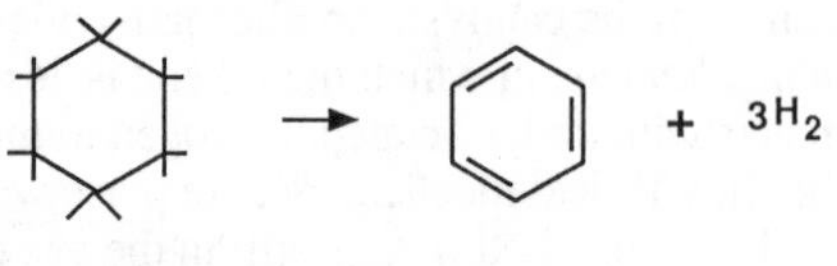

•Dehydroisomerization of cycloalkanes

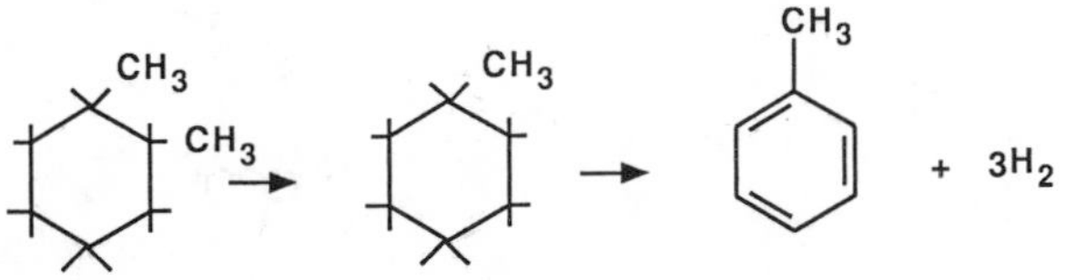

•Dehydrocyclization of alkanes

$$H_3C-CH_2-CH_2-CH_2-CH_2-CH_3 \longrightarrow + 4H_2$$

•Isomerization of alkanes

$$H_3C-CH_2-CH_2-CH_2-CH_2-CH_3 \longrightarrow H_3C-\underset{\underset{CH_3}{|}}{\overset{\overset{CH_3}{|}}{C}}H-CH-CH_3$$

•Hydrocracking of alkanes

$$CH_3-CH_2-CH_2-CH_2-CH_2-CH_2-CH_2-CH_2-CH_2-CH_3 + H_2 \longrightarrow$$

$$CH_3-CH_2-CH_2-CH_2-CH_3 + CH_3-CH_2-\underset{\underset{CH_3}{|}}{C}H-CH_3$$

•Hydrodesulfurization

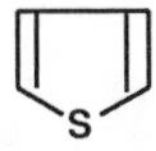

$$ + 4H_2 \longrightarrow CH_3-CH_2-CH_2-CH_3 + H_2S$$

Each of these reactions in its own way improves the product. Aromatic compounds generally have very high octane numbers. The formation of branched chain alkanes also raises the octane number. Loss of sulfur is always an improvement. Both the hydrocracking and hydrodesulfurization processes involve the addition of hydrogen to the molecule. For this reason reforming is sometimes called *hydroforming*.

Examples of reforming processes are illustrated in Figures 9.9 through 9.11. Isomerization of hexane involves three steps (not including the adsorption of hexane onto the catalyst surface or the desorption of the 2-methylpentane from the catalyst): a) on the metal surface a hexane molecule loses hydrogen to the metal to form 1-hexene; b) on the acidic oxide surface the 1-hexene rearranges to 2-methyl-1-pentene; and finally c) on the metal surface the 2-methyl-1-pentene regains hydrogen to form the desired

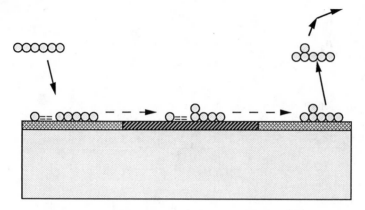

Fig. 9.9 The catalytic reforming of hexane occurs with adsorption of hexane onto a metallic surface (indicated by crosshatching), its dehydrogenation to hexene, migration to the acidic surface (indicated by dark shading) followed by rearrangement to 2-methyl-1-pentene, migration of the alkene to a metal surface where it is hydrogenated to 2-methylpentane, and finally desorption of the product 2-methylpentane into the process stream.

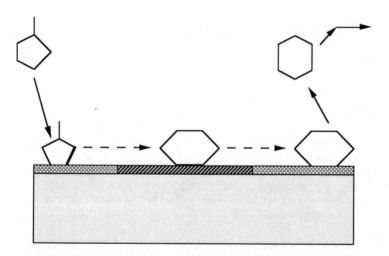

Fig. 9.10 The reforming of methylcyclopentane begins with adsorption onto a metal surface and dehydrogenation to a methylcyclopentene, followed by ring expansion on the acidic surface to cyclohexene, migration of the cyclohexene to a metallic surface where it is hydrogenated to cyclohexane, and finally desorption of the cyclohexane into the process stream.

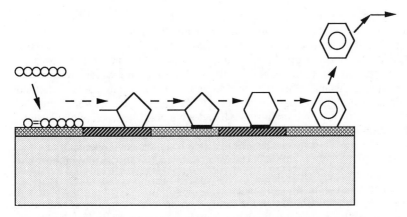

Fig. 9.11 Conversion of hexane to benzene during reforming involves some
of the steps shown in Figs. 9.9 and 9.10. In this case, the hexene is isomerized
to methylcyclopentane, and the cyclohexene is further dehydrogenated to benzene.

product, 2-methylpentane. Other reactions involve a similar sequence of loss
of hydrogen to the metallic component, formation and rearrangement of the
carbocation on the acidic component, and further gain or loss of hydrogen on
the metallic surface. The only one of the reforming processes in which the
dual functionality of the catalyst is not important is the dehydrogenation of
cycloalkanes to aromatics, which is entirely a loss of hydrogen. Using
cyclohexane as an example, the mechanism of dehydrogenation involves an
initial adsorption of cyclohexane onto the catalyst. The loss of six hydrogen
atoms is very rapid, and their loss may even occur simultaneously. The
resulting benzene is held on the catalyst surface by the interactions of the π
electrons in benzene with the d orbitals on the metal atoms.

   Isomerization of alkanes is more difficult than isomerization of
alkenes, because the former process requires hydride abstraction to generate
the carbocation, while the latter depends on protonation of a double bond.
Protonation of a double bond is easier than abstraction of a hydride. When
isomerization reactions are carried out on the surface of an acidic catalyst,
isomerization of alkanes can be facilitated by adding small amounts of the
related alkenes. The carbocations formed by protonation of the alkenes are
able to effect intermolecular hydride transfer reactions with the alkanes,
thereby increasing the rate of the process. Ideally, the isomerization of
alkanes should then be preceded by a dehydrogenation step to produce
alkenes. However, the use of a bifunctional catalyst eliminates the need for a
separate processing step, because the dehydrogenation can be effected on the
metal surface.

   Isomerization of alkanes can also occur on the metal surface. Consider
first the possibility that an alkane adsorbs on the metal surface at two adjacent

carbon atom positions. If each of these carbon atoms loses a hydrogen to the catalyst surface and no re-hydrogenation occurs, the resulting alkene can desorb into the product stream. In this case no isomerization is involved. Now consider what might happen if an alkane adsorbs on the catalyst not through two adjacent carbons, but, say, through two carbon atoms which are four or five positions distant from each other along the chain. Again, the two affected carbon atoms might lose hydrogen atoms to the catalyst, but now if a new C-C bond is formed, it will not contribute to a double bond between two adjacent carbon atoms, but rather will be the C-C bond closing a five- or six-membered ring. Thus the product of this step is an alkylcyclopentane or alkylcyclohexane. As mentioned above, a cyclohexane adsorbed on the catalyst can readily dehydrogenate, so that the eventual product is an alkylbenzene. In contrast, cyclopentanes do not readily lose hydrogen. However, their readsorption on the catalyst could lead to rupture of one of the C-C bonds in the ring; if this bond breaking is accompanied by rehydrogenation from the catalyst, the product would be a branched-chain alkane. Thus depending on the exact sequence of events, alkanes can be converted to branched chain alkanes, cycloalkanes, or aromatics.

Once an alkene has been formed, cyclization can be regarded as a self-alkylation process, analogous to the alkylation reactions discussed previously, except that now both the carbocation and the double bond are in the same molecule rather than being in different molecules.

This process can also be regarded as the reverse of a β-bond scission reaction.

A cyclic carbocation can transfer a proton to an alkene. The cyclic product is a cycloalkene. If the cycloalkene loses a hydride ion, a cycloalkenyl carbocation is formed. The continuing sequence of reactions leads eventually to the formation of an aromatic compound (next page).

Isomerization is used to convert pentane and hexane into their branched chain isomers. The reaction takes place in the presence of a bifunctional catalyst of nickel or platinum deposited on a SiO_2-Al_2O_3 support. The reaction pressure is 1.4-2.8 MPa; some hydrogen is mixed with the vapor stream.

Some reforming feedstocks may contain alkenes. During reforming alkenes may undergo a variety of reactions, including hydrogenation to alkanes, isomerization and hydrogenation to branched chain alkanes, or cyclization to cycloalkanes. High concentrations of alkenes are undesirable, because these compounds can consume large amounts of hydrogen during

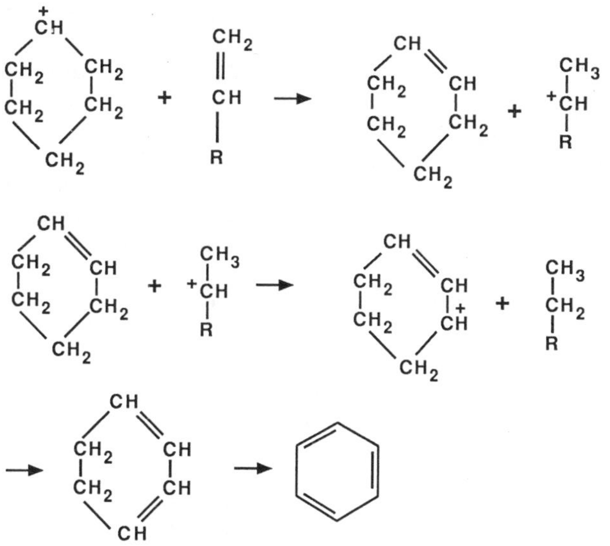

$$\underset{\text{CH}_2}{\overset{+}{\underset{\text{CH}_2}{\overset{\text{CH}}{\diagdown}}}}\ +\ \underset{\text{R}}{\overset{\text{CH}_2}{\underset{\text{CH}}{\parallel}}}\ \longrightarrow\ \text{(cyclohexene)}\ +\ \underset{\text{R}}{\overset{\text{CH}_3}{\underset{+}{\text{CH}}}}$$

$$\text{(cyclohexene)}\ +\ \underset{\text{R}}{\overset{\text{CH}_3}{\underset{+}{\text{CH}}}}\ \longrightarrow\ \text{(cyclohexenyl cation)}\ +\ \underset{\text{R}}{\overset{\text{CH}_3}{\underset{\text{CH}_2}{}}}$$

$$\longrightarrow\ \text{(cyclohexadiene)}\ \longrightarrow\ \text{(benzene)}$$

reforming or can dehydrogenate to the highly unsaturated compounds which are the precursors of coke. If a reforming feedstock does contain high concentrations of alkenes, a hydrogenation step to convert alkenes to alkanes may be carried out before the reforming process. Feedstocks may also contain up to 50% cyclopentane and cyclohexane. Dehydrogenation of the cyclohexanes produces aromatic compounds. Cyclopentanes also yield aromatic compounds, but first must undergo a ring expansion to form the six-carbon ring.

In hydrocracking, two adjacent carbon atoms must adsorb on the catalyst surface, with subsequent breaking of C-H bonds on these two carbon atoms. (This process also occurs during the isomerization reactions carried out on the metal surface.) For hydrocracking to proceed, further loss of hydrogen from the two affected carbon atoms is necessary. This additional loss of hydrogen results in the formation of carbon-metal double or triple bonds. The formation of strong, multiple carbon-metal bonds ruptures the C-C bond between the two adsorbed atoms. Subsequent addition of hydrogen to the two fragments results in the formation of two smaller alkane molecules. The hydrocracking can be represented as the formation of methane and a new alkane shorter by one carbon atom, as for example in the hydrocracking of pentane:

$$CH_3CH_2CH_2CH_2CH_3 + H_2 \rightarrow CH_3CH_2CH_2CH_3 + CH_4$$

The actual position along the chain at which hydrocracking occurs depends on the nature of the catalyst; nickel favors hydrocracking to methane as shown in the example above, whereas platinum favors hydrocracking near the middle of the chain. Hydrocracking generally requires high temperatures and metals which form very strong carbon-metal bonds. Both straight-chain and branched-chain compounds can undergo hydrocracking.

Hydrocracking can also occur on the acidic sites of the oxide portion of the catalyst. Again, hydrocracking is like isomeration in that it is facilitated if the alkane is first dehydrogenated to an alkene on the metal surface. An important characteristic of hydrocracking is that the rate increases very rapidly as a function of the molecular weight of the reacting species. For example, hexadecane cracks three times as fast as dodecane. This strong dependence of rate on molecular weight means that the undesirable long-chain alkanes are "cracked out" of the feedstock while the more desirable shorter chain compounds undergo hydrocracking much more slowly and therefore tend to survive into the product stream.

Considered overall, reforming is not a hydrogenation process, because dehydrogenation, dehydroisomerization, and dehydrocyclization all produce hydrogen, even though the hydrocracking and hydrode-sulfurization consume hydrogen. Thus in some cases the total reforming process could result in a net loss of hydrogen from the species undergoing reforming. In another sense, the reforming process can produce hydrogen, but removing it from some of the species being reformed.

Catalytic reforming in practice

Normally the feedstocks for catalytic reforming are straight run gasolines or low octane number liquids from thermal cracking of naphtha. As in catalytic cracking, high concentrations of alkenes are undesirable. Both nitrogen and sulfur are undesirable components because of their tendencies to poison catalysts. In addition, sulfur compounds might cause corrosion problems with the metal components of the reforming reactors.

The essence of catalytic reforming is that the very desirable branched chain and cyclic structures can be produced without further cracking of the molecules. Typical temperatures of reforming are on the order of 450 - 500°. Often hydrogen is added to the reforming reactor in order to suppress coke formation and "over-aromatization", the formation of too high a proportion of aromatic compounds. One of the best reforming catalysts now in commercial practice is an alloy of platinum and rhenium supported on aluminum oxide. The catalyst is partially sulfided prior to use, and it appears that both the sulfur and the rhenium have some role in reducing coke formation.

Typical reaction conditions for the catalytic reforming in the presence of a bifunctional catalyst are 475° and pressures of 1-2 MPa. A block flow diagram of a catalytic reforming process is shown in Figure 9.12.

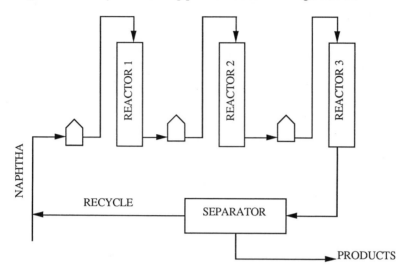

Fig. 9.12 A schematic flow diagram of a catalytic reforming process using three reactors. The small pentagons in the diagram represent heaters used to adjust the temperature of the process stream for each reactor. The performance of such a system in the reforming of naphtha is summarized in Table 9.4.

Generally reforming is used to treat straight-run naphthas, materials which contain up to 75% straight chain alkanes, and have octane numbers <50. Reforming is endothermic. If the temperature drops too low, the reaction rates will become unacceptably slow. In addition, the formation of aromatics from cycloalkanes is reduced by a shift in equilibrium at low temperatures. Because of these considerations, the reforming process in a typical refinery operation will be carried out in a sequence of reactors, with external heating supplied between each reactor. With three stages of reactors, a naphtha having an octane number of 38 can be converted to 90-octane gasoline. The process performance is summarized in Table 9.4.

Since some reforming reactions involve addition of hydrogen to molecules and others involve a loss, the partial pressure of hydrogen in the process is important for determining the nature of the products. At low hydrogen partial pressures, equilibrium between cycloalkanes and aromatics favors the formation of aromatics. Similarly, low hydrogen partial pressures will favor coke formation. In the operation of a reforming unit, the

Table 9.4 Three-stage catalytic reforming of naphtha

	Stage		
	1	2	3
Inlet temperature	500°	500°	500°
Outlet temperature	430°	470°	495°
Octane number of product	66	80	90
Percent of total amount of catalyst in this stage	15	35	50
Reactions occurring*	A,B	A,B,C,D	C,D

*A is dehydrogenation, B dehydroisomerization, C hydrocracking, and D dehydrocyclization.

specification of the hydrogen partial pressure will then determine the equilibrium between the cycloalkanes and the aromatics. Usually a high conversion of the cycloalkanes to aromatics is desired, because of the contribution of the aromatics to raising the octane number of the product. Since the equilibrium constant of a reaction varies with temperature, the desired aromatics production for a given hydrogen partial pressure then immediately establishes the minimum temperature to be used in the refining unit. At the same time, the values of the hydrogen partial pressure and the temperature establish the rate at which coke will form on, and deactivate the catalyst. Knowing how soon the catalyst will be deactivated allows the operators to plan the periodic regeneration of the catalyst. A high hydrogen partial pressure will favor hydrocracking and decrease the rate of coke formation, but will also decrease the yield of aromatics. Thus the choice of hydrogen partial pressure is obtained by striking a compromise among hydrocracking, aromatics production and catalyst deactivation. Typical pressures used in practice are ~2 MPa. High temperature operation leads to rapid deactivation of the catalyst, but low temperature operations give low reaction rates and a shift of equilibrium to favor cycloalkanes rather than aromatics. Thus it is important to insure that the temperature in the unit does not vary greatly during operation; the temperature range is usually held within ±20°.

As in catalytic cracking, coke formation can be a problem in reforming. Since coke formation is associated more with carbocation formation than with hydrogenation and dehydrogenation, in a bifunctional reforming catalyst coke formation is more likely to occur on the oxide surfaces than on the metal surface. A catalyst on which coke has formed can be regenerated by carefully burning the coke from the catalyst surface; for example, the regeneration may be run at 450° in an atmosphere containing about 1% oxygen.

Applications of hydrogen in refining

We have seen some reactions of compounds with hydrogen occurring during catalytic reforming, although hydrogenation is not the principal intent of the process. Specialized operations exist which take advantage of the deliberate reaction of hydrogen with various distillation cuts. Collectively, these operations are referred to as *hydroprocessing,* and involve the reaction of the hydrocarbon stream with hydrogen at elevated pressures in the presence of a catalyst. Among the reasons for which hydroprocessing might be considered are a) the pretreatment of the feedstock to remove impurities, such as potential catalyst poisons, before processing; b) to improve the characteristics of a product (*e.g.,* by reducing the sulfur content) to meet market or performance specifications; and c) to crack large molecules into the gasoline or diesel ranges. In examples (a) and (b) a large extent of cracking is not needed, but rather only the interaction of hydrogen with specific components of the feed, such as sulfur compounds or alkenes.

Hydrodesulfurization

To meet air pollution regulations for oil combustion systems used in electric generating plants or oil-fired heating installations, the sulfur content of the fuel should be <1%. However, many fuel oil cuts may have sulfur contents of >2%. The reduction of sulfur by reaction with hydrogen is the process known as *hydrodesulfurization.* This process has been introduced as one of the reactions occurring during catalytic reforming. However, hydrodesulfurization can be carried out as a separate operation its being coupled to a reforming process. The hydrodesulfurization of thiophene is a good example of the reaction

$$\text{[thiophene]} + 4\,H_2 \longrightarrow CH_3-CH_2-CH_2-CH_3 + H_2S$$

This reaction begins with the cleavage of the C-S bond to form 1,3-butadiene, which subsequently undergoes further hydrogenation. It is not necessary first to reduce the aromatic ring system in order to accomplish the cleavage of a C-S bond.

Many classes of organosulfur compounds can occur in crude oils or petroleum fractions. Each functional group has a characteristic reactivity toward hydrogen, so that the ease of hydrodesulfurization is in the order thiols > disulfides > sulfides > thiophenes > benzothiophenes > dibenzothiophenes > larger aromatic heterocyclics (*e.g.,*

249

benzonaphthothiophenes). Generally low molecular weight compounds are more readily desulfurized than compounds of high molecular weight.

In general, the reaction we seek to carry out during hydrodesulfurization is the breakage of a C-S bond. Using methanethiol as an example, the desired reaction would be

$$CH_3SH + H_2 \rightarrow CH_4 + H_2S$$

Inevitably the hydrodesulfurization reactions are accompanied by hydrocracking (*i.e.*, cleavage of C-C bonds), and by reduction of aromatic ring systems. In other words, the hydrodesulfurization catalyst not only catalyzes the breakage of C-S bonds, but it is also a reasonably good hydrogenation catalyst. Many of the metals used in catalytic reforming processes, such as platinum, are poisoned by sulfur compounds, and their catalytic activity is destroyed. Such is not the case for the hydrodesulfurization catalysts. Coking of the catalyst can also occur during hydrodesulfurization. As in the case of reforming, catalyst deactivation by coking can be retarded by increased hydrogen partial pressure.

Typical industrial conditions for hydrodesulfurization involve reactions at 300-450° and pressures as high as 20 MPa. Oxides of transition metals have been used as catalysts; however, these catalysts are converted, at least partially, to sulfides during operation. Some commercial catalysts are "pre-sulfided;" that is, already converted to sulfide form before being used. Sulfiding the catalyst involves both replacement of some oxide anions by sulfide and partial reduction of the metal. The two elements which are constituents of hydrodesulfurization catalysts are molybdenum or tungsten. Combinations of either of these elements with cobalt or nickel make especially effective hydrodesulfurization catalysts. However, neither cobalt nor nickel is, by itself, a good catalyst. Substances which have little or no catalytic activity themselves but which enhance the effectiveness of other, catalytically active materials, are called *promoters*.

During hydrodesulfurization, hydrogen, hydrogen sulfide, and the organosulfur compound are all adsorbed on the catalyst surface. Using a cobalt-promoted molybdenum sulfide catalyst and thiophene reactant as an example, hydrogen dissociation

$$H_2 \rightarrow 2H$$

and subsequent reaction of the dissociated hydrogen and donated electrons with the organosulfur compound occur

$$C_4H_6S + 2H + 2e^- \rightarrow CH_2=CHCH=CH_2 + S^{-2}$$

The electrons are donated by the molybdenum in a reaction such as

$$Mo^{+3} \rightarrow Mo^{+4} + e^-$$

The sulfide ion is then converted to hydrosulfide by *reductive adsorption*

$$Co^{+2} + H_2 + 2S^{-2} \rightarrow Co + 2\,SH^-$$

Alternatively, the sulfide ion reacts with adsorbed hydrogen

$$2H + S^{-2} \rightarrow H_2S + 2\,e^-$$

The overall hydrodesulfurization process therefore consists of two separate oxidation-reduction reactions. In one, electrons are donated, and, in the other, hydrogen atoms are donated.

Hydrodesulfurization of heavy oils is carried out at pressures of 3-10 MPa. Since some hydrocracking inevitably accompanies hydrodesulfurization, about 5-15% of the heavy oil is converted to material boiling below 350°. The hydrogen consumed in this process amounts to about 0.35 - 1.0% (by weight) of the heavy oil feed.

Hydrodenitrogenation

Nitrogen is an undesirable component of fuel for several reasons. First, nitrogen can react during combustion to produce various nitrogen oxides. The oxides of nitrogen are conveniently lumped together in the designation NO_x. The formation of NO_x is a concern because it can dissolve in water and be washed out of the atmosphere in rain. Since solutions of the nitrogen oxides in water are acidic (including the strong acid, nitric acid) NO_x contributes to the formation of acid rain. In addition, NO_x reacts with unburned hydrocarbons and carbon monoxide (both of which occur in the exhaust gases of automobiles, especially those in which the engine is not well tuned) through complex photochemical processes to form the noxious air pollution problem, smog. Concentrations of nitrogen >0.1% in gas oil deactivate the cracking catalysts. Since most organic nitrogen compounds are basic, they react strongly with the acidic sites on bifunctional catalysts, preventing the desired hydrogenation or carbocation formation reactions. Nitrogen compounds in lube oil are unstable at high temperatures. They may decompose, thus disappearing from the oil and reducing the amount of oil available for lubrication, or, even worse, they may polymerize to solid deposits which provide no lubrication.

Hydrodenitrogenation is analogous to hydrodesulfurization in that we seek to cleave a C-N bond, just as in the previous case the objective was to cleave a C-S bond. A simple example would be the reaction of aminomethane

$$CH_3NH_2 + H_2 \rightarrow CH_4 + NH_3$$

Hydrodenitrogenation reactions may take place in steps, as illustrated for the reactions of pyridine

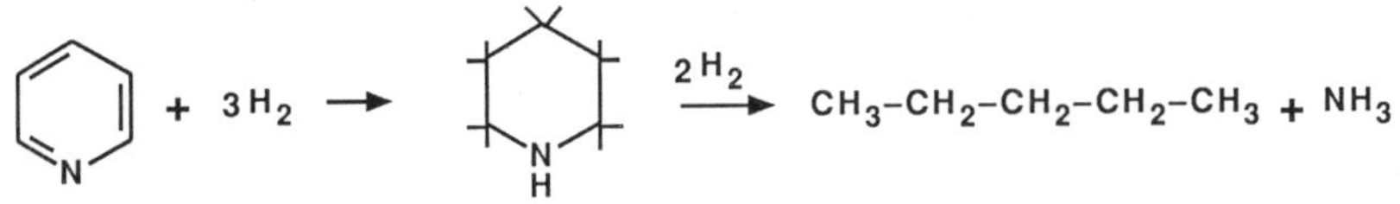

This reaction sequence highlights a difference between hydrodesulfur-ization and hydrodenitrogenation. In the hydrodesulfurization of aromatic sulfur-containing compounds a complete reduction of the ring was not necessary to effect cleavage of the C-S bond. In comparison, hydrode-nitrogenation does require a reduction prior to cleavage of the C-N bond. Thus pyridine is first converted to piperidine, whereas the analogous reaction of thiophene to tetrahydrothiophene is not a preliminary step for removal of sulfur from thiophene. Hydrodenitrogenation may sometimes be used as an alternative to clay treatment for lube oils.

Both oxide and sulfide catalysts can be effective for hydrode-nitrogenation. Hydrodenitrogenation catalysts therefore do not need to be pre-sulfided before use.

Hydrofining

Two types of reactions occur during hydrofining: hydrogenation and hydrocracking. The hydrogenation reactions are illustrated by the hydrogenation of alkenes

$$CH_3CH_2CH_2CH_2CH_2CH=CH_2 + H_2 \rightarrow CH_3CH_2CH_2CH_2CH_2CH_2CH_3$$

and reduction of aromatics to cycloalkanes

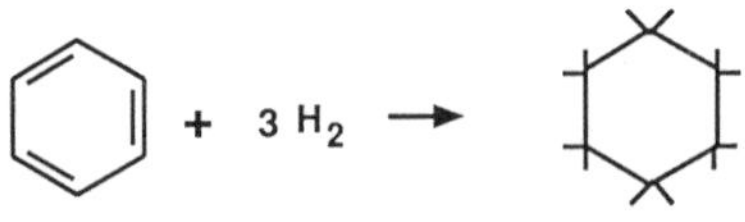

The hydrofining of aromatics to cycloalkanes is essentially the reverse of dehydrogenation of cycloalkanes. Hydrofining is especially useful for kerosene and jet fuel, because aromatic compounds in these fuels give rise to soot during combustion.

Hydrocracking reactions include the breakdown of long chain alkanes

252

$$CH_3-CH_2-CH_2-CH_2-CH_2-CH_2-CH_2-CH_2CH_2-CH_3 \; + \; H_2 \longrightarrow$$

$$CH_3-CH_2-CH_2-CH_2-CH_3 \; + \; CH_3-CH_2-\overset{\overset{\displaystyle CH_3}{|}}{CH}-CH_3$$

and the removal of long side chains from alkylated aromatic molecules

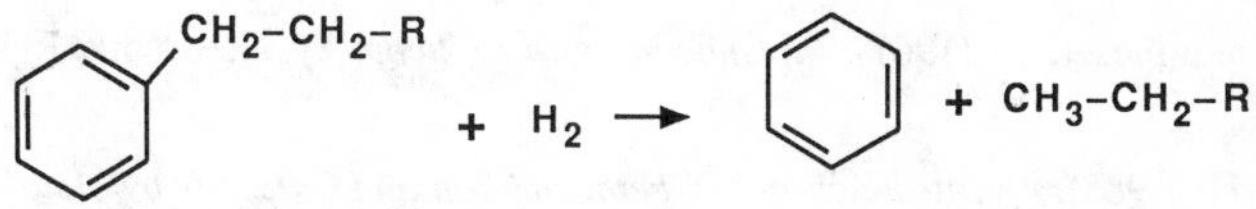

$$\text{(benzene ring)}-CH_2-CH_2-R \; + \; H_2 \longrightarrow \text{(benzene ring)} \; + \; CH_3-CH_2-R$$

In general reactions occurring during hydrocracking are very similar to those already discussed for catalytic cracking, and proceed via similar mechanisms. In hydrocracking the high partial pressure of hydrogen prevents the condensation of alkenes which leads to coke formation. Polycyclic aromatic hydrocarbons, which are also coke precursors, may undergo partial hydrogenation during hydrocracking, followed by a breakdown of the molecule into smaller or less condensed aromatic structures. The suppression of coke formation provides very long on-stream times for the catalysts without the need for regeneration.

Hydrocracking operations are carried out at 10-20 MPa. The conversion of heavy oil to products boiling below 350° is related to the amount of hydrogen consumed in the process. Consumption of about 1% (by weight) of hydrogen results in about 30% conversion of heavy oil to lighter materials. Conversions in the range of 90-100% can be achieved with hydrogen consumption of 5 - 5.5%.

A comparison of some of the aspects of hydrocracking with catalytic cracking (*i.e.*, in the absence of hydrogen) is shown in Table 9.5.

Table 9.5 Comparison of catalytic cracking and hydrocracking

	Catalytic cracking	*Hydrocracking*
Boiling range of feed	230-600°	50-540°
Hydrogen consumption, wt. %	0	1.5 - 5
Operating pressure, MPa	0.1-0.2	10.5-21
Temperature	480-540°	260-430°
Conversion to gasoline, wt. %	50-80	30-100
Major products	Gasoline and fuel oil	LPG, naphtha, jet fuel, and lube oils

The catalytic cracking process is more selective in that a narrower spectrum of products is produced. In hydrocracking, no alkenes are present in the gasoline, and coke formation is significantly lower than in catalytic cracking.

253

Further reading

Bland, W. F. and Davidson, R. L. (eds.) (1967). *Petroleum Processing Handbook*. New York: McGraw-Hill; Section 3.

Bond, G. C. (1987). *Heterogeneous Catalysis: Principles and Applications*. Oxford: Clarendon Press; Chapter 1.

Campbell, I. M. (1988). *Catalysis at Surfaces*. London: Chapman and Hall; Chapters 1,6.

Fieser, L. F., and Fieser, M. (1961). *Advanced Organic Chemistry*. New York: Reinhold; Chapter 7.

Gasser, R. P. H. (1985). *An Introduction to Chemisorption and Catalysis by Metals.* Oxford: Clarendon Press; Chapter 8.

Gates, B.C., Katzer, J. R., and Schuit, G. C. A. (1979). *Chemistry of Catalytic Processes*. New York: McGraw-Hill; Chapters 1 and 5.

Grasselli, R. K. and Brazdil, J. F. (eds.) (1985). *Solid State Chemistry in Catalysis*. Washington: American Chemical Society; Chapters 1,13,16.

Loudon, G. M. (1988). *Organic Chemistry*, 2d edn. Menlo Park, CA: Benjamin/ Cummings; Chapter 5.

Meyers, R. A. (ed.) (1986). *Handbook of Petroleum Refining Processes*. New York: McGraw-Hill; Parts 2,3,6.

Occelli, M. L. (ed.) (1988). *Fluid Catalytic Cracking: Role in Modern Refining*. Washington: American Chemical Society.

Royal Dutch Shell (1983). *The Petroleum Handbook*, 6th edn. Amsterdam: Elsevier; Chapter 5.

Satterfield, C. N. (1980). *Heterogeneous Catalysis in Practice*. New York: McGraw-Hill; Chapter 9.

Venuto, P. B. and Habib, E. T. (1979). *Fluid Catalytic Cracking with Zeolite Catalysts*. New York: Marcel Dekker.

Chapter 10

Coal combustion

By far the greatest amount of coal used in the world is consumed in various combustion applications: raising steam in boilers of electric generating plants, process steam or heating applications in industry, and the heating of commercial or residential buildings. This chapter introduces the basic principles of coal combustion. In many cases, coal as mined is not suited for direct introduction into some combustion device; consequently, various unit operations are employed to prepare the coal for use. The principal operations are size reduction and removal of some of the mineral matter. These operations are introduced before the discussion of combustion behavior.

Coal preparation

The discipline of *coal preparation* includes the operations necessary to convert coal from its as-mined condition into forms more suited for the eventual use, almost always either combustion or the manufacture of metallurgical coke. Broadly, coal preparation covers a variety of processes, including drying, grinding or other size reduction operations; and cleaning, to remove or reduce the amount of mineral matter. Some of these operations increase the calorific value of the coal, and those that do are sometime referred to as *coal beneficiation*. Although in strict terms coal beneficiation processes are a category in the broader area of coal preparation, the two terms are often used loosely as synonyms. This section provides an introduction to coal grinding and coal cleaning.

Hardness and grindability

Mining procedures generally produce large lumps of coal. Most processes which use coal will require fairly small sizes. For example, the feed to metallurgical coke ovens (Chapter 12) requires pieces of about 5 cm; combustion processes may use coals ranging in size, depending on the specific equipment, from 6mm to <74 μm. An important step in coal preparation is therefore the reduction of particle size by crushing or grinding operations.

The hardness of coal, and the susceptibility of coal to breakage, are important both in size reduction and in transportation and handling. It is important to know how easy or difficult it will be to effect a size reduction of the coal during preparation. It is also important that during transportation or handling the coal not be so easily broken that it produces excessive amounts of fine powder that is lost during handling or that is so small as to be undesirable for use.

The tendency of coal to break is determined by its *friability*. Various approaches are used to measure friability, one being the *drop shatter test*. The average particle size of the coal, X, is measured before the test. The sample is then allowed to fall 2 m onto a steel plate, and then the average particle size after the test, Y, is measured. The size stability of the coal, S, is calculated as

$$S = 100 \, (Y / X)$$

and the friability is then calculated as

$$F = 100 - S$$

There are also many tests to determine the *grindability*. One such is the Hardgrove test. A weighed sample of known size is subjected to grinding under a known weight for a known period of time. At the end of the test the weight of <74 μm particles is measured and used to determine the *Hardgrove grindability index*, I. For a "good" (*i.e.,* relatively easy to grind) bituminous coal, I = 100. The Hardgrove grindability index varies with rank in a way similar to many other coal properties, in that a distinct slope change is observed in the bituminous coal region when grindability is plotted as a function of rank. The rank variation is traditionally shown as a plot of I as a function of the maf volatile matter content (Fig. 10.1). The value of I for a given coal can be affected by the moisture content of the sample being tested and by the temperature at which the test is run. Thus there is usually a range of I values for a given rank of coal.

Until the late 1950's virtually all large-scale use of pulverized coal in the United States involved the consumption of bituminous coal and anthracite. It was usually assumed that I was a linear function of rank, especially since

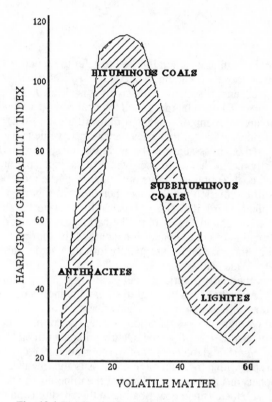

Fig. 10.1 The Hardgrove index passes through a maximum in the bituminous coal region and drops for both anthracites and low-rank coals. For most coals the Hardgrove index will fall in the region depicted by the shaded band, although the measured value can be very sensitive to the moisture content of the sample tested.

many lignites appear to be very friable. When the first lignite-fired power plants were constructed, it was found that the grinding mills were badly undersized. That discovery led to the detailed investigation of the grindability of low-rank coals and indeed that the value of I decreases in the lower ranks relative to bituminous coals.

Coal cleaning

Introduction

The term coal cleaning is generally used to refer to reduction of inorganic constituents or, loosely speaking, reduction of ash. Whether a coal is cleaned before it is used, and, if so, how extensively it is cleaned, depend on market specifications which must be met to sell the coal competitively.

There are three major reasons for cleaning coal. First, removal or reduction of mineral matter increases the calorific value and carbon content when expressed on an as-analyzed basis. In the transportation of cleaned coal, this essentially means that more energy is being shipped per unit weight of fuel. Second, coals containing large quantities of pyrite must be cleaned to help meet sulfur emission regulations when the coal is burned in a electric generating plant. Third, if coal is to be used for making metallurgical coke, the inorganic components may dilute the carbonaceous material, interfering with the formation of coke. In addition, some of the inorganic materials might be carried into the coke and subsequently introduce undesirable impurities into the metal made using that coke.

As commercially practiced, coal cleaning relies upon differences in physical properties between the coal itself and the inorganic constituents. The physical property most commonly taken advantage of is density. Depending on rank and petrographic composition, the density of the carbonaceous portion of the coal is 1.2-1.6 g/cm^3. Quartz and clay minerals have densities of about 2.6 g/cm^3, and the density of pyrite is about 5 g/cm^3. In principle, treating the coal with a medium having a density intermediate between those of the carbonaceous portion of coal and of the minerals should allow a sharp separation, with the cleaned coal floating in this medium and the minerals sinking. In practice, it is never possible to effect a perfect separation. The minerals that formed by precipitation reactions in cracks or fissures in the coal seam can be liberated from the coal by crushing or grinding. A density separation of coal and mineral matter in the manner just described is then straight-forward. However, other minerals may have formed as tiny grains in coalified or carbonized plant cell cavities (*e.g.,* as in fusinites) or may have become intimately mixed with the carbonaceous material by being washed into the swamp as the coal is forming. Generally it is not economically feasible to grind coal finely enough to liberate all of the minerals incorporated in these ways. Furthermore, no possible amount of grinding could liberate the cations bonded to the carboxyl groups in low-rank coals.

Float-sink testing and washability curves

We saw in Chapter 6 that coals contain a wide variety of inorganic constituents, that the ash content generally does not correlate with rank, and that ash composition correlates only roughly with rank. The extent to which

258

crushing or grinding liberates minerals for subsequent removal depends on how they are held in the coal. Thus there is often no way to predict in advance how a given coal will respond in coal cleaning. Each individual coal has to be evaluated in the laboratory.

One approach to laboratory testing is the *float-sink test*. In this experiment the pulverized coal is shaken with a liquid of known density (the liquid could be an organic compound or an aqueous solution of a salt such as calcium chloride). A density separation should occur (Fig. 10.2):

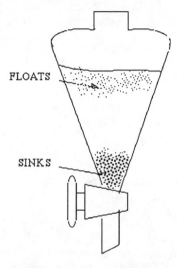

FLOATS

SINKS

Fig. 10.2 A laboratory float-sink test conducted in a separatory funnel. The fluid is a liquid of density between the coal and its mineral constituents; coal will float while the minerals will sink.

The carbon-rich cleaned coal is called the *floats*; the inorganic-rich mineral matter and very dirty coal make up the *sinks*. A preliminary evaluation is often made by plotting the yield of floats as a function of the ash in the floats (as determined by proximate analysis). A hypothetical example is shown as Figure 10.3. In this curve the individual data points are obtained by using liquids of increasing density. Thus the point at (5.5, 25) might have been obtained using a liquid of density 1.3 g/cm^3, and the point at (7.5, 71) might have been obtained by using a liquid of density 1.4 g/cm^3. At a 100% floats yield we have recovered all the coal in the floats. Thus the intersection of the curve with the abscissa represents the ash content of the original coal. For many coals the curve will approach the ordinate asymptotically. The implication of this behavior is that it is impossible to clean the coal completely (at least by density separation) to 0% ash. The amount of ash in the floats

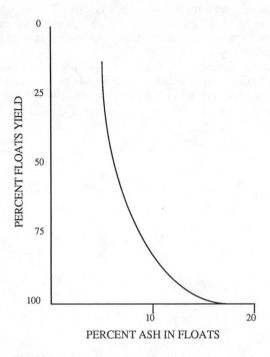

Fig. 10.3 An example of a preliminary evaluation of a coal, plotting the floats yield from the float-sink test as a function of the ash content of the floats.

at the point at which the curve essentially becomes vertical, sometimes called the residual ash, is a result of the mineral particles which are too small to be liberated by grinding, or of the inorganic components chemically bonded to the coal, or both. The curve provides an estimate of the "penalty" involved in cleaning coal. In this example, a reduction of ash from 17% to 6% would correspond to a floats yield of 50%; in other words, half the original coal would be discarded.

The separation in a float-sink tests is sometimes not as sharp as in the sketch shown above. Sometimes there will be some material that remains in suspension in the liquid; this material is known as the *middlings*. The middlings represent a potential problem in that they do not appear with the floats and, in an actual plant operation, they might be difficult to separate from the cleaning liquid.

A more detailed appraisal of coal cleaning can be obtained from a *washability curve*, Fig. 10.4.

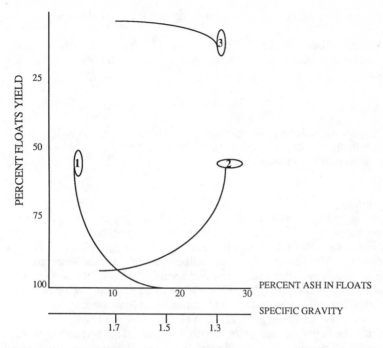

Fig. 10.4 A washability curve for a coal sample. The way in which the curve is read is explained in the text below.

These curves can be used in several ways. As one example, suppose that it is desired to determine the cleaning performance of a liquid of given density, say, 1.5 g/cm³. In this case, one would first read from the specific gravity axis at 1.5 to curve ②; finding the value of this intersection on the floats yield axis gives a predicted floats yield of 80%. Then, reading from 80% on the floats yield axis to curve ① and finding the value of this intersection on the ash axis, one determines that there would be about 7.5% ash in the floats. Finally, reading from the 1.5 g/cm³ position on the density axis to curve ③, one finds that the middlings yield would be 9%. A second example of the use of the washability curve is the determination of the density necessary to achieve a desired ash level in the cleaned coal, and the consequent performance of the cleaning process. In this case, suppose that it is desired to produce a coal having 6% ash. By reading from 6% on the ash axis to curve ① and finding the value of the intersection on the floats yield axis one finds that the floats yield for a 6% ash cleaned coal would be 75%. Now reading from 75% on the floats yield axis to curve ② and finding the density equivalent to the intersection point, indicates a liquid density of 1.4 g/cm³. Then finally reading from 1.4 on the density axis to curve ③, one finds a middlings yield of 13%.

The performance of a separation at a particular specific gravity is sometimes expressed as the *distribution coefficient* **c**, which is given by the expression

$$c_x = 100 \, F / (F + S)$$

where the subscript x denotes the specific gravity for which **c** is being specified, F is the weight percent of the coal found in the floats and S is the weight percent of the coal found in the sinks.

Froth Flotation
Coal cleaning by froth flotation takes advantage of both the physical (density) differences between coal and the inorganic constituents and the chemical differences between them. Consider for example a finely pulverized bituminous coal. The pulverized mixture will consist of particles which are coal-rich and other particles which are mineral-rich. As a rule, the surface of a bituminous coal is composed largely of hydrocarbon macromolecules which have no attraction for water. In fact, such surfaces are hydrophobic. In contrast, surfaces of common minerals such as clays and quartz are hydrophilic. The high concentrations of oxygen atoms on the surfaces of these atoms provide numerous sites for hydrogen bonding of water molecules. The flotation process involves putting the crushed coal in water and blowing air through the mixture. The dense, hydrophilic mineral particles will be wetted by the water and will sink. However, the hydrophobic coal particles will not be wetted. Instead, air bubbles can attach to the surfaces of the coal particles and carry them to the surface of the liquid (Fig. 10.5). In practice, a flotation cell can be represented by the schematic diagram of Figure 10.6.

The separation of coal from minerals is never complete because not all minerals are liberated by grinding, some minerals may be hydrophobic and thus attach to air bubbles and float, and some coal particles may be hydrophilic (especially if oxygen functional groups are available on the surface). Low-rank coals contain a large variety of oxygen groups (Chapter 6) and usually have hydrophilic surfaces. In addition, many of the minerals are of extremely small size and cannot be liberated by ordinary grinding. Therefore the low-rank coals are seldom amenable to cleaning by traditional processes.

Chemical coal cleaning
Some inorganic components of coal may be chemically bonded to the macromolecular structure of the coal. For example, transition elements may

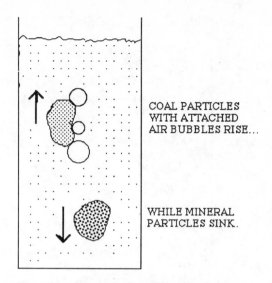

COAL PARTICLES
WITH ATTACHED
AIR BUBBLES RISE...

WHILE MINERAL
PARTICLES SINK.

Fig. 10.5 A schematic view of the operation
of a floatation process. The key is the difference
in attachment of air bubbles to hydrophobic
rather than hydrophilic surfaces.

form coordinate covalent bonds with oxygen, nitrogen, or sulfur atoms; and
alkali or alkaline earth elements can be incorporated as cations associated
withcarboxylic acid groups. In addition, a portion of the sulfur is part of the
macromolecular structure of the coal. The removal or reduction of these
elements cannot be effected by processes relying on differences in physical
properties; rather, some kind of chemical bond cleavage must be performed.
A variety of processes have been developed in laboratories and tested at pilot
scale for removal of organic sulfur or the chemically bound metallic
elements. None of these processes has proven to be economically feasible;
consequently, none are in commercial use.

Combustion processes

Chemical aspects of combustion

The combustion of coal begins with pyrolysis. During the heating of a coal
particle to combustion temperatures, thermal cracking reactions occurring
inthe macromolecular structure of the coal result in the evolution of a variety

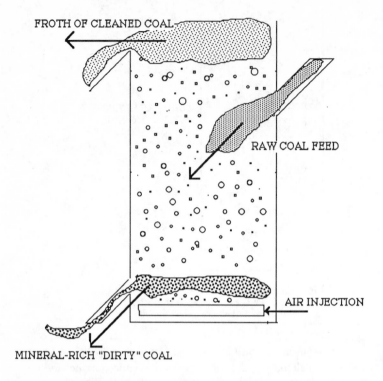

FROTH OF CLEANED COAL

RAW COAL FEED

AIR INJECTION

MINERAL-RICH "DIRTY" COAL

Fig. 10.6 A froth flotation cell in operation, with separation of the raw coal into a stream with a high concentration of minerals and some coal; and a second stream with cleaned coal containing a reduced concentration of minerals.

of volatile compounds as gases, vapors, or tars. The extent to which this evolution occurs, as well as the kinds of compounds evolved, depends on the rank and petrographic composition of the coal, the heating rate in the particular combustion device in use, and the particle size of the coal being burned.

Because most of the compounds evolved as a result of the pyrolytic breakdown of the coal structure are hydrocarbons, they are of course combustible. The evolved volatiles ignite and burn homogeneously in the gas phase in the combustion device. The loss of the volatiles leaves behind a residual carbonaceous char. After the cloud of evolved volatiles has burned, the char then ignites and burns in a heterogeneous reaction. The heterogeneous char combustion can be an order of magnitude slower than homogeneous gas phase combustion of the volatiles. Furthermore, the rate of char combustion will be dependent on the chemical and physical structure of the char itself; in turn, the char structure derives from the structure of the original coal, as affected by the pyrolysis processes which led to char formation.

Lignites and subbituminous coals tend to produce chars which are more reactive than those from bituminous coals. We have seen in Chapter 6 that the low-rank coals have the open structure, which is characterized by a high degree of isotropy in the chemical structure and a large degree of porosity. With the increasing structural ordering which accompanies a change in rank, reactivity decreases. (Even with essentially pure carbons, the highly ordered graphite is much less reactive than an amorphous carbon.) In addition, the caking coals of bituminous rank pass through a fluid stage which allows further alignment of the polycyclic aromatic structural units and a consequent increase in structural anisotropy. After the volatiles have escaped and the material has resolidified, the char which remains is highly anisotropic.

Depending on the rank of coal being burned, the amount of char remaining after the initial combustion of the volatiles may amount from ~50% to, in extreme cases, >90% of the weight of carbon in the original coal. Thus the heterogeneous char burnout is critical to complete and efficient combustion. The char burnout process is said to occur in three *zones* of combustion.

Zone I reaction is the situation in which the rate of reaction at the char surface is controlling the rate of the combustion. In this zone, the rate of diffusion of oxygen to the char surface and of combustion products away from the burning char is much faster than the rate of the chemical reaction at the surface. Typically Zone I processes occur at relatively low temperatures. At high temperatures, the situation is reversed; the rate of chemical reaction is now so fast that the overall process is limited by the rate of diffusion of oxygen molecules from the bulk of the gas to the char surface. This situation represents Zone III. In this case the reaction rate is essentially determined by the number of moles of oxygen arriving per unit external surface area of In Zone III the reaction rate will increase as the gas flow around the char particle increases, but will decrease as the size of the char particle increases. For these reasons, some practical combustion devices such as pulverized coal fired boilers (discussed in the following section), use fuel pulverized to small particle sizes and injected into highly turbulent flow patterns inside the combustor. Zone II represents an intermediate case. The oxygen still arrives rapidly at the char surface, but the diffusion of oxygen molecules through the pore system into the interior of the char particle is slow.

Practical aspects of coal combustion

There are three general types of processes in which coal is burned: fixed-bed combustors, pulverized coal combustors (also known as pc combustors or as suspension fired combustors) and fluidized bed combustors.

Fixed-bed combustors

All fixed bed combustors derive in some way from the obvious, albeit primitive, approach to burning coal by heaping up a pile of coal lumps and setting the pile afire. A variety of designs has been developed over many years for improving the efficiency of this primitive approach to coal combustion. The factors which essentially all fixed-bed combustors share are 1) a method of introducing more coal into the bed as combustion proceeds; 2) some device, generally a grate, to allow the introduction of air into the burning coal bed; and 3) a way of removing the ash from the system. Those combustors which are used to generate hot water or steam for heating or for electricity generation also share a fourth feature, a means of heat transfer from the combustion gases to water or steam.

Fixed-bed combustors have a variety of applications, from simple domestic heating units to sophisticated and automated combustors for industrial process heating and electricity generation. The coal properties which must be taken into account in fixed-bed combustion are the tar production during the pyrolysis which occurs before combustion actually begins, and the caking properties of bituminous coals. Many of the bituminous coals used in fixed-bed combustors may contain ~30% volatile matter. As the temperature of the coal particles rises, after the coal is put on the fire, thermal decomposition of the coal to form tar will begin around 400°. The temperature at which tars are liberated is below their ignition temperature. In many of simplest designs of fixed-bed combustors, it is not possible to obtain adequate mixing of air with the liberated trars to insure their complete combustion. The unburned tars consequently contribute to the formation of a greyish-yellow smoke. The unburned tars contribute significantly to smoke and soot formation, which are undesirable on both aesthetic and environmental grounds.

The most straightforward approach to eliminating much of the smoke production from fixed-bed combustors is to alter the pattern of air flow. In the case of a simple coal fire, as in the grate of a fireplace or stove, the air usually enters beneath the burning coal and sweeps upward through the coal bed, a design known as the *updraft combustor*. If instead air is forced in through the top of the coal bed, in a *downdraft combustor*, the volatiles emitted from pyrolyzing coal and transported into the region of red-hot burning coal particles, where the volatiles can be consumed effectively. Although the downdraft combustor design offers potential advantages for destroying volatiles, it is still necessary to insure an adequate supply of oxygen and to sufficient reaction time in the region of the burning coal to allow for complete reaction of the volatiles.

Caking coals can potentially form large, fused agglomerates which could impede the flow of air through the bed and which could interfere with the operation of any mechanical devices in the bed. Two approaches have been

266

employed to deal with caking coals. The caking properties of coals can be destroyed, or significantly reduced, by mild oxidation of the coal before it is fed to the combustor. Alternatively, the internal parts of the combustor can be designed so that, for example, the mechanical movement of the grate helps to break up any agglomerates which might form.

The smallest fixed-bed combustors, such as many domestic furnaces, are fed coal by hand. In larger devices, automatic feeding of coal is accomplished by using mechanical stokers. A variety of stoker designs have been developed over the years, to cope with increasing size of the combustors, to add coal in ways to suppress smoke formation, and to deal with the inevitable ash. The major disadvantage of fixed-bed combustors is that the upper limit of practical size (about 10^7 kJ/m^2), even with mechanical stoker feeding, is too small for many modern, large-scale applications, such as electric power stations.

Pulverized coal firing
Pulverized coal firing is the predominant method of coal combustion when heating rates greater than those attainable from fixed-bed combustors are needed. The principal use of pulverized coal combustors is the raising of steam for electric power generation. The coal used for pc-fired units is ground so that 80% of the particles are less than 74 μm. The pulverized coal is entrained in a stream of air blown into the combustion chamber. The hot (~1400°), highly turbulent gas stream inside the combustion chamber promotes rapid ignition of the coal particles. Since reaction rates increase with temperature, most types of coals can be burned in a pc-fired combustor. However, it is important to realize that although the pc-firing concept can be adopted to most coals, it is still critical to design a specific combustion device for the specific coal to be burned. A pc-fired combustor cannot operate efficiently if radical changes in the feed coal characteristics are made. The high temperature inside the combustion chamber, and the consequently large temperature gradient between the combustion gases and water or steam contribute to very good heat transfer.

The pulverized coal particles burn first by emitting volatiles, which burn in a cloud around the particle, and then the remaining char burns in a heterogeneous reaction. The char combustion is controlled by diffusion of oxygen inside the pore system of the char, being a case of the Zone II reaction described earlier. The time required for complete char burnout in Zone II varies by about a single order of magnitude between the most reactive and least reactive chars. The particle implication of this situation is that essentially the same fundamental design of a pc-fired combustor can be used for the combustion of all ranks of coal. However, because there are still differences in char burnout rates, ash content, calorific values, and moisture contents of various coals, the specific design details of a given pc-fired unit must be matched reasonably well with the specific coal to be burned. If the reactions in a pc-fired combustor were governed by Zone I kinetics, the reactivities of chars from various ranks of coal would have a much greater impact on the overall process, and it is likely that the char burnout in that case might vary by three or four orders of magnitude between the least and the most reactive chars. If Zone I reactions did predominate, it seems unlikely that a single basic design on combustor could be applied to all ranks of coal.

The high temperatures do, however, have some disadvantages. High temperatures promote the oxidation of nitrogen, both in the fuel and in the air supplied for combustion, to various oxides of nitrogen. Emissions of nitrogen oxides are of concern for air quality. Further, the high combustion temperatures can promote sintering, or in extreme cases melting, of the inorganic components of the coal. Partially molten ash can stick to the heat transfer surfaces inside the combustor, severely reducing heat transfer to the water or steam and restricting the flow of combustion gases. In the most severe cases, deposits of sintered ash inside large boilers in electric power stations can grow to the size of a compact automobile. When such deposits eventually slough away from the boiler walls, their fall through the boiler can cause extensive, and expensive, damage.

Fluidized bed combustion
Fluidized bed combustion is a technology which came into increased prominence during the 1980's. Air is blown upward through a bed of coal, ash, and an inert material such as sand. By adjusting the air velocity upward to exceed the terminal settling velocity of the particles in the bed, the particles remain suspended. The coal fed to the bed is crushed to the 2-6 mm size range. Coal particles added to a hot bed of fluidized sand, or other inert inorganic material, burn rapidly and establish a fairly uniform temperature distribution throughout the bed. The rapid, efficient combustion makes it possible to sustain combustion in a fluidized bed with less than 5% carbon in the bed.

An advantage of being able to operate with very low carbon inventories in the bed is that fluidized bed combustors can operate with material of low carbon content, such as the refuse from coal cleaning operations. Such material may contain 60-70% mineral matter, and would be of little or no use in most other combustion devices. The fluidized bed combustor provides a way of utilizing the energy values of materials which would otherwise be discarded as waste.

In normal operation the concentration of coal in the bed is low. Provided that one can respond by varying the feed rate, it is possible to accomodate wide variations in coal quality in the feed. This is particularly true for burning a coal in which the ash content varies greatly; the concern then must be that the coal feed rate is adjusted to maintain a steady output of heat from the combustion process. (As the ash content increases, the calorific value of the coal on an as-fired basis will drop; to maintain a constant heat output, more coal must be fed.)

Most fluidized bed combustors operate at temperatures of 750-1000°. These lower temperatures, relative to pc-fired units, significantly reduce the sintering and partial melting of ash. Furthermore, these temperatures are below the decomposition temperature of calcium sulfate. The consequence of this fact is that calcium-based materials, such as limestone or dolomite, can be added to the bed to capture sulfur oxides produced during the coal combustion. This in-bed capture of sulfur reduces or eliminates the need for pollution control equipment to remove sulfur oxides from flue gases, as is needed on pc-fired combustors. In addition, the comparatively low temperatures also substantially reduce formation of nitrogen oxides.

Ash behavior during combustion
Unlike natural gas and fuel oil, coal contains a significant quantity of incombustible materials which form ash during the combustion process. The ash must be collected in some way, and then disposed of in manners consistent with environmental regulations. If there were no other factors of concern, ash would at least have to be collected and disposed. However, in some situations the ash can be much more troublesome than merely representing an inert material that needs some handling for collection, removal, and disposal.

Coals contain a variety of minerals, with quartz, clays, and pyrite predominating. Low-rank coals also contain cations of the alkali and alkaline earth elements bonded to carboxyl groups in the coal structure. Some elements may also be incorporated by coordinate covalent bonding to oxygen, nitrogen, or sulfur functional groups. As the temperature of the coal is raised in the early stages of combustion, a complex sequence of phase changes and reactions occurs. At some point the temperature may become high enough to allow those inorganic components with relatively low melting points to melt. These components are often aluminosilicates, formed by thermal alteration of the clay minerals, and, in low-rank coals, reaction of altered clay minerals with cations liberated when the carboxyl groups are decomposed.

The liquid phase formed by melting a portion of the ash can act as a "glue" that facilitates the vitreous sintering of the remaining, unmelted ash particles. In fixed-bed combustors, the sintering of ash can result in the formation of *clinkers*, hard aggregates of ash particles that may be troublesome for operation of the ash handling and collection equipment. In fluidized beds, sintering of the ash, possibly in combination with the bed material (such as sand), can form aggregates which are too large to be fluidized by the air stream passing through the bed. Again, this can lead to serious operational problems in the combustor.

In pc-fired combustors a portion of the ash particles is entrained in the rapidly flowing stream of combustion gases. These particles can impact, and adhere to, heat exchange surfaces in the boiler. The ash deposit, sometimes referred to as *fouling*, reduces heat transfer to the water or steam. In order to maintain the desired steam temperature, it becomes necessary to increase the firing rate, thus increasing temperatures inside the combustor. This increase of temperature only exacerbates the ash fouling problem, because the temperature in the interior of the combustor is increased, greater opportunities exist for sintering the ash deposit. Further, a partially molten surface of ash provides an efficient means of trapping and retaining other ash particles as they continue to impact the surface. The ash deposit continues to grow, setting off a cycle of increasing firing rates and growth of deposit. If the ash deposits cannot be removed by blowers built into the combustor, it is necessary to shut the unit down periodically for cleaning. In serious cases, the ash deposits can grow so large that eventually they break away from the boiler surfaces, and can cause serious damage as they fall through the boiler.

The extreme case of ash behavior is *slagging*, which is the formation of a liquid, free flowing stream of molten ash. Some coal combustion systems have been designed to operate in the so-called slagging mode, in which the ash is deliberately melted, so that it can be collected as the fluid stream flows out of the combustor. In that case it is critical that the temperatures be kept high enough to insure that the flowing slag has a sufficiently low viscosity to flow without solidifying inside the combustor or the slag handling system. On the other hand, most pc-fired combustors are designed to cope with deliberate slagging operation. In those cases, slagging represents a serious problem which can damage the boiler or require unscheduled shutdowns for maintenance and cleaning.

Further reading

Barnard, J. A. and Bradley, J. N. (1985). *Flame and Combustion*, 2d edn. London: Chapman and Hall; Chapters 8 and 12.

Berkowitz, N. (1979). *An Introduction to Coal Technology*. New York: Academic Press; Chapter 10.

de Lorenzi, O. (ed.) (1947). *Combustion Engineering*. New York: Combustion Engineering, Inc.; Chapter 9.

Pitt, G. J. and Millward, G. R. (eds) (1979). *Coal and Modern Coal Processing*. London: Academic Press; Chapter 6.

Singer, J. G. (ed.) (1981). *Combustion: Fossil Power Systems*. Windsor, CT: Combustion Engineering, Inc. ; Chapters 3, 4,5,12.

Chapter 11

Production and use of synthesis gas

Synthesis gas, a mixture of carbon monoxide and hydrogen, is one of the most useful materials in the chemical industry. Its usefulness derives from its versatility in producing a diversity of products, depending on the selection of temperature , pressure, CO/H2 ratio, and catalyst. A further aspect of the versatility of synthesis gas is its ability to be manufactured from any of the hydrocarbon fuels - indeed, in principle, almost any source of carbon - and the ease of altering the composition via the water-gas shift reaction.

The production of synthesis gas

Steam reforming of natural gas

We have discussed the steam reforming reaction briefly in Chapter 7.

$$CH_4 + H_2O \rightarrow CO + 3\,H_2$$

Steam reforming of natural gas is the major industrial route to synthesis gas today. It is also one of the major processes for the manufacturing of hydrogen, accounting for about 90% of the world's supply.

Steam reforming is endothermic, and therefore is carried out in externally fired furnaces. The heat required for this process is obtained by burning a portion of the gas stream.

The process uses nickel catalysts. The catalyst is poisoned by sulfur, so it critical that the feedstock be sweetened before reforming. The nickel metal is normally dispersed on aluminum oxide or magnesium oxide as a support. The concept of a *catalyst support* derives from the fact that catalytic reactions occur at surfaces (Fig. 11.1)

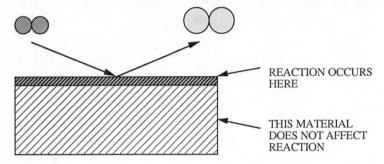

REACTION OCCURS HERE

THIS MATERIAL DOES NOT AFFECT REACTION

Fig. 11.1 The action of a catalyst occurs at the surface; the bulk of the material, which supports the surface, does not have a role in the reaction.

The nature of the surface is critical to catalysis of the process, but in essence the bulk of the material has no effect on the chemistry of the process. When the catalyst is a relatively inexpensive material, the fact that the bulk is not participating in catalysis of the process is unimportant. However, if the catalyst is quite expensive, such as the platinum used in platforming, it is not economically sound to use large pieces of the catalytically active material. Since the reaction only occurs at the surface, then from the point of view of the process chemistry the composition of the bulk of the catalyst particle is immaterial. The catalyst support provides the physical foundation to support a surface layer of the catalytically active material. A good catalyst support should have the following properties: low cost; the ability to withstand the thermal and mechanical conditions of the process; have enough interaction with the catalyst material to provide good physical dispersion and mechanical support; and should itself not cause any undesirable side reactions.

Two problems are encountered in steam reforming. First, the high temperatures of the reaction can cause sintering of the nickel. As the particles of nickel grow, their surface area decreases, and hence the catalytic activity drops. Second, there is a potential side reaction of the hydrogen and carbon monoxide

$$CO + H_2 \rightarrow H_2O + C$$

In this case the carbon deposits on the catalyst, decreasing the effective surface area to which the reacting gases can have access, and again decreasing the activity of the catalyst.

An extension of the steam reforming of natural gas is to use LPG or naphtha as feedstock. For example,

$$C_4H_{10} + 4\,H_2O \rightarrow 4\,CO + 9\,H_2$$

The same general considerations apply. The feedstock must be desulfurized. A nickel catalyst is used on an aluminum or magnesium oxide support, at high temperatures and pressures.

A potentially serious problem with the steam reforming of natural gas, LPG, or naphtha is that these feedstocks may themselves be in short supply by the early 21st century. In fact, a major thrust of research into synthetic fuel production during the so-called energy crises of the late 1970's and early 1980's was to make methane and liquid hydrocarbons from other sources. Over the long term, measured in decades, it is important to find alternative routes to synthesis gas which do not rely on natural gas or petroleum derivatives as feedstocks.

Coal gasification

The key feature of steam reforming of natural gas, LPG, or naphtha is the reaction of a carbon compound with steam. Indeed, this is a fairly general reaction which could be written as

$$C + H_2O \rightarrow CO + H_2$$

where the symbol C is taken to mean any source of carbon in a carbonaceous material, to distinguish from the use of the symbol C to mean the element carbon. In essence, almost any carbon-containing material could be used in this process. Among the candidate feedstocks which have been considered are agricultural and forest product wastes, peat, petroleum resids, and tar sands. However, the most abundant, and one of the least expensive, sources of carbon is coal. Therefore we can consider the prospect of converting coal to synthesis gas.

The technology of converting coal to a gaseous fuel of any sort is known as *coal gasification*. In this chapter we will focus mainly on gasification to produce synthesis gas. Other gasification processes for the manufacture of fuel gases will be discussed in Chapter 12.

Carbon is a sufficiently powerful reducing agent to reduce steam in the carbon-steam reaction. If steam is passed over a bed of red-hot coal, the gaseous product should in principle be a 1:1 mixture of carbon monoxide and hydrogen. In practice, some leaks into the system are inevitable, and the gaseous product has the approximate composition 50% H_2, 40% CO and 5% each CO_2 and N_2. This product is *water gas*. Because the carbon-steam reaction is endothermic, the coal bed quickly cools to a point at which the reaction stops. Water gas production could be made continuous by supplying external heat to the reactor, as was done for the steam reforming of natural gas. However, the external combustion of coal to heat the reactor essentially wastes the carbon of the coal, because it is converted to carbon dioxide and lost. If instead we recognize that the incomplete combustion of coal produces carbon monoxide

$$C + 0.5\ O_2 \rightarrow CO$$

then the reaction of coal with a mixture of steam and air, or of steam and oxygen, will produce carbon monoxide via an exothermic reaction, and a carbon monoxide - hydrogen mixture via an endothermic reaction. If a balance is achieved between these two reactions so that the heat liberated from the partial combustion of coal is sufficient to drive the endothermic carbon-steam reaction, then the gasification process can run continuously.

All of the approaches to commercial coal gasification processes developed in the last sixty years represent different designs for carrying out the carbon-steam-oxygen (or carbon-steam-air) reactions and appropriately balancing the various endothermic and exothermic reactions. At least five reactions are likely to occur inside a gasifier, as summarized in Table 11.1.

Table 11.1 Reactions occurring during coal gasification

Name	Equation	Enthalpy change
Combustion	$C + O_2 \rightarrow CO_2$	Exothermic
Boudouard reaction	$C + CO_2 \rightarrow 2\ CO$	Endothermic
Carbon-steam reaction	$C + H_2O \rightarrow CO + H_2$	Endothermic
Water-gas shift reaction	$CO + H_2O \rightarrow CO_2 + H_2$	Exothermic
Hydrogenation	$C + 2\ H_2 \rightarrow CH_4$	Exothermic

The final composition of the gas actually produced in a gasification reactor will depend on three factors, all of which relate to the specific design and operating procedures for the reactor: the balance among the five reactions shown in Table 11.1 (*i.e.*, the extent to which each occurs); whether air or pure oxygen is used as the combustion gas (use of air means that the product will be diluted with nitrogen); and whether or not the coal is heated slowly through Stage II thermal decomposition, which would add an assortment of hydrocarbon species to the product.

Lurgi gasification

The most successful commercial gasifier is the Lurgi (Fig. 11.2). The design is sometimes referred to as a *fixed-bed gasifier*. Despite the implication of the name, the bed of coal is actually moving continuously down the gasifier. Since fresh coal is continuously added to the top, it is the height of the bed which actually remains more-or-less fixed, not specific pieces of coal. For this reason the design is also known as a moving bed gasifier.

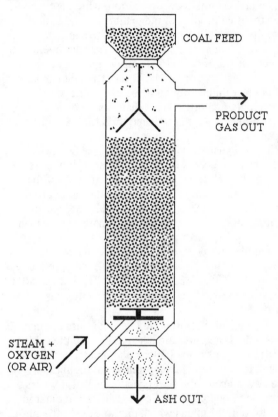

Fig. 11.2 In the Lurgi gasifier, coal introduced at the top settles downward through the gasifier vessel against a counter-current flow of steam, oxygen, and gasification products rising through the vessel.

The Lurgi gasifier typically operates at pressures of 3-4 MPa. Because the reactor vessel is operating at elevated pressures, the coal must be introduced, and ash withdrawn, via special locks (analogous to the air locks on spaceships). The distributor smooths the coal so that the height of the bed is even across the gasifier. The temperature near the grate is 900-1025°. Despite the added mechanical complexity of the locks necessary to add or withdraw materials at high pressure, the pressurized operation provides several advantages: a greater throughput rate of coal for a given size of vessel, compared to operation at ambient pressure; elimination or reduction of the need for compressors to raise the gas pressure for distribution in pipelines; and a shift of the equilibrium in the hydrogenation reaction to the right, thus increasing methane formation, which is especially desirable when the final product of the plant is to be substitute natural gas. In addition to the mechanical complexity of the equipment needed for pressure operation, the gasifier and its ancillary equipment will have a greater cost than equipment of comparable size operating at atmospheric pressure. The use of oxygen instead of air offers the advantage that the product gas will not be diluted with incombustible nitrogen and therefore will have a much higher calorific value than gas produced using air. On the other hand, extra plant investment is required for the equipment to manufacture the oxygen.

The coal entering a Lurgi gasifier will be heated by the hot gases ascending from the grate area. The coal is heated slowly and undergoes some thermal decomposition before being gasified. The products of the gasification reactions themselves, mainly carbon monoxide and hydrogen with some carbon dioxide and methane, will be mixed with products of thermal decomposition, including carbon dioxide, hydrocarbon gases, and tars.

Carbon dioxide is readily removed from the product gases by washing with a reagent such as monoethanolamine. The cleaned product gas might have the composition 35% CO, 50% H_2, and 15% CH_4, with small amounts of C_2-C_4 hydrocarbon gases. The calorific value of this gas is about 17.7 GJ/m^3.

The disadvantages of the Lurgi gasifier are that it will not operate with strongly caking coals unless measures are taken to destroy the caking properties before gasification, or to stir the bed, or both; and that the by-product tars must be treated to comply with environmental regulations. However, the undeniable advantage of the Lurgi gasifier is that it works. Lurgi gasifiers are used in the only commercial coal gasification plant in the United States, and are used in the plants which produce over 90% of the liquid fuels used in South Africa.

Koppers-Totzek gasification
The Koppers-Totzek (K-T) gasifier avoids some of the problems of tar formation and limited use of caking coals associated with the Lurgi gasifier by blowing finely pulverized coal into the vessel with steam and oxygen (Fig. 11.3).

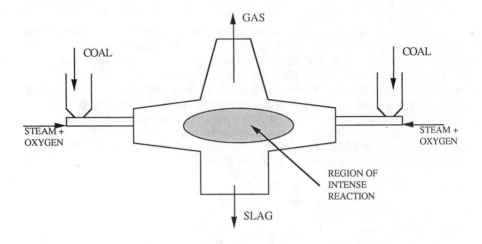

Fig. 11.3 In the Koppers-Totzek gasifier finely pulverized coal is blown, with steam and oxygen, into the vessel, where very rapid, intense reaction takes place.

The design shown in Figure 11.3, with two injection systems diametrically opposed, is called a two-headed K-T gasifier. Four-headed designs, in which four systems are arrayed at 90° angles around the circumference, are also in use.

The reaction temperatures in a K-T gasifier can exceed 1600°. The residence time of a coal particle is ~1 s. The coal does not have a chance to undergo thermal decomposition to tar before it is gasified. Consequently there are no by-product tars, and no chance for caking to occur. Therefore virtually any rank of coal can be used as a feedstock. The very high reaction temperatures cause the mineral matter in the coal to melt to a liquid slag. The gas produced in a K-T gasifier is typically 55-60% carbon monoxide, 30% hydrogen and 10% carbon dioxide.

The advantages of the K-T gasifier are that it will operate on virtually any coal, that the high temperatures provide very high reaction rates, and that ~99% of the carbon in the coal becomes carbon compounds in the gas, rather than unwanted by-products. The disadvantages are that operation at atmospheric pressure operation requires the gas to be compressed at a later stage; the units tend to have very high oxygen consumption; and that the gas leaves the gasifier at 1350-1400° C, so that much of the sensible heat of the product gas is wasted.

K-T gasifiers are in commercial use in several countries. A major application of the K-T gasifier is the manufacture of hydrogen for use in ammonia fertilizer production.

Texaco gasification

In several respects the Texaco gasifier combines the main advantages of the K-T gasifier - high reaction rates with no by-product tars - with the advantage of pressure operation provided by the Lurgi gasifier. The mechanical problems of the pressure locks in the Lurgi gasifier are eliminated in the Texaco gasifier by feeding the coal as a slurry in water. It is comparatively easy to raise the pressure of a fluid by pumping, the injection of a pressurized fluid into a vessel at high pressure being much simpler than injecting or withdrawing solids.

The Texaco gasifier (Fig. 11.4) operates in the range 2-9 MPa with reaction temperatures of 1100-1350°.

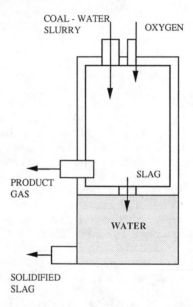

Fig. 11.4 Schematic diagram of a Texaco gasifier.
A coal-water slurry is injected along with oxygen;
reactions occur as the mixture moves downward through
the gasifier.

A typical product gas has the composition 45-50% CO, 35% H_2, and 15% CO_2. One Texaco gasifier is in commercial use in the United States, producing synthesis gas for the manufacture of chemicals.

The water-gas shift reaction

Each of the processes discussed in the previous section provides a route to the production of a mixture of carbon monoxide and hydrogen. The actual H_2/CO ratio varies considerably, from 3 in the steam reforming of natural gas to ~1 in water gas. When a carbon monoxide - hydrogen mixture is used in a specific chemical process, a specific H_2/CO ratio, dictated by the stoichiometry of the desired reaction, will be required. Before the synthesis gas can be used, its composition must be adjusted to the appropriate H_2/CO ratio needed to make the desired product. Thus an intermediate step is needed between the production of synthesis gas and its ultimate use.

Changing the H_2/CO ratio of synthesis gas is accomplished by the water-gas shift reaction

$$CO + H_2O \rightarrow CO_2 + H_2$$

The reaction of synthesis gas with steam will convert some carbon monoxide to carbon dioxide, which may be removed by absorption in monoethanolamine. In this way the H_2/CO ratio is increased. Since the water-gas shift is an equilibrium, reacting synthesis gas with carbon dioxide will convert some of the hydrogen to water, which can be removed by condensation. In this case the H_2/CO ratio is reduced. By appropriate choice of reacting the synthesis gas with steam or carbon dioxide, and by selecting reaction conditions which favor one side or the other of the equilibrium, one can start with synthesis gas of virtually any composition and "shift" its composition to any desired value from pure carbon monoxide to pure hydrogen.

Suppose we have a synthesis gas with H_2/CO of x_0, and that we desire to produce a gas having H_2/CO of x. To change the composition requires that some fraction, f, of the gas must be shifted. Initially we can presume that the gas has x_0 moles of H_2 and 1 mole of CO. After shifting, the gas will have $1-f$ moles of CO and x_0+f moles of H_2. Since the desired H_2/CO ratio of the shifted gas is x, we can say that

$$x = (x_0 + f) / (1 - f)$$

so that f can be found by straightforward algebra,

$$f = (x - x_0) / (x + 1).$$

As written (*i.e.*, with carbon dioxide and hydrogen on the right hand side), the reaction is exothermic. Therefore the equilibrium shifts to the right as the temperature is lowered. Examples of the effect of temperature are shown in Table 11.2.

Table 11.2 Effect of temperature on composition (expressed as mole fractions) during water gas shift

Temperature	CO_2	H_2	CO	H_2O
427°	0.37	0.37	0.13	0.13
812°	0.25	0.25	0.25	0.25
1027°	0.13	0.13	0.37	0.37

Because there are the same number of moles on each side of the equation, the equilibrium is independent of pressure. Usually the pressure in a shift reactor will be dictated by the pressure requirements of the process in which the shifted gas is to be used.

Initial water-gas shift technology relied on catalysts of Cr_2O_3 and Fe_2O_3, which operate most effectively at 350-475°. Current catalyst development focuses on ZnO-CuO catalysts that can operate at 200-250°. The lower temperature increases the equilibrium constant by a factor of 10. The current technology uses reactor pressures of 3 MPa with a gas-catalyst contact time of 1 s. The catalyst lifetime is ~3 years.

The uses of synthesis gas

Since synthesis gas is composed mainly of two highly combustible gases, it can be used as a fuel. Both water gas and the gaseous products of oxygen-blown gasifiers have calorific values in the range 11-19 GJ/m³. The fuel use of this gas will be discussed in more detail in Chapter 12. However, there are two concerns associated with the use of water gas or synthesis gas as a fuel. First, the calorific value is only a third to half that of natural gas. Second, carbon monoxide is a highly toxic gas, and is an undesirable component of a fuel, especially for domestic use.

Methanation

The projected lifetimes of remaining reserves of coal and natural gas suggest that supplies of natural gas will be depleted long before coal. The demand for gaseous fuels may have to be met by producing them from coal. Since none of the principal gasification systems produce substantial quantities of methane, it is necessary to convert the product synthesis gas to methane.

Methanation of synthesis gas is the exact reverse of steam reforming of natural gas:

$$CO + 3 H_2 \rightarrow CH_4 + H_2O$$

The product is called *substitute natural gas* (SNG). Currently one plant in the United States is producing SNG from lignite, making the synthesis gas in Lurgi gasifiers.

The position of the equilibrium among methane and steam *vs.* carbon monoxide and hydrogen is affected by temperature and pressure. The equilibrium constant of a reaction is a function of temperature. In this instance low temperatures favor methanation, *i.e.,* formation of methane at the expense of carbon monoxide and hydrogen. High temperatures therefore favor steam reforming. Since methanation involves a reduction in the number of moles of gaseous species, from four to two, high pressures favor methanation. Thus a methanation reactor would be run at high pressures and low temperature, whereas the steam reforming of natural gas would be favored at low pressures and high temperatures. Methanation is highly exothermic, as should be expected since steam reforming is endothermic. The reaction is so exothermic that precautions must be taken to avoid runaway temperatures which could sinter the catalyst or cause methane to decompose to carbon black.

Before methanation, the synthesis gas must be shifted to achieve the necessary H_2/CO ratio and must be purified. The principal impurities are carbon dioxide, ammonia, and hydrogen sulfide. The latter two gases are formed from nitrogen and sulfur in the coal, and are present in small amounts in the product gas from a gasifier. The removal of impurities is carried out in processes analogous to the purification of natural gas. Typical methanation processes operate at 260-370° and 1.5-7 MPa in the presence of nickel catalysts supported on aluminum oxide. A special form of very porous, unsupported nickel is also used. When an alloy of nickel and aluminum is treated with a strongly basic solution (*e.g.,* sodium hydroxide), the aluminum dissolves in the base, leaving a highly porous form of nickel known as *Raney nickel.*

Methanation of synthesis gas removes the disadvantages of using synthesis gas directly as a fuel. Specifically, the carbon monoxide is destroyed, and the calorific value is increased to ~37 GJ/m^3. Of course, methanation adds an additional processing step, and therefore additional costs, to a process converting coals or other carbon sources to gaseous fuels.

Methanol synthesis

The production of methanol occurs according to the reaction

$$CO + 2H_2 \rightarrow CH_3OH$$

This reaction was discovered in 1913. Until World War II, methanol produced from synthesis gas used coal as the starting material for the manufacture of the synthesis gas. Since then, the synthesis gas has been made by steam reforming natural gas, LPG, or naphtha.

Like methanation, the production of methanol is exothermic and occurs with a decrease in number of moles. Therefore the reaction is driven to the right by low temperatures and high pressures. Typical process conditions use ZnO catalysts at 400° and 30-38 MPa. With these reaction conditions a 62% conversion to methanol is obtained at equilibrium. Current studies on catalyst development focus on copper catalysts that can operate at 260° and 5 MPa. This very substantial reduction in process severity could result in a significant improvement in process economics (both the initial cost and maintenance of reactor vessels operating at 400° and 35 MPa are much higher than for comparable equipment operating at 260° and 5 MPa). This potential improvement in economics can justify a substantial research and development effort on catalyst formulation and performance. The choice of catalyst for this process is critical, since the potential exists for forming a wide variety of other organic compounds from synthesis gas via reactions of the type to be discussed in the following section. Whatever catalyst is chosen must have a high selectivity to methanol.

A crucial intermediate in the synthesis of methanol (or methane) appears to be the HCOH species formed on the catalyst surface

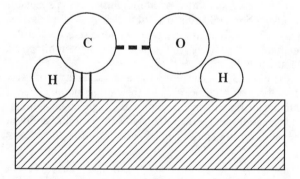

Fig. 11.5 The behavior of HCOH on the catalyst is critical in the synthesis of methanol. If the C-O bond remains intact the product is methanol, whereas if it is broken, the product is methane.

If the C-O bond is broken, the HC fragment will continue to add hydrogen to form methane. On the other hand, desorption and hydrogenation of the HCOH fragment with the C-O bond intact leads to the desired methanol. Both zinc and copper are very effective for hydrogenation of HCOH without C-O bond cleavage.

Methanol has been suggested as a liquid transportation fuel to replace gasoline. One commercial scale plant is now operating, in New Zealand, to convert methanol into gasoline (Chapter 13). Large scale use of methanol as a fuel may be realized in the future; however, methanol currently has a number of important uses. The largest use of methanol is the conversion to formaldehyde by catalytic oxidation

$$2\ CH_3OH + O_2 \rightarrow 2\ H_2C{=}O + 2\ H_2O$$

Formaldehyde is used in the production of many condensation polymers such as phenol-formaldehyde resins, or urea-formaldehyde foams. Some of these uses are discussed in more detail in Chapter 14.

The Fischer-Tropsch synthesis

The Fischer-Tropsch reaction was developed in Germany during the 1920's and 30's. Synthesis gas was reacted in the presence of a supported cobalt catalyst at ~200° and 0.7 - 1 MPa. The product was a complex mixture of hydrocarbons up to about C_{40}. The heavier compounds were useful as synthetic diesel fuel and lubricating oil. After the Second World War, American and British scientific teams investigating German wartime technology became enthusiastic about the remarkable use of coal in Germany to produce synthetic fuels and chemicals. However, it soon became clear that processes that had to be run in the necessity of wartime conditions with minimal petroleum supplies were not economically feasible in the postwar world with abundant, inexpensive petroleum. Research was carried out by the United States Bureau of Mines using the very inexpensive mill scale (a mixture of iron oxides and carbonates that forms on steel as it is processed in rolling mills), showing that reaction at 300-375° and 1.5-3.5 MPa produced a mixture of branched chain compounds having octane numbers >70, a potentially useful synthetic gasoline, as well as oxygenated compounds. Nevertheless the process was not economically viable in the face of competition from cheap petroleum products.

The Fischer-Tropsch (F-T) synthesis is the major application for synthesis gas produced from coal. During World War II, much of the aviation and armored vehicle fuel used by the German military was produced from coal, via gasification to synthesis gas followed by the

Fischer-Tropsch process. Today, over 90% of the liquid fuels used in South Africa are made in much the same way from coal. The viability of the process derives from a combination of the lack of indigenous petroleum in South Africa but abundant local coal supplies, South African difficulties of purchasing petroleum on open world markets because of trade embargoes, and the availability of a large supply of cheap labor. As we will see, the F-T synthesis is far more versatile than only being a source of substitute liquid fuels.

The chemistry of the F-T syntheses can be written in the form of generic equations for the major products. For straight-chain alkanes

$$(2n+1) \ H_2 + n \ CO \rightarrow C_nH_{(2n+2)} + n \ H_2O$$

which has a heat of reaction of -231 kJ/mol, and

$$(n+1) \ H_2 + 2n \ CO \rightarrow C_nH_{(2n+2)} + n \ CO_2$$

The second reaction is the *Kolbel reaction*. The heat of reaction is -228 kJ/mol. Straight-chain alkenes are formed by

$$2n \ H_2 + n \ CO \rightarrow C_nH_{2n} + n \ H_2O$$

The reactions leading to alcohols are

$$2n \ H_2 + n \ CO \rightarrow C_nH_{(2n+1)}OH + (n-1) \ H_2O$$

and

$$(n+1) \ H_2 + (2n-1) \ CO \rightarrow C_nH_{(2n+1)}OH + (n-1) \ CO_2$$

In addition to these principal products, some ketones, aldehydes, and carboxylic acids are produced. In some cases the F-T reactions are accompanied by isomerization, dehydrocyclization, or both. In essence, almost any hydrocarbon product that could be derived from petroleum can be made from synthesis gas via the F-T reactions. The exact mixture of products obtained depends on the specific conditions of catalyst, pressure, and temperature used in the reaction. The F-T synthesis leads in principle to a complex mixture of alkanes, alkenes, and oxygenated compounds, having up to about 40 carbon atoms. The process is operated over a wide range of conditions, 225-365° and 0.5-4 MPa, depending on the required product distribution. Catalysts based on iron, nickel, or cobalt are used. Low temperatures favor formation of higher molecular weight compounds, while temperatures at the higher end of the range are used for gasoline production. If the reactor temperature becomes too high, a high yield of methane is achieved.

There are four basic variations of the F-T synthesis. *Medium pressure synthesis* is run at 220-340° and 0.5-5 MPa in the presence of iron catalysts. The main products are gasoline, diesel oil, and some heavier alkanes. The *high pressure synthesis* operates at 100-150° but pressures of 5-100 MPa. The catalyst is ruthenium with an aluminum oxide support. The main product of the high pressure synthesis is waxes having melting points up to ~135°. Increasing the H_2/CO ratio in the feed encourages the formation of lower molecular weight compounds. In fact, if H_2/CO is increased to 4, about 96% of the product is methane. The *iso synthesis* uses thorium oxide or thorium and aluminum oxides as the catalyst with reaction conditions of 400-500° and 10-100 MPa. The major products are C_4-C_5 branched chain alkanes. About 75% of the product stream is 2-methylpropane and 25% is branched chain alkanes larger size. With reaction temperatures over 400°, production of aromatics along with branched chain alkanes is encouraged. If the temperature is kept below 400°, the formation of oxygenated compounds (*e.g.*, alcohols) is encouraged. Straight-chain alcohols are produced in the *synthol synthesis* at 400-450° and 14 MPa in the presence of an iron catalyst.

In most of these processes metallic catalysts are used. Metal catalysts chemisorb both hydrogen and carbon monoxide. (Recall, for example, the platforming catalyst.) These chemisorption properties are especially pronounced for the group VIII elements: Fe, Co, Ni, Ru, Rh, Pd, Os, Ir, and Pt. Iron and nickel are especially favored catalysts for methanation. The extent of dissociation of carbon monoxide, once it has chemisorbed on the catalyst, increases with increasing temperature.

A key to understanding the chemistry of synthesis gas conversion must be the events which take place on the catalyst surface. Consider the products from the methanation, methanol synthesis, and the various F-T processes: the hydrocarbon products are methane, C-C branched chain alkanes, gasoline, diesel oil, waxes, and some aromatics; the oxygenated products include methanol, higher straight-chain alcohols, and some ketones, aldehydes, and acids. In those products that are hydrocarbons all of the carbon monoxide must be destroyed in the process, whereas for the production of oxygenated products at least one C-O bond must be preserved per molecule. These products range from methane and methanol with one carbon atom each to waxes having high numbers of carbon atoms. This distribution suggests that some type of polymerization must occur on the catalyst surface.

When a chain of carbon atoms is growing on the surface of a catalyst, at any given step one of two events can occur: the growth can stop (termination) or one more carbon atom can be added (propagation). If we assume that the probability of propagation vs. termination is constant regardless of the chain length, the amount of some compound with chain length of x carbon atoms will be some fraction, α, of the amount of the compound with x-1 carbon atoms. The value of this fraction α will be governed by the relative rates of propagation and termination. In fact, α can be found from

$$\alpha = r_p / (r_p + r_t)$$

where r_p is the rate of propagation and r_t is the rate of termination. Now suppose that the weight fraction (in the product) of the compound containing x carbon atoms is M_x. This value is related to α by the expression

$$\log (M_x / x) = c + x \log \alpha$$

This expression indicates that a plot of $\log (M_x / x)$ as a function of x should be linear with a slope of $\log \alpha$. This relationship is called the *Schulz-Flory distribution*. In principle it allows us to predict the distribution of various sized molecules in the product, although in practice α is usually linear only for values of $x \geq 4$. In most F-T processes, α is usually 0.5-0.8. For the same conditions of temperature, pressure, and catalyst, α is independent of the number of carbon atoms and the nature of the surface species on the catalyst. Thus the product distribution can be described by using the single variable α. For example, the weight percent W of the species containing N carbon atoms in the product stream will be given by

$$W = 100 \, N \, (1 - \alpha)^2 \, \alpha^{(N-1)}$$

As a rule α increases with pressure and is inversely proportional to temperature. For example, on a ruthenium catalyst at 0.1 MPa and 200° the principal product is methane, but at 100 MPa the products include very high molecular weight greases and waxes and in some cases polymers with ~10,000 carbon atoms. Generally, low values of α and high reaction temperatures produce lighter products, while high values of α and low temperatures result in heavier products.

The fact that a distribution of products is obtained is a disadvanatge of the F-T process. For hydrocarbons in the C_5-C_{10} range typical of gasoline, the maximum yield is achieved at $\alpha = 0.76$, but the yield of compounds in this range is only 44%. The product distribution can be changed by changing the catalyst, the operation conditions, or both, but there is always a balance to be achieved. For example, to produce gasoline by the F-T process one must accept inevitable accompanying high yields of C_1-C_4 gaseous products. Because of the wide product distribution, a large number of subsequent unit operations are necessary to separate and refine the products. The need for additional separation and refining adds both to the capital investment for a plant and to operating costs. A major thrust of catalyst research has been to find catalysts which will break the wide distribution and provide better selectivity to a narrow range of products. The Dow Corporation has reported the use of a molybdenum catalyst on carbon support which provides a 70% yield of C_2-C_5 hydrocarbons, with the main products being ethane and propane.

The mechanisms of F-T reactions on catalyst surfaces are still under active investigation. Hydrocarbon synthesis can begin with the dissociation of chemisorbed carbon monoxide

$$
\begin{array}{c}
O \\
\parallel \\
C \\
\vert \\
\overline{/\!/\!/\!/\!/}
\end{array}
\quad \longrightarrow \quad
\begin{array}{c}
C \quad O \\
\parallel\!\parallel \quad \parallel \\
\overline{/\!/\!/\!/\!/}
\end{array}
$$

The carbon atom bonded to the catalyst surface can react with hydrogen atoms (from dissociated, chemisorbed hydrogen) to produce a methylene group; the methylene can react further with another hydrogen to form a methyl group on the catalyst surface.

$$
\begin{array}{c}
C \\
\parallel\!\parallel\!\parallel \\
\overline{/\!/\!/\!/\!/}
\end{array}
+ \ 2\ H\cdot \ \longrightarrow \
\begin{array}{c}
CH_2 \\
\parallel \\
\overline{/\!/\!/\!/\!/}
\end{array}
+ \ H\cdot \ \longrightarrow \
\begin{array}{c}
CH_3 \\
\vert \\
\overline{/\!/\!/\!/\!/}
\end{array}
$$

The methyl group on the catalyst can terminate by reaction with one additional hydrogen to form methane, which desorbs from the catalyst surface.

$$
\begin{array}{c}
CH_3 \\
\vert \\
\overline{/\!/\!/\!/\!/}
\end{array}
+ \ H\cdot \quad
\overline{/\!/\!/\!/\!/}
\quad + \ CH_4
$$

Alternatively, the carbon chain can grow as a result of *methylene insertion*

$$
\begin{array}{c}
CH_2 \ CH_3 \\
\parallel \quad \vert \\
\overline{/\!/\!/\!/\!/}
\end{array}
\quad \longrightarrow \quad
\begin{array}{c}
CH_2\!-\!CH_3 \\
\vert \\
\overline{/\!/\!/\!/\!/}
\end{array}
$$

If a methylene insertion reaction can occur, the ethyl group on the catalyst can also undergo one of two reactions: termination to form ethane by reaction with a hydrogen atom, or growth by methylene insertion to produce a propyl group

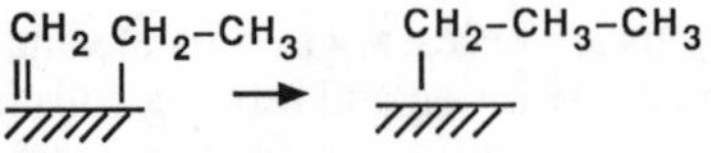

$$CH_2 = CH_2-CH_3 \quad \longrightarrow \quad CH_2-CH_3-CH_3$$

(surface-bound species)

At each step of the chain growth the options of termination or further growth by methylene insertion are available.

A problem with practical operation of the Fischer-Tropsch synthesis is that it is highly exothermic. If the temperature in the reactor becomes too high the principal product is methane, which is undesirable if the intent is to make liquid fuels.

The formation of oxygenated compounds requires that at some point a *carbon monoxide insertion* reaction occurs

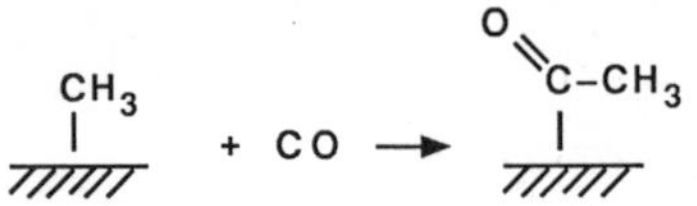

$$CH_3 \;\; + \;\; CO \;\; \longrightarrow \;\; \overset{O}{\underset{}{\backslash\!\!\!=}}C-CH_3$$

The formation of an alcohol occurs via subsequent reaction with hydrogen atoms

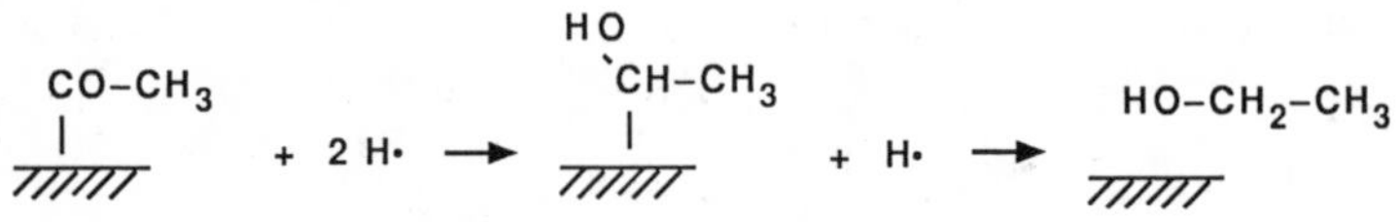

$$CO-CH_3 \;\; + \;\; 2\,H\cdot \;\; \longrightarrow \;\; HO\text{-}CH-CH_3 \;\; + \;\; H\cdot \;\; \longrightarrow \;\; HO-CH_2-CH_3$$

and the formation of longer chain alcohols is exactly analogous

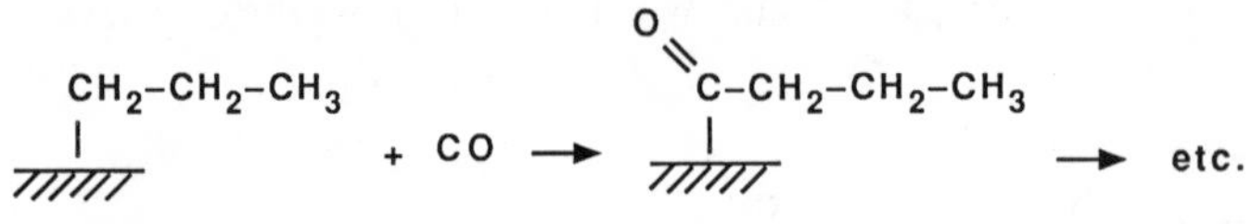

$$CH_2-CH_2-CH_3 \;\; + \;\; CO \;\; \longrightarrow \;\; \overset{O}{C}-CH_2-CH_2-CH_3 \;\; \longrightarrow \;\; etc.$$

Aldehydes may form as by-products

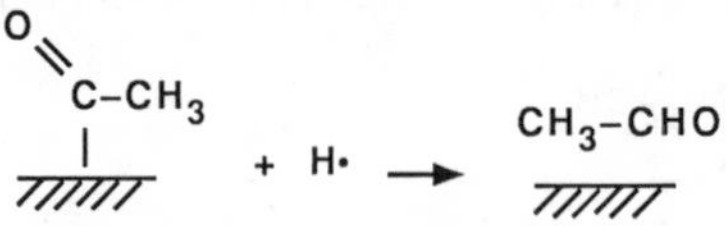

$$\overset{O}{C}-CH_3 \;\; + \;\; H\cdot \;\; \longrightarrow \;\; CH_3-CHO$$

As an example of the remarkable versatility of the F-T syntheses, consider the following sequence of reactions: First, coal is gasified to produce synthesis gas. A portion of the synthesis gas is reacted in the synthol process to form long chain alcohols; the remainder is reacted in the medium pressure synthesis to form alkanes. Mild oxidation of the alcohols produces long chain acids, the so-called fatty acids. Thermal cracking of the alkanes can be used to manufacture propene by β-bond scission processes. The propylene can be converted, via a series of reactions, into glycerol. The glycerol can be reacted with the fatty acids to produce fats

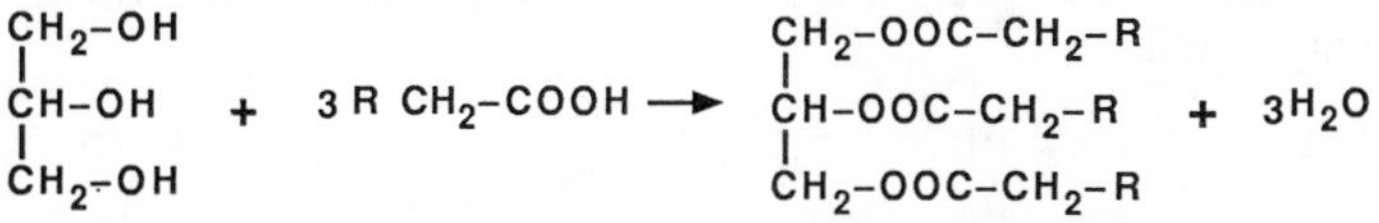

$$
\begin{array}{l}
CH_2-OH \\
CH-OH \\
CH_2-OH
\end{array}
\ +\ 3\,R\ CH_2-COOH \longrightarrow
\begin{array}{l}
CH_2-OOC-CH_2-R \\
CH-OOC-CH_2-R \\
CH_2-OOC-CH_2-R
\end{array}
\ +\ 3H_2O
$$

and the fats can be used for cooking or related kitchen purposes as synthetic margarine. Gasification followed by F-T reactions, as well as some other known processes such as thermal cracking, can be used to convert coal to synthetic foodstuffs.

The Kolbel reaction is a variation of the F-T synthesis in which the product being rejected is carbon dioxide rather than water. Thus

$$2n\ CO + n\ H_2 \rightarrow (\text{-}CH_2\text{-})_n + n\ CO_2$$

Because the rejected material, carbon dioxide, contains carbon rather than hydrogen, the Kolbel reaction makes it possible to produce the desired hydrocarbon products with a lower H_2/CO ratio in the synthesis gas than is used in the customary F-T synthesis where water is the rejected by-product. Consequently, the Kolbel reaction offers the possibility of eliminating the water gas shift reaction to increase the H_2/CO ratio. An extreme case is represented by the *Kolbel-Engelhard reaction*, which uses a mixture of carbon monoxide and water:

$$3n\ CO + n\ H_2O \rightarrow (\text{-}CH_2\text{-})_n + 2n\ CO_2$$

The Oxo process

The Oxo process was developed in Germany during the 1930's for the preparation of higher alcohols. The original interest was for the synthesis of C_{12}-C_{18} alcohols for conversion in a subsequent step to sulfate esters that are used in detergents. Carbon monoxide and hydrogen are added to an alkene, in a reaction known as *hydroformylation*, and the resulting aldehyde is subsequently reduced to the alcohol. An example is the hydroformylation of propene:

$$CH_3CH=CH_2 + CO + H_2 \rightarrow CH_3CH_2CH_2CH=O$$

The Oxo process is a variation of the Fischer-Tropsch synthesis, and originally used a Fischer-Tropsch catalyst at 110-150° and 15 MPa. The process is now performed with a homogeneous cobalt carbonyl catalyst at 75-200° and 10-30 MPa. The active form of the catalyst is $HCo(CO)_4$.

A disadvantage of the Oxo process is that the addition across the double bond is not specific, so that in the example above one also obtains isobutyraldehyde

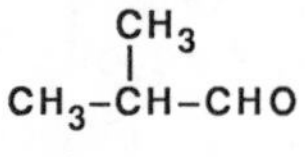

$$
\begin{array}{c}
CH_3 \\
| \\
CH_3-CH-CHO
\end{array}
$$

in about 40% yield.

The straight chain aldehyde is then reduced in a second step to the corresponding alcohol. The long chain alcohols are desirable are precursors to detergents. The shorter alcohols are esterified with phthalic acid to form plasticizers.

Further reading

Berkowitz, N. (1979). *An Introduction to Coal Technology*. New York: Academic Press; Chapter 12,.

Gates, B. C., Katzer, J. R., and Schuit, G. C. A. (1979). *Chemistry of Catalytic Processes*. New York: McGraw-Hill; Chapter 3.

Merrick, D. (1984). *Coal Combustion and Conversion Technology*. New York: Elsevier; Chapter 4.

Probstein, R. F. and Hicks, R. E. (1982). *Synthetic Fuels*. New York: McGraw-Hill; Chapter 3.

Satterfield, C. N. (1980). *Heterogeneous Catalysis in Practice*. New York: McGraw-Hill; Chapter 10.

Chapter 12

Alternate fuels from coal

The estimates of the remaining lifetimes of petroleum and natural gas suggest that these resources may be depleted well before society has made a major change to non-fossil energy sources as solar, geothermal, or nuclear energy. Some source of hydrocarbon fuels will be necessary as a "bridge" between the petroleum and gas based energy economy of the second half of the twentieth century and a new energy economy for the twenty-first century, based on renewable energy sources. That "bridge" is most likely to be coal. This chapter discusses strategies for manufacturing gaseous and liquid fuels from coal. The conversion of coal to an alternative solid fuel form, coke, is also discussed.

Coal carbonization

Carbonization is the heating of coal (or other materials) in the absence of air to effect chemical changes by thermal decomposition reactions. We have already seen that thermal decomposition results in the evolution of relatively hydrogen-rich gases, vapors, and liquids, and leaves behind a carbon-rich solid. Carbonization is another example of the hydrogen distribution processes discussed previously in connection with, for example, catagenesis and coking of catalysts. Three broad categories of carbonization processes have been developed, depending on whether the desired principal product is a gas, a liquid, or a solid.

Low-temperature carbonization occurs in the range 450-700°. In this temperature range the yield of tar is at a maximum; the tar can be separated through a variety of distillation and extraction processes into a large number of useful chemical products. The gas produced during low-temperature carbonization is useful as a fuel. The remaining char is highly reactive and is also of use as a fuel.

Medium-temperature carbonization is run at 700-900°. The gas yield is maximized in this temperature range. Medium-temperature carbonization is a potential route to the production of a gaseous fuel from coal. The char is also fairly reactive and is of potential use as a fuel. In particular, fixed-bed combustors (Chapter 10) can produce significant amounts of smoke as a result of incomplete combustion of the tars liberated as the coal is slowly heated. The smoke can contribute to severe air pollution problems, especially in cities where many households and small commercial or industrial establishments burn coal in fixed-bed combustion devices. Although the problem of air pollution from coal smoke is at least 600 years old, the cities of London and Pittsburgh in the early years of the 20th century are notorious for poor air quality. Several processes have been developed, especially in the United Kingdom, for producing a smokeless coal-derived fuel by a mild carbonization process that drives most of the tar out of the coal. The carbonized coal char can then be burned in, for example, domestic fires with a considerable reduction in the amount of smoke formed.

High-temperature carbonization occurs at 900-1200°. The principal product is a hard coke of value for use in blast furnaces in the metallurgical industry. High-temperature carbonization is the most important of the three types of processes and is discussed in detail in the following section.

Coke production

Chemistry of coke formation

Caking coals produce a fluid phase when passing through the severe degradation of structure accompanying Stage 2 thermal decomposition. A special subset of the caking coals resolidify to produce coke, the special material used as a fuel in metallurgical processes. Such coals are known as *coking coals*. Before discussing some of the aspects of coke production on an industrial scale, it is useful to treat in greater detail the distinctions between non-caking, caking, and coking coals, and particularly to examine the molecular basis for these distinctions.

Regardless of whether a coal is a caking coal, a carbonaceous solid will remain at the end of the thermal decomposition process (assuming of course that the process is carried out in inert and not a reactive atmosphere). These solid products of *carbonization* are of two types: a *char* is the product of carbonization of a material that did not pass through a fluid stage during the carbonization process, whereas a *coke* is the product of a carbonization process that at some point passed through a liquid or liquid crystal phase. These definitions of chars and cokes are general definitions, not restricted to materials derived from coals, but indeed are applicable to materials derived from any natural source of carbon (including wood and petroleum) or synthetic organic substances such as polymers. Chars and cokes differ from each other in a second fundamental way: a coke, if heated to extreme temperatures (*e.g.*, 3000°) will form graphite, whereas a char, even if heated to the same temperatures, will not form graphite. This distinction is expressed by saying that chars are *non-graphitizable* but that cokes are *graphitizable*.

The structure of graphite is

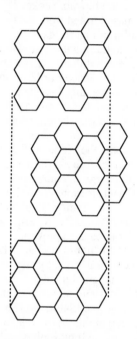

Graphite is highly anisotropic. Within one plane of aromatic rings the carbon-carbon bonds are strong, covalent bonds. From one plane to another the structural interactions are relatively weak electrostatic forces between the π electron systems in the aromatic rings (Fig. 7.4). Thus there are large differences in many physical properties of graphite, depending on whether the measurement is made parallel or perpendicular to the planes of the aromatic rings. A metallurgical coke derived from coal is also anisotropic; for example, if the surface of a coke sample is examined by optical microscopy with polarized light, the coke is found to be optically active. Chars, on the other hand, are isotropic, and are not optically active. In essence the situation is

Coke (anisotroptic) → Graphite (anisotropic)

Char (isotropic) -**X**-> Graphite.

Further heat treatment of an anisotropic coke leads to the anisotropic graphite structure, whereas heat treatment of an isotropic char does not.

To work backwards one additional step, it is necessary to ask why a caking coal produces a fluid phase which will lead eventually to a non-graphitizable, isotropic char, while a coking coal also produces a fluid phase, but one which forms an anisotropic, graphitizable coke. The distinction in this case lies in the nature of the fluid phase formed from the two types of coals. The generation of an anisotropic solid from a fluid phase is greatly facilitated if the arrangement of molecules in the fluid is itself anisotropic. Liquids are ordinarily thought of as being isotropic, but there exists a special class of liquids - *liquid crystals* - in which the molecular order is anisotropic. The existence of liquid crystals represents a balance between intermolecular forces (in the case of aromatic molecules, these would be interactions of the π electron systems) and the vibrational and translational energies of the molecules. In a solid crystal,the intermolecular or interatomic interactions provide sufficient cohesion to overcome the vibrational or translational energy which would allow the molecules free movement. In an ordinary liquid, the molecules have sufficient energy to overcome the intermolecular forces of the solid and to have sufficient freedom of motion so that the properties of the liquid are isotropic. In a liquid crystal, substantial intermolecular forces still exist (thus making the material crystal-like in this respect), yet the forces are not strong enough to prevent flow (thus making the material liquid-like); hence the term "liquid crystal." The anisotropic liquid crystal fluid which is the precursor to coke formation is the *mesophase*.

The production of an anisotropic, graphitizable coke from coal requires formation of an anisotropic mesophase, whereas if instead the fluid formed during carbonization is an isotropic liquid, the resulting solid product will be an isotropic char. The final step in this analysis is to inquire what features of structure favor the formation of a mesophase relative to an isotropic liquid. We have seen that Stage 2 thermal decomposition of coal leads to the evolution of a variety of products, including various hydrocarbons produced by thermally cleaving bridging groups or alkyl side chains from aromatic ring systems, and molecules such as CO, H_2S, and NH_3 which are produced by thermal disruption of oxygen, nitrogen, or sulfur functional groups. The reactions occurring during the thermal decomposition of coal are free radical processes. We have seen that one approach to the stabilization of free radicals is hydrogen capping or hydrogen abstraction. The liquid phase that forms initially on the thermal breakdown of the macromolecular structure of the coal is isotropic. The reactivity of this liquid phase will be determined to some extent by the total concentration of free radicals in the liquid. Suppose first of all that the redistribution of hydrogen within the liquid is difficult. Rather than being capped, the highly reactive free radicals will enter into recombination reactions with other nearby radicals. If we presume the existence of a random, isotropic distribution of radicals within the isotropic liquid, the formation of new carbon-carbon bonds by radical recombination reactions will proceed in random directions. Consequently, the structure of the solid that eventuates from this radical recombination process will be essentially isotropic because of the formation of bonds between the randomly oriented liquid molecules has taken place in all possible directions (Fig. 12.1)

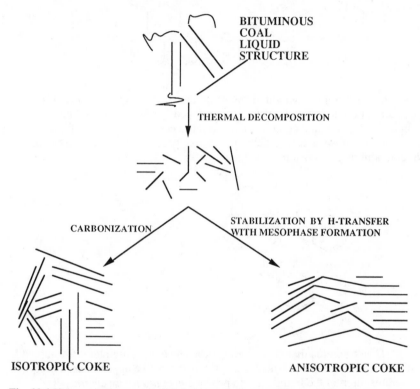

BITUMINOUS
COAL
LIQUID
STRUCTURE

THERMAL DECOMPOSITION

CARBONIZATION

STABILIZATION BY H-TRANSFER
WITH MESOPHASE FORMATION

ISOTROPIC COKE

ANISOTROPIC COKE

Fig. 12.1 If the mesophase produced during coal decomposition undergoes only further carbonization, the product is likely to be an isotropic coke. On the other hand, the stabilization of the mesophase by internal hydrogen transfer can lead to formation of an anisotropic coke.

Consider a situation in which the isotropic liquid first formed from the coal is able to undergo internal hydrogen redistribution to allow the capping of some of the free radicals. Capping reduces the total population of radicals in the liquid, and, as a consequence, this liquid has less overall reactivity than that in the previous example. In addition, the less reactive liquid phase tends to be of lower viscosity (*i.e.,* higher fluidity). The higher fluidity allows molecules in the liquid the mobility to align; the lower overall reactivity gives molecules the chance to realign before they are consumed in a recombination reaction. In this situation the potential exists for *dehydrogenative polymerization* reactions of a type illustrated below for anthracene.

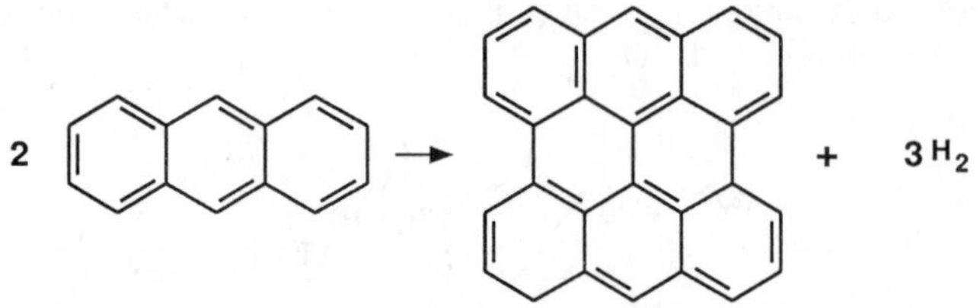

$$2 \quad \text{[anthracene]} \quad \longrightarrow \quad \text{[polycyclic aromatic structure]} \quad + \quad 3\,H_2$$

The structures of the compounds present in the actual liquid produced from the thermal breakdown of coal are more varied (a coal tar may contain hundreds of individual components) and more complex than the simple example of anthracene. The dehydrogenative polymerization of the mixture of compounds gives rise to structures such as

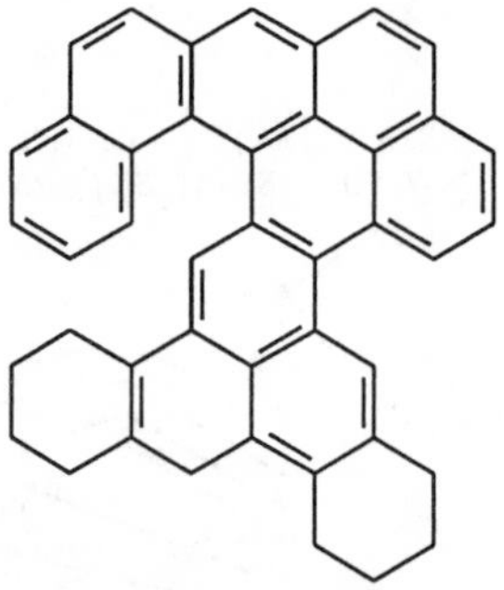

[polycyclic aromatic hydrocarbon structure]

The structures that arise from anthracene have a high length-to-breadth ratio and, at least in two dimensions, present a rod-like appearance. These structures align in the liquid state to produce *nematic liquid crystals*. The ordering in a nematic liquid crystal is shown in Figure 12.2, where the molecules are viewed from the side, in a manner similar to the convention introduced in Chapter 6 of representing an aromatic ring system viewed edge-on by a straight line. There is a distinction between the relatively rod-like structures produced by the dehydrogenative polymerization of anthracene and those structures produced from a mixture of compounds typical of coal tars. In the latter case (Fig. 12.2), the molecule is flat but there is no significant difference between the length and breadth of the molecule. As structures of this type continue to grow, they become almost circular. To

296

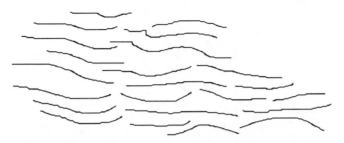

Fig. 12.2 A schematic view of a nematic liquid crystal, showing significant structural ordering of the molecules in the liquid phase. In this example the molecules are aromatic systems viewed edge-on.

distinguish this structure from the anthracene polymers, the circular-shaped structure is said to form *discotic nematic liquid crystals*, where the term "discotic" means disc-like. The discotic molecules that can participate in the formation of the discotic nematic liquid crystal phase, which in fact is the mesophase, are sometimes referred to as *mesogens*.

In systems in which the overall reactivity has been reduced by hydrogen capping of some of the radicals, the growth of molecules is sufficiently slow so that increase of molecular size is not significant until high temperatures are reached. But since the fluidity of the liquid increases with increasing temperature, the point at which molecular growth becomes significant is at or near the point at which fluidity is high. Thus the mesophase is able to continue to accommodate further molecules of mesogens, and, therefore, the mesophase continues to grow.

The relationship of fluidity, as measured by the Gieseler test, to coal rank is shown in Figure 12.3. This relationship is traditionally expressed using volatile matter expressed on a dry, mineral-matter-free basis (dmmf) as an indicator of rank. The coals which have the highest fluidity (point A in Fig. 12.3) are not ideal coking coals. With ~30% dmmf volatile matter, these coals are high and medium volatile bituminous and, recalling Fig. 6.5, these coals contain appreciable amounts of oxygen. The destruction of oxygen functional groups during Stage 2 thermal decomposition generations large numbers of radicals, consequently resulting in a very reactive liquid phase. Furthermore, the aromaticity and ring condensation of these coals is also low, and it would be necessary to have extensive growth of the aromatic ring systems to form the discotic mesophase. Coals at point B in Figure 12.3 are highly aromatic and may have large condensed ring systems, but the fluidity of the liquid phase is too low to allow good their extensive alignment in the liquid. The best coking coals are therefore found at point C, which essentially represents a compromise among fluidity, aromaticity and ring condensation, and reactivity of the liquid.

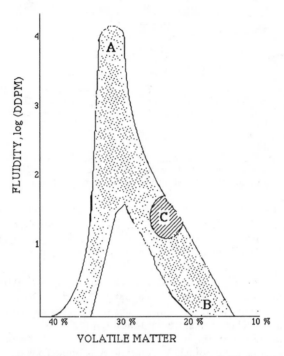

Fig. 12.3 The relationship between Gieseler fluidity and volatile matter content. The best coking coals occur in region **C**, indicated by the shaded region.

In order to form a coke that is satisfactory for metallurgical properties, a coal *must* go through a fluid state. The decomposition reactions occurring within the fluid mass give gas and other volatiles which force their way through the viscous fluid. As the fluid resolidifies the tracks left by the bubbles of volatiles passing through the fluid do not fill up with fluid. The property of passing through a fluid state with evolution of volatiles in this manner is characteristic of the vitrinites of bituminous rank, of about 78-89% carbon. Fusinites or micrinites do not become fluid at any level of rank, but experiments on blends show that up to 25% concentration of these macerals can increase the strength of the resulting coke, just as adding an aggregate to cement makes a strong concrete. Sporinite becomes fluid at a slightly higher temperature than vitrinite, but then becomes very fluid, loses much weight as volatiles and then resolidifies. When heated by itself sporinite behaves somewhat like popcorn, in blowing up to a very weak, thin walled bubble structure. In relatively small amounts in coal sporinite can contribute usefully to the fluidity in cases in which the vitrinite has relatively poor agglomerating properties.

Industrial production of coke

The discussion of petroleum products and petroleum refining in Chapters 8 and 9 showed how developments in technology and patterns of energy consumption caused fluctuations in demand for various crude oil cuts, for example the historical trends of demand of kerosene *vis-a-vis* gasoline. The limitations on the availability of straight-run gasoline could have limited the development and growth of the automobile industry, and this concern led to the increase in gasoline yield afforded by thermal cracking. Improvements in engine performance showed the need for improved octane numbers in gasoline, a need which was met in part by the development of catalytic cracking processes. The forces which led to the development of coke production provide another example of how limitations in availability of an energy supply led to new developments in fuel chemistry, which in turn allowed the further advancement of technology.

The production of a metal from its ore requires reaction of the ore with some kind of reducing agent. To keep the cost of the metal low, the reducing agent used must be inexpensive. One of the cheapest and best reducing agents is carbon. In the early days of the iron industry, before 1700, the carbon used for smelting iron ore was obtained in the form of charcoal. Charcoal is made be heating wood in the absence of air. Unfortunately, the requirement for wood was prodigious. Approximately three acres of trees needed to be felled per ton of iron produced. Eventually the cutting of trees produced such extensive deforestation that a law was enacted in Great Britain forbidding the installation of new iron-making furnaces. To find a substitute for charcoal, iron makers turned to coal. Coal, however, gives a poor grade of iron, because impurities released as the coal passes through Stage 2 thermal decomposition enter the metal. About the same time, brewers were experimenting with the use of coal to dry hops and malt used for making beer. The same problem was encountered, in that the volatile constituents of the coal entered the beer ingredients, ruining the flavor of the resulting beer. Brewers discovered that the heating of coal in the absence of air drove out the volatile constituents, leaving behind a non-volatile solid product that still had considerable value as a fuel. This solid is coke. Iron makers quickly learned that the coke served as an excellent fuel and reducing agent for the blast furnaces used for smelting iron ores.

In the 1700's, coke was produced by preparing a large stack of coal with pieces of wood running through it. The wood was set on fire, and the heat from the burning wood carbonized the coal. This process was unsatisfactory because it was impossible to control once the wood had been ignited, the progress of carbonization was highly dependent on prevailing weather conditions, and, consequently, the quality of the product varied widely.

The first major improvement was the *beehive coke oven*, a simple dome constructed of refractory material. Volatile components emitted from the coal were burned to provide the heat required for the carbonization. In the late 1800's organic chemists discovered that the tar produced from coal was a rich source of a wide variety of useful and valuable materials, including solvents, dyes, medicines, and fertilizers. Unfortunately, the very concept of operation of the beehive coke oven was that tar and other volatiles had to be burned to provide the heat for carbonization. The necessary destruction of a potentially valuable by-product led to the need for a new design of coke oven.

The new coke oven design, which is still in use, is the *by-product coke oven*, also known as the *slot-type coke oven* (Fig. 12.4).

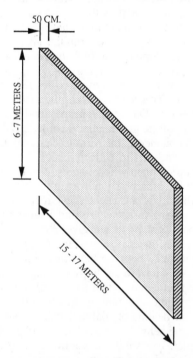

Fig. 12.4 Slot-type coke ovens are designed to be very narrow for effective heat transfer to the mass of coal being coked. Usually 20 - 100 of these individual units will be built together in a battery of coke ovens.

Individual ovens are built together in batteries of twenty to a hundred ovens. Between each pair of ovens is a flue in which gas is burned to provide heat for the coking reactions. The very narrow width of each oven, typically 50 cm, is governed by limitations on heat transfer to insure that heat can penetrate completely to the center of the charge.

Coal used for the production of coke needs a number of special qualities: high free swelling index (~6-7), high fluidity, large plastic range, and low mineral matter content (the mineral matter tends to dilute the metaplast, interfering with the growth of large aromatic ring systems). In addition, the mineral matter should be low in phosphorus, sulfur, and silicon, because if these elements were carried into the coke and thence into the metal, the quality of the metal would be impaired.

During operation of a slot-type oven, the charge is coked from the walls inward (Fig. 12.5).

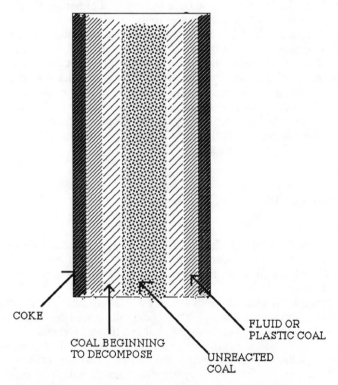

COKE

COAL BEGINNING
TO DECOMPOSE

UNREACTED
COAL

FLUID OR
PLASTIC COAL

Fig. 12.5 In a slot-type coke oven, coal cokes from the walls inward, producing distinct layers of coke, plastic coal, coal beginning to decompose,and unreacted coal. As coking proceeds, the layers of coke grow inward and eventually meet at the center.

The various layers - coke, partially decomposed coal, plastic coal - move toward each other from the walls, so that at the end the whole mass of the charge has been converted to coke. If the maximum fluidity of the charge is too great, or if the dilitation is too great (especially for a strongly euplastic coal), enough pressure can be built up to damage or destroy the oven walls. Therefore candidate coals are tested extensively in small-scale ovens before being used in large-scale operation.

The desired properties of coke relate to composition, reactivity, size, and strength. Ideally coke should have <1.5% volatile matter, <10% ash, <1.0% sulfur, and a calorific value of >34 MJ/kg. Coke should react readily with oxygen or with carbon dioxide to form carbon monoxide (which in fact is the active reducing agent in a blast furnace). Because the formation of carbon monoxide is a heterogeneous reaction, cokes should have a high surface area. Particle size needs to be large (>5 cm) and of narrow range to allow good permeability of the coke charge to the air blast in the furnace. The mechanical strength needs to be sufficient to resist crushing either during handling or when supporting the weight of the charge in the blast furnace.

The reduction of iron ore requires at least 0.5 tonne of carbon (coke) per tonne of iron produced, so a very large steelmaking complex may need to draw on the production of two or more coal mines. Individual coals that give excellent quality, strong, reactive coke are now very rare. For both of these reasons, it is common to blend coals to obtain the charge to a coke oven battery. Petrographic analysis of the coals is the most important technique for evaluating coal quality for coke making.

Coal gasification

We have introduced the topic of the conversion of coal to gaseous fuels in Chapter 11, with particular concern for the production of synthesis gas. In this chapter the discussion of gasification is continued, with focus on the production of gaseous fuels.

The least complicated approach to coal gasification is to take advantage of the fact that gas produced during high temperature thermal decomposition of coal is very combustible. By heating the coal into Stage 3 thermal decomposition, it is possible to produce a gas which is 40-50% hydrogen and 30-40% methane (the balance being nitrogen, carbon monoxide, carbon dioxide, and ethene). Any ammonia, hydrogen sulfide, or carbon dioxide produced can be removed from the gas by absorption processes analogous to those discussed in Chapter 7 for natural gas purification. The product of this process has various names: *coal gas, town gas,* or *illuminating gas.* The calorific value is about 24 GJ/m^3, compared with 37 GJ/m^3 for natural gas.

Coal gas collected as the by-product of coke production could be sold to generate extra revenue for the coke plant. However, if coal is carbonized only for the purpose of producing coal gas (and not at the same time manufacturing coke), only about 20% of the coal is actually converted to gas; a use or market must be found for the remaining 80% of the coal. The great disadvantage of the carbonization of coal - as a gasification process - is that only a fraction of the coal is converted to gas. It would be preferable to have a process that converts all of the carbon in the coal to a gaseous product.

The reaction of steam with a red-hot bed of coal produces *water gas*, a mixture of the approximate composition 50% H_2, 40% CO, 5% CO_2, and 5% N_2. Because the carbon-steam reaction is endothermic, the reaction would quickly stop in the absence of an additional source of heat. Burning of a portion of the coal to provide external heat to the reactor is undesirable because the carbon burned to carbon dioxide is lost as a potential fuel. Instead, heat is derived from incomplete combustion of coal to carbon monoxide.

If air is passed slowly over hot coal, incomplete combustion occurs

$$C + 0.5\ O_2 \rightarrow CO$$

and if any carbon dioxide does form, it would in principle by destroyed by the Boudouard reaction

$$C + CO_2 \rightarrow 2\ CO$$

(in practice some carbon dioxide inevitably winds up in the product). As the coal is heated in this incomplete combustion process, it passes through the Stage II thermal decomposition; consequently, some hydrocarbons are formed as by-products. The eventual gaseous product is 20-25% CO, 55-60% N_2, 2-8% CO_2, and 3-5% small hydrocarbons. This product is known as *producer gas*. Although the carbon monoxide and the hydrocarbons are good fuels, they are so diluted by the nitrogen and carbon dioxide that the calorific value of producer gas is quite low, about 5.2 GJ/m^3.

Despite its very low calorific value, producer gas has been a popular fuel gas because of its ease of manufacture. It has been used in applications where very large quantities of hot gas were needed, as in open hearth furnaces in steel mills, glass-making furnaces, and pottery kilns. Producer gas is not a desirable fuel for domestic use because of its low calorific value (about one-sixth that of natural gas) and because of the very toxic nature of carbon monoxide.

The formation of producer gas via the exothermic incomplete combustion reaction provides a strategy for converting all the carbon in the coal to gas. If the process is begun with the production of producer gas, the coal bed is heated by the exothermic reaction. When the coal bed is hot, the air is turned off and steam is switched on. Now water gas is generated until the coal cools to the point at which the steam-carbon reaction no longer occurs. At this point the steam is turned off and the air switched on. Producer gas is manufactured until the coal is once more hot enough to react with steam. This cycle of alternate air- and steam-blowing can be continued indefinitely.

A gas plant might make water gas and producer gas in a roughly 1:2 ratio. One of the advantages of water gas is its higher calorific value (compared to producer gas) of ~11 GJ/m³. When water gas and coal gas are made in the same plant, they can be blended for sale as *illuminating gas*. A block diagram of the process is shown as Figure 12.6.

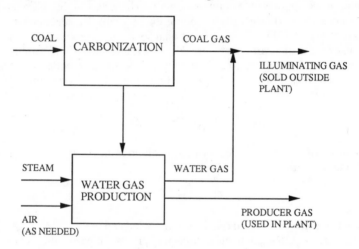

Fig. 12.6 Schematic diagram of a gas plant, the eventual products being illuminating gas for commercial use and producer gas for use as a heat source in the plant.

The cycling of air- and steam-blowing of the gas producer can be eliminated by injecting steam and air (or oxygen) simultaneously into the gasifier. With appropriate choice of conditions, the exothermic reactions balance the endothermic reactions and the process can run continuously for as long as desired. The simultaneous injection of air or oxygen and steam into a single vessel of coal is the basis of design of all modern coal gasification systems. The Lurgi, Koppers-Totzek, and Texaco gasifier designs have been discussed in Chapter 11 in connection with the manufacture of synthesis gas. Many other coal gasification schemes have been developed, some of which have been tried on the pilot plant scale, and a few which have seen some commercial use. However, the Lurgi and Koppers-Totzek are the principal gasifier designs being used for large-scale coal gasification, and the Texaco design appears to have promise for commercial manufacture of chemicals as well as for use in combined-cycle power plants.

Coal liquefaction

If the known reserves of petroleum are divided by the annual rate of consumption, the lifetime of the reserve, at present consumption rates, is about 30 years. Although there is interest in developing engines which would run on pulverized coal (as the early diesel), steam, or hydrogen, there is nevertheless an immense investment world-wide in combustion devices (vehicle engines as well as stationary systems such as furnaces and boilers) using liquid fuels. Neither the economic strength nor the manufacturing capacity exists to convert or replace all such liquid-fuel-fired systems in a short time. Consequently it is important to consider the prospects for manufacturing liquid, petroleum-like fuels from other sources. In addition, most of today's organic chemical industry is based on petroleum. Thus as remaining supplies of petroleum dwindle, it may be more important to society to divert the petroleum to the chemical industry rather than to continue burning it as a fuel.

Indirect liquefaction

We have discussed in Chapter 11 a useful route to the production of gasoline, diesel oil, and other hydrocarbons via F-T processes. If the synthesis gas used as the feed for the F-T synthesis is made from coal, then it is possible to manufacture liquid fuels from coal by the two steps that are illustrated by the "equations" below

$$Coal \rightarrow CO + H_2$$
$$CO + H_2 \rightarrow Gasoline, diesel oil, etc.$$

The formation of synthesis gas is an intermediate step in the production of liquid fuels from coal by this scheme. Because of this, the overall process is known as *indirect liquefaction*.

The general process scheme for indirect liquefaction involves the gasification of coal to produce synthesis gas, the purification and shifting of the gas, and then the reaction of the shifted gas via F-T processes to produce liquid fuels. This process scheme is being used today in South Africa for the manufacture of liquid fuels from coal. Recall that one of the equations describing the F-T synthesis of alkanes is

$$(2n+1)\,H_2 + n\,CO \rightarrow C_nH_{(2n+2)} + n\,H_2O$$

If it were desired to synthesize octane (n = 8) the required H_2/CO ratio of the feed would be 17/8, or ~2.1. If the process were begun with the production of water gas, $H_2/CO \sim 1$. Recall from Chapter 11 that the fraction of gas to be shifted is given by

$$f = (x - x_0) / (x + 1)$$

where in this example $x_0 = 1$ and $x = 2.1$. Then about one-third of the total gas stream must be shifted to achieve the desired H_2/CO ratio. The appropriate F-T processes have been introduced in Chapter 11, where we have seen that the choice of reaction conditions such as temperature, pressure, catalyst and H_2/CO ratio allow the production of a variety of liquid, petroleum-like materials from synthesis gas. Although gasification, the water-gas shift, and F-T syntheses are all proven technology in commercial use, we must recognize, as the name *indirect* implies, that this route to liquid fuels is the long way around, because the coal must be gasified before liquids can be produced.

The Fischer-Tropsch process was used in Germany during World War II for the production of liquid fuels. At the high point of production, about 750 million liters was produced annually. Unfortunately, most of the compounds in the gasoline were straight chain alkanes, so the octane number of the untreated gasoline was about 40. Thus reforming, addition of tetraethyllead, or both, were required to increase the octane number to acceptable levels. The preponderance of straight-chain compounds however made the higher boiling material a superb fuel for diesel engines, with cetane numbers as high as 85.

Direct liquefaction

The conversion of coal to liquid fuels without the intermediate formation of synthesis gas, *e.g.*,

$$Coal \rightarrow Gasoline, diesel, etc.$$

is *direct liquefaction*. The process requirements for direct liquefaction can be inferred from a comparison of the H/C ratios in coals and petroleum products. A typical coal has an H/C ~0.75, while for many petroleum products H/C ~2. Thus the principal requirement in the direct conversion of coal to petroleum-like liquid fuels is to increase H/C by adding hydrogen, in some form, to the coal.

In gasification, whether for the production of a fuel gas or of synthesis gas for indirect liquefaction, the macromolecular structure of the coal is broken into products containing single carbon atoms, such as carbon monoxide and methane. Since this transformation completely destroys all features of the macromolecular structure of the coal, in principle any rank of coal could be used as a feedstock for gasification. (In practice, some specific ranks of coal may be unsuitable for certain process configurations, as, for example, the use of highly caking coals in fixed-bed gasifiers.) However, in direct liquefaction processing, the desired liquid fuels may contain, say, 5 to 20 carbon atoms in their molecules. For example, kerosenes may contain C_{12}-C_{14} alkanes, alkylated cycloalkanes and multicyclic cycloalkanes, and alkylated aromatics. Thus some of the structural features of the coal must be preserved in the liquid products.

A major problem of the chemistry of direct liquefaction is that the coal is principally aromatic in structure. The good stability of aromatic systems toward addition reactions makes hydrogenation of coal difficult. A consequence is that direct liquefaction processes typically operate at high temperatures (400-500°) and high pressures (up to 70 MPa). Equipment which operates at such high pressures and temperatures is not inexpensive and may have high maintenance costs. The investment in plant equipment and maintenance is one reason why coal-derived liquid fuels are not economically competetive with petroleum-based fuels today.

In the hydrogenation reactions that might be used to increase the H/C ratio of the coal, aromatics tend to be less reactive than alkanes or alkenes. For example, under mild conditions styrene reacts with hydrogen to produce ethylbenzene; the double bond but not the aromatic ring is hydrogenated. Under more drastic conditions, the ethylbenzene would react to form toluene and methane, again preserving the aromatic structure. (This is not to imply that it is impossible to reduce aromatic ring systems, because such reductions can in fact be carried out with appropriate catalyst and reaction conditions. The important point is that aromatic rings are less vulnerable to attack than are other parts of the coal structure.) The relatively low reactivity of aromatic rings relative to aliphatic carbons implies that coals are likely to undergo reactions initially at the aliphatic crosslinks, rather than in the aromatic systems.

Recalling the generalizations of coal structure from Chapter 6, the open structure is highly crosslinked and very porous; the liquid structure is less porous and less crosslinked that the open structure; and the anthracite structure, although somewhat porous, is almost completely aromatic with virtually no crosslinks. The porosity of a coal will affect the access of reagents to the internal surface area and the egress of reaction products.

Preferred feedstocks for direct liquefaction are coals of lignite through high volatile bituminous in rank. These are coals with abundant crosslinks or other reactive functional groups, high porosity for reagent access, and small aromatic ring systems. Some success has been had in liquefaction experiments with medium volatile bituminous coal. It is generally agreed that, with today's understanding of coal chemistry, low volatile bituminous coals and anthracites are virtually impossible to process in direct liquefaction.

The mechanism of direct liquefaction is not fully understood. There is general agreement that liquefaction proceeds via free radical processes which are initiated by bond cleavage reactions, such as the rupture of an aliphatic crosslink:

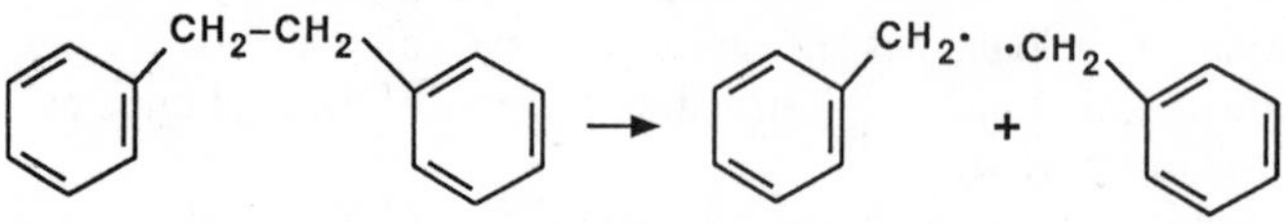

At stringent reaction conditions, it is possible to reduce aromatic rings or even to open rings:

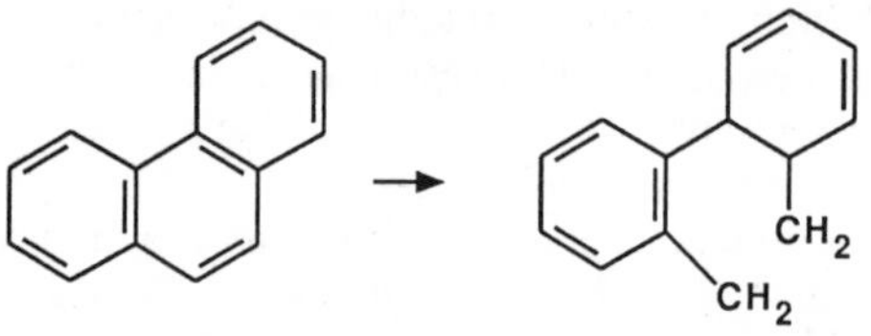

Of the many free radical reactions discussed earlier, those of particular concern to direct liquefaction are hydrogen abstraction, hydrogen capping, and recombination. The stabilization of radicals by hydrogen capping preserves molecular fragments that are smaller in size than the original coal macromolecule. Repeated bond cleavage and hydrogen capping eventually leads to the production of molecules small enough to be liquids.

Hydrogen abstraction in principle will also stabilize the radicals against recombination, and probably helps to break apart the coal macromolecule. If, however, the hydrogen is being abstracted from some other part of the coal structure, the hydrogen abstraction itself is generating a new radical which might be able to react further. If this is the case, there is no net addition of hydrogen to the system. Rather, the coal will transform as would be expected from the hydrogen redistribution scheme introduced in Chapter 3, forming relatively hydrogen-rich liquids, but leaving behind an increasingly carbon-rich, unreactive solid.

Recombination of radicals is very undesirable. Not only does recombination not generate a useful liquid product, but in addition the carbon-rich solid product of recombination is highly resistant to further reaction. (In a sense, this accumulation of unreactive carbonaceous solid is analogous to coking accompanying carbocation reactions.) In order to maximize the formation of useful liquid products, and at the same time minimize recombination, an enormous amount of ingenuity has been devoted to designing catalysts and hydrogen donor molecules (discussed below) and to selecting appropriate reaction conditions to insure that a reactive form of hydrogen is available at or close to the sites at which free radicals are being generated.

The most straightforward approach to liquefaction is the *Bergius process*. A block flow diagram of this process is given in Figure 12.7. The coal is slurried in oil to facilitate charging to pressure vessels (recall the same strategy used in the Texaco gasification process). The coal/oil slurry is reacted with hydrogen at 475-485°. For low-rank coals the reactor pressure is 25-32 MPa; with bituminous coals the pressure exceeds 70 MPa. The main product of the reaction is *middle oil*. This material is somewhat similar to No.2 fuel oil. Once a liquid product of some sort has been formed, it is then able to be treated by many of the petroleum refining processes we have discussed in Chapters 8 and 9.

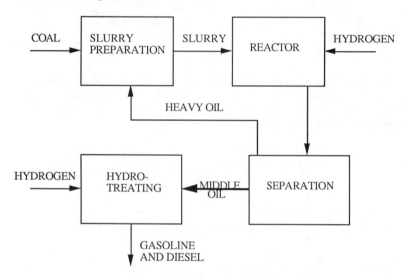

Fig. 12.7 The Bergius process involves reaction of a coal-oil slurry with hydrogen, followed by a separate hydrotreating of the coal-derived liquids to produce gasoline and diesel oil.

In the Bergius process, the middle oil is vaporized and reacted further with hydrogen. This process is called *hydrotreating*, also sometimes known as hydrocracking or hydrofining. Hydrotreating the Bergius middle oil provides the desired gasoline and diesel oil. About 50-55% of the coal (on a moisture-and-ash-free basis) is converted to middle oil. Another 40% is converted to *heavy oil*, which is recycled to provide the liquid vehicle for the slurry. The gasoline produced from the Bergius process was typically 75% paraffins, 20% aromatics, and 5% olefins. It had octane numbers in the range 75-80.

The chemistry of the Bergius process is basically an attack by hydrogen on the macromoleuclar structure of the coal. It is generally agreed to proceed via free radicals

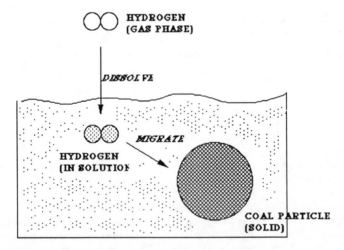

The Bergius process involves a three-phase reaction. Gaseous hydrogen must dissolve in the oil; the dissolved hydrogen must then migrate through the oil to the coal surface; and finally the hydrogen must dissociate and react at the solid coal surface (Fig. 12.8).

Fig. 12.8 Unless a strategy is available for transferring hydrogen from components of the liquid vehicle, direct liquefaction must occur as a three-phase process in which gaseous hydrogen dissolves in the vehicle, migrates through the liquid, and finally reacts with the solid coal.

This need for dissolution and migration of hydrogen accounts in part for the very high pressure requirements of the Bergius process. We have seen previously that high temperatures, high pressures, or especially the combination of high temperatures and pressures are not desirable from an economic point of view. The more stringent the reaction conditions, the greater the performance requirements for the equipment, and in turn more severe equipment requirements result in higher initial costs and, often, higher maintenance costs. Much of the research and developmental work on direct liquefaction in the past fifty years has been devoted to finding strategies to reduce the temperature and pressure requirements of the process.

Hydrogenation of coal can be facilitated by the presence in the liquid phase of organic molecules which are able to transfer hydrogen to the coal. These molecules are then themselves re-hydrogenated by gaseous hydrogen or hydrogen dissolved in the liquid medium. Compounds which act in this way are called *hydrogen donors* or sometimes H-donors. The classic example (though by no means the best) hydrogen donor is 1,2,3,4-tetrahydronaphthalene, generally known by its trivial name, tetralin.

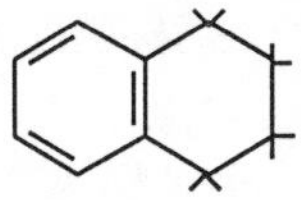

The reaction which can be envisioned is

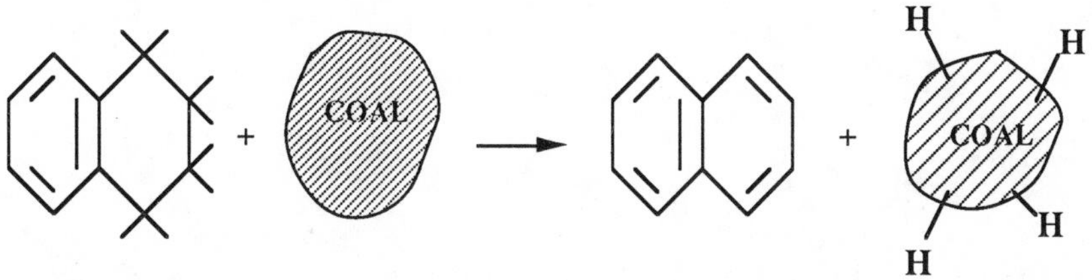

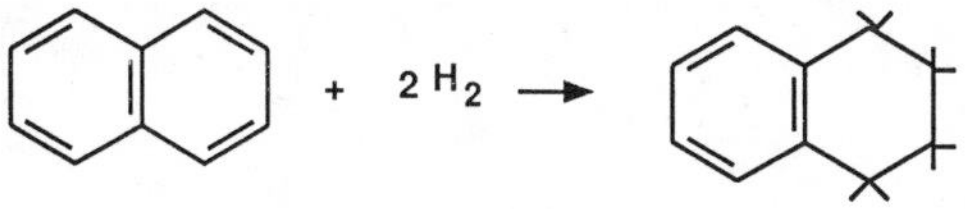

which may proceed sequentially through 1,2-dihydronaphthalene. The hydrogen donated by the tetralin is used to cap radicals, thus facilitating coal depolymerization and liquids production. In a subsequent reaction the tetralin is regenerated

so that the net process is

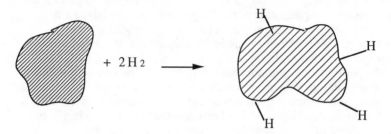

$$+ 2H_2 \longrightarrow$$

Further reading

Berkowitz, N. (1979). *An Introduction to Coal Technology*. New York: Academic Press; Chapters 11,12,13.

Grainger, L. and Gibson, J. (1981). *Coal Utilisation: Technology, Economics, and Policy*. London: Graham and Trotman; Chapters 7,8,9.

Marsh, H. (ed.) (1989). *Introduction to Carbon Science*. London: Butterworths; Chapters 2,9.

Probstein, R. N. and Hicks, R. E. (1982). *Synthetic Fuels*. New York: McGraw-Hill; Chapters 4,6.

Chapter 13

Oxygenated fuels

Our use of hydrocarbon fuels is tempered by two considerations. First, there are environmental consequences of the combustion of these compounds, consequences which are well-known, if not necessarily well understood, as acid rain and the greenhouse effect. Second, the remaining reserves of these fuels, especially petroleum and natural gas, are limited and future needs for liquid fuels may have to be met from other sources. This chapter discusses oxygenated compounds, particularly alcohols and ethers, which have promise as fuels or fuel additives.

A major environmental concern, especially in highly populated areas, is the emission of nitrogen oxides (collectively known as NO_x) from automobile exhausts. NO_x is one of the chief contributors to smog formation, which is a serious pollution problem in such cities as Denver and Los Angeles. NO_x has two principal sources: nitrogen chemically incorporated in the fuel reacting at combustion temperatures, called *fuel NO_x* ; and the reaction of nitrogen in the air inside the combustion chamber (*e.g.* the cylinder of an automobile engine) reacting at the high combustion temperatures with oxygen to form *thermal NO_x*. One approach to limiting NO_x emissions is the use of *catalytic converters* in the exhaust system. In the converter the carbon monoxide, NO_x, and unburned hydrocarbons react over a platinum catalyst to form carbon dioxide and nitrogen. The catalysts used in these converters are poisoned by lead, which enters the exhaust system when gasoline containing tetraethyllead is used as a fuel. The poisoning of catalysts is one of the major reasons for phasing out the use of leaded gasolines (the other reason being concerns for the effects of lead emissions on the environment and public health). A consequence of the phasing out of leaded gasoline is the need to develop alternative fuel additives for improving the octane rating of gasoline.

Attention has focused on the use of oxygenated compounds, mainly ethers or alcohols, as fuel additives. The alcohols in fact are receiving serious consideration not only as an additive, but also for the complete replacement of petroleum-derived fuels. The two alcohols considered as leading candidates for alternate fuels are methanol and ethanol. Methanol is made from synthesis gas. Ethanol is made from the fermentation of such materials as grain, agricultural wastes, and forest by-products.

The heat of combustion of an oxygenated compound is less than that of the corresponding hydrocarbon. Therefore the energy available from the combustion of a gasoline/alcohol blend is less than that from gasoline alone. This difference adversely affects the fuel economy of the vehicle. Generally, however, the other performance of gasoline-alcohol blends is good. The effects of ethanol addition on the knock-limited compression ratio and thermal efficiency are shown in Table 13.1.

Table 13.1 Effects of blending ethanol on knock-limited compression ratio and thermal efficiency of gasoline

% Ethanol in blend	Knock-limited compression ratio	Thermal efficiency
0	6	0.32
10	6.2	0.33
25	8	0.36
50	10	0.38
100	>10	0.38

For comparison, the addition of 3 mL tetraethyllead per gallon of gasoline gives a fuel having a knock-limited compression ratio of 7.5 and a thermal efficiency of 0.34. The specific fuel consumption, SFC, is inversely proportional to the calorific value of the fuel on a weight basis:

$$SFC = 2545 / \eta\, Q$$

where η is the thermal efficiency ands Q is the calorific value. A blend of ethanol and gasoline has been sold commercially as *gasohol*.

t-Butyl alcohol and ether

In addition to providing good combustion performance with reduced NO_x emissions, a second benefit of oxygenated fuels is their reduction of carbon monoxide emissions. Denver has the worst carbon monoxide emission problem in the United States, and the Colorado state legislature has recently mandated the use of oxygenated fuels during the winter driving season

(November to March). A compound of interest as an additive in this application is methyl t-butyl ether, also known as MTBE.

The synthesis of MTBE is performed by the reaction of isobutylene with methanol in the presence of a strong Bronsted acid catalyst, usually an acidic ion exchange resin, at 40-90° and 1 MPa. The use of an acid catalyst is an indication that the reaction proceeds via a carbocation mechanism. In the presence of a Bronsted acid catalyst, isobutylene is converted to the 3° carbocation:

$$(CH_3)_2C=CH_2 + H^+ \rightarrow (CH_3)_3C^+$$

This carbocation then attacks a molecule of methanol to form the oxonium ion

$$(CH_3)_3C^+ + HOCH_3 \rightarrow (CH_3)_3CO^+CH_3$$
$$H$$

The intermediate oxonium ion loses the proton, regenerating the catalyst, and forming the desired methyl t-butyl ether:

$$(CH_3)_3CO^+CH_3 \rightarrow (CH_3)_3COCH_3 + H^+$$
$$H$$

The synthesis of methyl t-butyl ether is carried out as a liquid phase reaction, typically at 40-90° and 1 MPa.

A second route to MTBE is production from propylene. The oxidation of propylene with t-butyl hydroperoxide, $(CH_3)_3COOH$ (note that this is *not* a carboxylic acid!) in a presence of a molybdenum catalyst produces propylene oxide and t-butyl alcohol as a by-product. Air oxidation of isobutane is done at 120-150° and 3 MPa, the product being the hydroperoxide:

$$(CH_3)_3CH + O_2 \rightarrow (CH_3)_3COOH$$

The hydroperoxide is subsequently reacted with propylene to form the desired alcohol product and propylene oxide.

$$(CH_3)_3COOH + CH_3CH=CH_2 \rightarrow (CH_3)_3COH + \overset{O}{CH_3CHCH_2}$$

The methyl ether is then produced from the alcohol. The principal source of propylene is the thermal cracking of naphtha (Chapter 9). Propylene oxide is mainly used for the production of propylene glycol (propane-1,2-diol), which in turn is the starting point for making polyester resins. The t-butyl alcohol is an alternative to the MTBE as a fuel additive. t-Butyl alcohol has an octane number of 108.

Ethanol

We have seen that some aspects of engine performance with gasoline/alcohol blends are quite favorable. Since the alcohols themselves are combustible liquids, one can also run vehicle engines on alcohols alone. Both ethanol and methanol have been considered in this connection. Existing SI engine designs can be run on ethanol with minimal modification to the engine. Ethanol provides good resistance to knocking and low levels of emissions, particularly NO_x, from the engine. Unfortunately, CI engines do not operate on ethanol or methanol.

There are two main sources of ethanol. The first is the hydration of ethylene:

$$CH_2=CH_2 + H_2O \rightarrow CH_3CH_2OH$$

usually carried out in the presence of an acid catalyst. The ethylene is produced by thermal cracking of petroleum derivatives. The other ethanol source is fermentation of biomass (agricultural or forest products). Since the principal source of ethylene is the petroleum industry, production of ethanol by this route is not likely to be feasible in times of petroleum shortages, The more likely approach is the fermentation of biomass, such as grain, sugar cane, sugar beets, or potatoes.

A block diagram of the fermentation process is shown in Figure 13.1. Grain, agriculutral wastes, or other forms of biomass could potentially be used as feedstocks. The fermenters are often of enormous size, being in the range of 10^6 liters in capacity. A concern for this process is that the external heat requirements for the distillation step could use so much heat that more energy is consumed in the process, to separate ethanol from water, than is manufactured in the form of the heat available from burning the product ethanol. Process research is focused on approaches for reducing the energy requirements of the distillation step or for finding alternate methods for removing the water, such as by absorbing the water on zeolites.

A comparison of the heats of combustion of ethanol and 2,2,4-trimethylpentane shows the advantage, a volumetric basis, to the latter. The heat of combustion of ethanol is 23.5 MJ/L while that of 2,2,4-trimethylpentane is 33.1 MJ/L. Ethanol has 71% of the calorific value of 2,2,4-trimethylpentane when the fuels are compared on the basis of equal volumes. If all other factors were equal, the distance driven per volume of fuel for an ethanol-fuelled automobile would be ~70% of that for a gasoline-fueled vehicle.

Despite the disadvantage of the lower heat of combustion per volume, ethanol can be a useful vehicle fuel when no other options are attractive or available. The most successful use of ethanol in automobiles has come in

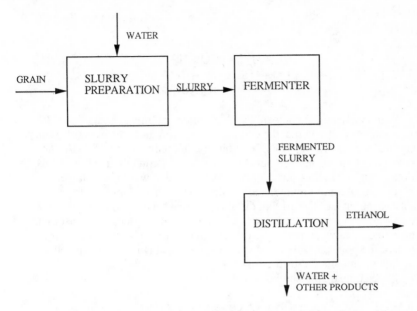

WATER

GRAIN → SLURRY PREPARATION —SLURRY→ FERMENTER

FERMENTED SLURRY

DISTILLATION —ETHANOL→

WATER + OTHER PRODUCTS

Fig. 13.1 Schematic diagram of a fermentation process for production of ethanol from grain or biomass.

Brazil, which has virtually no domestic petroleum reserves but abundant supplies of biomass such as sugar cane waste. In Brazil about 3.5 million automobiles are running on a 97% ethanol - 3% gasoline blend (the gasoline is not added for its fuel value, but rather to render the ethanol unfit to drink), and another 8 million use a 22% ethanol - 78% gasoline blend.

Methanol

Methanol is produced from synthesis gas by the method discussed in Chapter 11. Currently the synthesis gas is produced from the steam reforming of natural gas, followed by shifting to reduce the H_2/CO ratio:

$$CH_4 + H_2O \rightarrow CO + 3 H_2$$
$$H_2 + CO_2 \rightarrow CO + H_2O$$
$$CO + 2 H_2 \rightarrow CH_3OH$$

In the event of shortages of natural gas, an alternative approach would be the manufacture of synthesis gas by the steam-carbon reaction in coal gasification, followed by shifting (in the opposite direction) to raise the H_2/CO ratio:

$$C + H_2O \rightarrow CO + H_2$$
$$CO + H_2O \rightarrow H_2 + CO_2$$
$$CO + 2 H_2 \rightarrow CH_3OH$$

When methanol is compared with gasoline in terms of heat of combustion per unit volume, the comparison is even more disfavorable than in the case of ethanol. The heat of combustion of methanol is 17.8 MJ/L, about half the value of a gasoline component such as 2,2,4-trimethylpentane. Thus we obtain roughly half the energy from methanol that would be obtained from gasoline per unit volume, and would expect, under comparable conditions, about half the driving distance per tankful of fuel. It would be necessary to stop at a "methanol station" about twice as often as for gasoline. To be economically competitive with gasoline, methanol should therefore be about half the price per volume, since we would need twice as much methanol for an equivalent amount of driving.

If we compare the combustion equations

$$CH_3OH + 1.5 O_2 \rightarrow CO_2 + 2 H_2O$$
$$C_8H_{18} + 12.5 O_2 \rightarrow 8 CO_2 + 9 H_2O$$

and recognize the difference in the number of moles per liter (249 mol/L of methanol and 60.7 mol/L of an eight-carbon alkane) we find that the combustion of a liter of gasoline requires about twice as much air as combustion of a liter of methanol. Thus if an IC engine set up for burning gasoline were to be converted to methanol, changes in the fuel injector or carburetor system would be needed to obtain the appropriate air/fuel ratio.

Methanol is infinitely miscible with water. The potential exists for methanol to mix with water (for example, during long-term storage) to reduce the calorific value or, in extreme cases, essentially destroy the value of the methanol as a fuel. Methanol can cause blindness when ingested, absorbed through the skin, or even inhaled in large amounts. Ingestion of 100-250 mL can be fatal. Therefore methanol is a potential hazard to health if improperly handled.

Despite these disadvantages of methanol as a potential automotive fuel, there are several advantages in its favor. First, there already exists a industrial infrastructure for manufacturing methanol (large quantities are made each year for use as an industrial solvent). In comparison, the development of a coal liquefaction industry would require the construction of facilities which do not exist today. Second, methanol is made from synthesis gas. We have seen previously that synthesis gas can be made in principle from the steam gasification of almost any carbon source, including petroleum resids, coals, and biomass. Thus many potential sources of the synthesis gas are available, and of which could be used for methanol production with appropriate application of the water-gas shift reaction as an intermediate step.

Methanol has a large endothermic heat of vaporization. Because heat must be used to vaporize the fuel, and because the heat of combustion is lower than a hydrocarbon fuel, the flame temperatures are lower and the engine is said to "run cooler." The beneficial consequences of this are reduced thermal stress on engine components and lower production of thermal NO_x.

With growing concern for the possible global warming - the greenhouse effect - caused by a build-up of carbon dioxide in the atmosphere, methanol also provides the advantage of reduced CO_2 emissions per unit of energy liberated during combustion. Gasoline combustion produces 1.47 moles of CO_2 per MJ of energy, whereas for methanol combustion the corresponding figure is 1.40 mol/MJ. While at first sight this may seem like a very small difference, when these numbers are applied to the enormous annual consumption of liquid fuels in vehicles, a significant reduction in CO_2 emissions is obtained by the use of methanol.

To sum up, two major advantages are achieved by the use of methanol or ethanol in blends with gasoline or as a substitute for gasoline: First, a higher octane number is attainable, typically in the range 106-108. (Other oxygenated fuels used as octane boosters include t-butyl alcohol and methyl t-butyl ether, the latter compound being a methanol derivative.) Second, the fuels are cleaner burning than gasoline, with a 30-50% reduction in NO_x emissions attainable as a result of the lower flame temperature. On the other hand, there are also several disadvantages: Alcohols, especially methanol, produce aldehydes as by-products during combustion. Since aldehydes are not major components of gasoline emissions, their emission from combustion sources is not regulated. Aldehydes can be highly irritating to the mucous membranes of the respiratory tract, and thus are undesirable in air. This problem can be remedied by using catalytic converters in the exhaust system to insure the complete oxidation of the aldehydes to carbon dioxide and water. The "energy density" (heat of combustion per volume) of alcohols is about half that of gasoline. Thus the fuel economy measured on a volume basis (kilometers per liter or miles per gallon) is much lower than for gasoline. At temperatures below about -10° methanol cannot vaporize sufficiently to form a flammable methanol vapor - air mixture. This behavior presents problems in starting engines in cold climates. Gasoline, in comparison, is sufficiently volatile to form flammable gasoline vapor - air mixtures at -30°. Alcohols have a tendency to absorb moisture, forming solutions which are moderately good conductors of electricity. These mixtures promote electrochemical effects in corrosion processes and thus exacerbate corrosion of the materials used in the fuel system. This problem may be overcome by changing the material specifications.

Methanol to gasoline

Concerns about the potential disadvantages of methanol as a vehicle fuel, such as the need for conversion of the fuel systems in existing engines to a new air/fuel ratio and the possible health effects of methanol, can be eliminated by converting the methanol to gasoline. Such a process exists and is in commercial operation in New Zealand. It is known as the *MTG process* (Methanol to Gasoline). The key to successful operation of the process is a special zeolite catalyst, *ZSM-5* (for Zeolite-Sucony-Mobil, named for the companies involved in its development).

Zeolite catalysts were introduced in Chapter 9 in connection with catalytic cracking. The ZSM-5 catalyst has the special property of converting oxygenated compounds to hydrocarbons. Like all zeolites, ZSM-5 will only allow molecules of a certain size or shape to enter or exit the pores where catalyzed reactions take place. The pore openings in ZSM-5 are about 0.6 nm in diameter. Only C_{10} compounds or smaller can move through the structure. Molecules larger than C_{10} which form in the catalyst will crack to smaller molecules. Therefore a narrow distribution of products is obtained, with about 80% of the product being gasoline in the C_5 - C_{10} range.

As currently practiced, the MTG process begins with natural gas. The gas is sweetened by reaction with zinc oxide:

$$ZnO + H_2S \rightarrow ZnS + H_2O$$

The sweetened gas is then steam reformed, and the synthesis gas is converted to methanol. Because the MTG process is based on manufacture of methanol from synthesis gas, the potential exists to obtain the synthesis gas from hydrocarbon sources other than natural gas. The methanol is converted to ethylene via dimethyl ether:

$$2\ CH_3OH \rightarrow CH_3OCH_3 + H_2O$$
$$CH_3OCH_3 \rightarrow CH_2{=}CH_2 + H_2O$$

The formation of dimethyl ether proceeds at 300-325° and 2.3 MPa. The ethylene reacts further with methanol to form propylene:

$$CH_2{=}CH_2 + CH_3OH \rightarrow CH_3CH{=}CH_2 + H_2O$$

The formation of alkenes in the presence of an acidic catalyst (the zeolite) provides the opportunity for "polymerization" reactions to occur, similar to the reactions discussed in Chapter 9 regarding gasoline production.

$$CH_3CH=CH_2 + H^+ \rightarrow CH_3C^+HCH_3$$
$$CH_3C^+HCH_3 + CH_3CH=CH_2 \rightarrow (CH_3)_2CHCH_2C^+HCH_3$$
$$(CH_3)_2CHCH_2C^+HCH_3 \rightarrow H^+ + (CH_3)_2CHCH=CHCH_3$$

The 2-methyl-3-pentene is able to react further to cycloalkenes, in the processes we have discussed for reactions on bifunctional catalysts

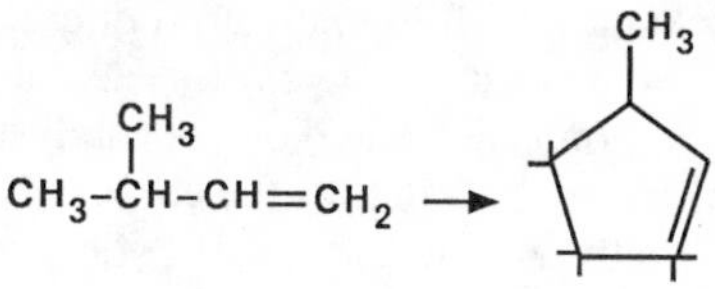

The formation of 1-methylcyclopentene can be followed by ring expansion to cyclohexene

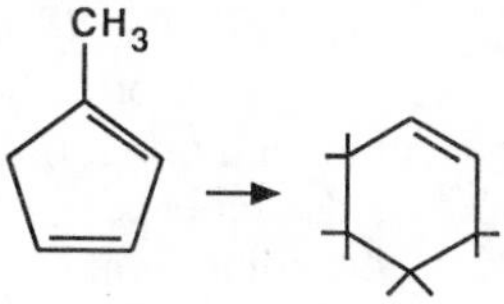

and subsequent aromatization of the cyclohexene to benzene

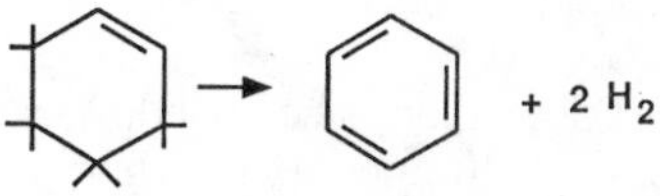

Aromatic compounds are valuable products because they boost octane numbers of the gasoline. Aromatization can also occur by hydrogen transfer

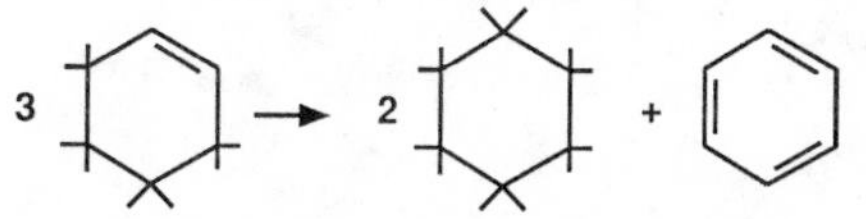

Hydrogen transfer is also possible between alkenes

$$(CH_3)_2CHCH=CHCH_3 + (CH_3)_2CHCH=CHCH_3 \rightarrow$$
$$CH=C(CH_3)CH=CHCH_3 + (CH_3)_2CHCH_2CH_2CH_3$$

Formation of compounds containing multiple double bonds is not desirable because they are the precursors for coke formation on the catalyst.

321

Conversion of the dimethyl ether to gasoline occurs at 330-400° and 2.2 MPa pressure in the presence of the ZSM-5 catalyst. Gasoline formation is exothermic (ΔH is -75 kJ/mol at 700 K). Exothermic reactions on acidic catalysts can lead to excessive coking of the catalyst, particularly if fixed-bed reactors are used. In the MTG process the problem is solved in part by carrying out the synthesis in two stages. In the second stage reactor the temperature increase is controlled by recycling some of the product through the reactor. In this case the recycle ratio is 6-9 moles of product per mole of feed recycled, the high recycle ratio resulting in a very expensive process. Despite these precautions, some carbon is deposited on the catalyst, necessitating a periodic regeneration by burning off the coke.

The size of the "windows" in the zeolite catalyst is critical for the formation of products in the molecular size range typical of gasolines. Small windows favor only ethylene formation. Medium-sized windows favor gasoline formation; the ZSM-5 catalyst has been designed with windows of appropriate size to favor products in the C_7-C_9 range. With the ZSM-5 catalyst the mixture of products is 2% C_6, 16% C_7, 39% C_8, 28% C_9, and 13% C_{10}. With larger windows in the catalyst the formation of heavy ($>C_{10}$) aromatics is favored. An example of such a compound is 1,2,4,5-tetramethylbenzene (durene). This product is undesirable because it has a high melting point (79°) and is easy to crystallize from solution in gasoline. The formation of solid crystals by deposition from solution during cold weather is called *carburetor icing* and is undesirable because it blocks the flow of fuel to the engine. Another problem with the formation of large aromatics is that the process, once begun, is difficult to stop, and therefore may lead to coke formation which deactivates the catalyst.

Even with the ZSM-5 catalyst of medium-sized windows some light alkenes and some heavy aromatics will be formed. In fact there is a high concentration of durene in the gasoline produced in New Zealand by the MTG process. The light alkenes, however, are very valuable by-products for use in the chemical industry. Propylene, for example, can be converted into solvents, antifreeze, and the monomers for the manufacture of acrylic fibers and resins and polyester resins.

Further reading

Barnard, J. A. and Bradley, J. N. (1985). *Flame and Combustion,* 2d edn. London: Chapman and Hall; Chapter 11.

Chang, R. and Tikkanen, W. (1988). *The Top Fifty Industrial Chemicals*. New York: Random House; Chapters 4,10,22,41.

Colucci, J. M. and Gallopoulos, N. E. (1977). *Future Automotive Fuels*. New York: Plenum; Session III.

Loudon, G. M. (1984). *Organic Chemistry*, 1st edn. Reading, MA: Addison-Wesley; Chapter 20.

Sperling, D. (1988). *New Transportation Fuels*. Berkeley, CA: University of California; Chapters II, III.

Chapter 14

Chemicals from fuels

Natural gas, petroleum, and coal are commonly thought of as fuels - that is, substances which are burned as sources of energy. Such a view, however, neglects the full scope of the potential of these materials. Despite their diverse characteristics, natural gas, petroleum, and coal are all organic materials containing abundant amounts of carbon. As we consider the best ways of using these materials for the benefit of humanity, it is important to recognize that natural gas, petroleum, and coal in fact represent immense stores of organic compounds, which the skillful industrial organic chemist can transform into wide array of non-fuel products. If concerns for global warming caused by combustion of fossil fuels accelerates the evolution to an energy economy based on nuclear, solar, or other sources, such a change by no means represents an end to research on the chemistry of natural gas, petroleum, and coal, but rather is the opening of an exciting new era.

Coal tar chemicals

We have seen previously that coal, when heated slowly, undergoes three stages of thermal decomposition. The second stage, 350-550°, is marked by the evolution of a mixture of hydrocarbons which condense to a dark colored, viscous liquid known as *coal tar*. The tar can be treated by a range of chemical processes to yield an enormous array of organic compounds which are useful materials in their own right or which are precursors for the industrial syntheses of other materials. The products manufactured from coal tar include solvents, dyes, pharmaceuticals, artificial flavorings, perfume, plastics, explosives, and fertilizers.

From about 1865 to 1940 the organic chemical industry was based on chemicals from coal tar. Because the compounds in tar are mainly aromatics, much of the research in organic chemistry during those years was devoted to the structure and reactions of organic compounds. Since World War II the organic chemical industry has been based on petroleum. In addition to the availability of abundant and inexpensive (at least until the 1970's) petroleum, there are several other reasons for the decline of the coal tar industry.

One reason is environmental concerns. The production of coal tar (which was mainly obtained on a large scale as a by-product of the metallurgical coke industry) is a messy, smelly business which today requires vigorous efforts to ensure compliance with environmental regulations. Coal tar is a very complex material chemically. Coal tar consists of several hundred compounds, most of which occur in concentrations less than 1%. To separate the tar into individual compounds requires a lengthy sequence of unit operations; the alternative is to be satisfied with using mixtures of materials or impure compounds. The production of chemicals as a by-product of the coke industry was not able to keep up with the market demand for organic chemicals. For example, years ago 95% of the phenol consumed annually was obtained from coal tar. Today the amount of phenol from coal tar is about the same in absolute numbers, but the annual consumption of phenol has increased so much that coal-derived phenol amounts at best to 5% of the total market. Furthermore, the changes which are occurring in the steel industry are indicative of a continual shrinkage of traditional technology of reduction of iron ore by coke in a blast furnace. The overall decline of the steel industry and the changing technology both contribute to a reduction in demand for coke, and, as a direct consequence, a reduced production of coal tar. Indeed, some industry analysts believe that the last coke plant to be built in the United States has already been built.

Processing of coal tar recovered as a by-product from coke manufacturing begins with distillation. The major product streams from distillation include *light oil*, which contains mainly benzene, toluene, and the xylenes; naphtha; naphthalene; *tar acids*, which include phenols, cresols, and the xylenols; and *tar bases*, which include pyridine, methylated derivatives of pyridine such as the lutidines, and quinoline. Higher-boiling materials, which can be recovered by further distillation, include large condensed-ring aromatics and heterocyclic compounds, such as *anthracene oil*, which contains such compounds as anthracene, phenanthrene, and carbazole.

Usually more than half of the tar is distilled to *pitch*, a very viscous material used mainly in the manufacture of carbon electrodes for electro-metallurgical processes, such as the smelting of aluminum. Pitch is also used as a roofing material. *Creosote oil* is used as a wood preservative, such as in treating the wooden poles used for electric power or telephone lines, and wooden railway ties.

Petrochemicals

Today's organic chemical industry is based almost entirely on chemicals derived from petroleum. The major starting points are methane; ethane, propane, and butane (obtained from natural gas as discussed in Chapter 7); and the naphtha fraction of crude oil. Each of these substances is converted to one or more which are either of direct use in their own right or can serve as intermediates for the production of other useful chemicals. The intermediates are methanol, from methane; ethylene and propylene, which can be made from ethane, propane, or naphthas; and benzene, toluene and xylene, which are also made from naphtha.

Methanol

We have already seen how methanol can be produced from synthesis gas, which in turn is produced by the steam reforming of natural gas. That sequence of operations is the main synthetic route to methanol today. However, the versatility of the methanol production lies in the fact that synthesis gas can in principle be produced from a variety of carbon-containing feedstocks, most notably, of course, coal. If the availability of natural gas decreases in the future, the gasification of coal, wood, or petroleum resids in steam can supply the synthesis gas needed for methanol production. Methanol is a useful solvent for a variety of organic materials. However, most of the methanol produced at present is used for conversion to other compounds, particularly formaldehyde, acetic acid, and methyl t-butyl ether.

Formaldehyde is produced by mild oxidation of methanol. A mixture of air and methanol is reacted at 100 - 300° in the presence of a catalyst to form formaldehyde. Since methanol boils at 65°, the reaction occurs in the vapor phase. Various metal oxides have been used as the catalyst for this reaction, particularly oxides of iron, molybdenum, or vanadium. The gaseous products are contacted with a water spray, which both reduces the temperature and dissolves the formaldehyde, along with unreacted methanol. Distillation of the aqueous solution removes the methanol, leaving a solution of formaldehyde that is often called formalin. Formaldehyde is perhaps most commonly known as a preservative, particularly for specimens used in biological or medical laboratories. This application, however, is a relatively minor use of formaldehyde. The most important application is the production of resins by reaction of formaldehyde with phenol or urea. Reaction of formaldehyde with phenol produces a resin widely used as the glue in the manufacture of plywood from thin sheets of wood, and in the manufacture of other wooden articles, such as furniture. These resins are also used as adhesives in fiberglass fabrication. The resins that are produced by reaction of formaldehyde with urea are good thermal and electrical insulators. At one time, formaldehyde-urea resins were quite popular for wall and ceiling insulation in homes. However, some of the resins may contain small amounts of unreacted formaldehyde. Formaldehyde has come under suspicion as a possible carcinogen, and for this reason the use of formaldehyde-urea resins in home insulation was banned in the United States in 1983.

Many processes exist for the production of acetic acid, especially since this compound is one of the oldest organic compounds produced by mankind. Acetic acid is perhaps best known as the component of vinegar responsible for its characteristic sour taste and odor. The production of acetic acid from methanol is one of several options for the industrial synthesis. The reaction of methanol with carbon monoxide is performed at 150-200°.

$$CH_3OH + CO \rightarrow CH_3COOH$$

The insertion of carbon monoxide into another organic molecule (in this case, methanol), is called *carbonylation*. The reaction is carried out in the liquid phase; since the reaction temperatures are well above the normal boiling point of methanol, high pressures must be used to insure that the methanol is liquid. Typical reaction pressures are 3-4 MPa. The most important uses of acetic acid are for the production of the monomers vinyl acetate and cellulose acetate. These compounds in turn are polymerized to produce useful materials used as synthetic fabrics.

Ethylene

Ethylene is by far the most important organic compound in the chemical industry. It is made from two sources: the predominant source is the cracking of ethane to ethylene, and the secondary source is the cracking of naphtha. We have seen in Chapter 9 how naphtha can "unzip" by successive β-bond scissions to produce ethylene. Ethylene is the industrial starting point for the production of a wide variety of compounds, some of which are used directly and others of which are further intermediates in the synthesis of useful products. Examples of the industrial utility of ethylene are discussed in this section, but it should be recognized that this discussion by no means gives a complete review of the uses of this very versatile compound.

A major use of ethylene is its polymerization to produce polyethylene. It is one of the plastic materials which are virtually ubiquitous in everyday life; its uses are as diverse as molded plastic items (such as bottles), plastic food wrap, and electrical insulation. Roughly 50% of all the ethylene produced is converted directly to polyethylene. Two general types of polymerization are used in the production of polyethylene. Free radical polymerization is run at temperatures above 100° and pressures of about 100

MPa. The formation of radicals is initiated by the presence of small amounts of oxygen or peroxides. A chain reaction proceeds, with the product containing about 1000 monomer units. The processes responsible for termination of polymerization are the same as those introduced in Chapter 3: disproportionation and recombination. The polyethylene made in this way contains some branching between the chains of ethylene monomers. This branching resulted from a hydrogen abstraction somewhere along the chain forming a new radical site in the interior of the chain, rather than at the ends, with a new C-C bond being formed by recombination reactions. The polymer formed in this process is *low-density polyethylene*. The alternative process is an ionic reaction carried out in the presence of catalyst. The catalyst is formed by the reaction of triethylaluminum, $(CH_3CH_2)_3Al$, with titanium (III) chloride, $TiCl_3$. The product is an example of a *Ziegler-Natta catalyst*, a family of compounds developed in the early 1950's which introduced major changes in polymer production. Neither the exact structure of Ziegler-Natta catalysts nor the mechanism of their action is fully understood. The reaction appears to proceed via formation of an organotitanium compound, containing a (CH_3CH_2)-Ti bond; coordination of an ethylene molecule via its π bond to the titanium atom, and then insertion of the coordinated ethylene into the Ti-C bond. Polyethylene produced using a Ziegler-Natta catalyst is called *high-density polyethylene*. This polymer contains 500 to 1000 monomer units in a linear, rather than branched, structure. High-density polyethylene is superior in both mechanical strength and heat resistance to the low-denisty form, and is the polyethylene used in the many applications of this polymer in household articles.

Ethylene is converted to vinyl chloride in a two-stage process. In the first step, ethylene is reacted with hydrogen chloride in the presence of oxygen to produce 1,2-dichloroethane (ethylene dichloride). This reaction is run at about 300°. A catalyst of copper(II) chloride, supported on potassium chloride, is used. This reaction is known as *oxychlorination*, although no oxygen is introduced into the product. In the second step, the 1,2-dichloroethane is heated at higher temperature, typically 500°. Thermal decomposition of the 1,2-dichloroethane produces the desired vinyl chloride and hydrogen chloride. (The hydrogen chloride is captured and recycled to the first stage oxychlorination process.) By far the major use of vinyl chloride is its polymerization to form another virtually ubiquitous plastic, poly(vinyl chloride). Poly(vinyl chloride) is widely known as *PVC*. Many of important objects in everyday life are, or could be, made of PVC, including the siding, water piping, and floor covering in a house; some clothing articles; and many types of plastic toys.

A small fraction, about 10%, of the ethylene dichloride is used to manufacture chlorinated solvents which are used in industry to dissolve oils, greases, and lubricants; and are used for the dry cleaning of clothing. An example of a solvent used as a "degreasing" solvent in industry is 1,1,1-

trichloroethane; the dry cleaning industry has employed solvents such as 1,1,2,2-tetrachloroethene, commonly known as perchloroethylene.

The formation of a carbocation from ethylene allows it to react with benzene to form ethylbenzene. The mechanism is slightly different from the formation of methyl t-butyl ether discussed in Chapter 13 or the alkylation and polymerization reactions in Chapter 9. The synthesis of ethylbenzene proceeds via a carbocation formed from a Lewis acid, rather than a Bronsted acid. The Lewis acid catalyst frequently used is anhydrous aluminum chloride, and the reactive intermediate is formed by

$$CH_2=CH_2 + AlCl_3 \rightarrow (CH_2CH_2)^+(AlCl_3)^-$$

This intermediate attacks the benzene ring to form a second carbocation intermediate which has the ion on a ring carbon

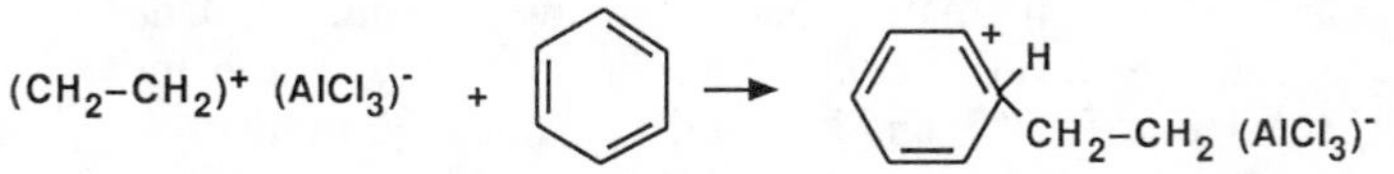

and then the loss of aluminum chloride generates the desired ethylbenzene

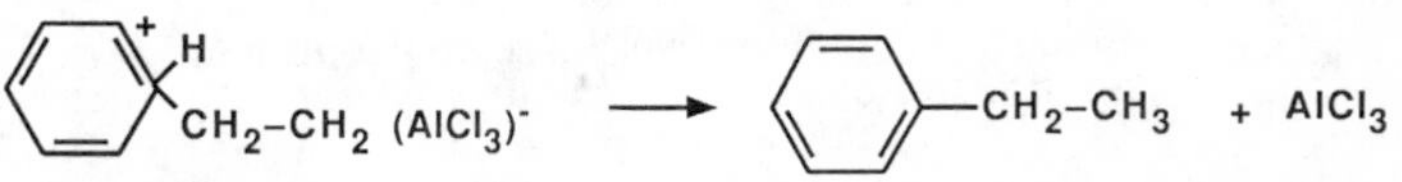

This reaction is an example of a very versatile reaction used in synthetic organic chemistry, the *Freidel-Crafts alkylation*. A tiny fraction of all the ethylbenzene produced is used as a solvent. About 99% of the ethylbenzene is converted directly to styrene by dehydrogenation of the ethyl side chain. The reaction occurs in the vapor phase at temperatures of 600-650° in the presence of a zinc oxide catalyst supported on aluminum oxide.

$$C_6H_5CH_2CH_3 \rightarrow C_6H_5CH=CH_2 + H_2$$

About half of the styrene produced is converted to polystyrene, which, along with polyethylene and PVC is a very common plastic in everyday life. We have seen that in principle many parts of a house could be made of PVC; polystyrene is used for many food containers in grocery stores (the styrofoam egg cartons are an example) and fast food restaurants, beverage cups, and even plastic eating utensils. The second major use of styrene is for the production of synthetic rubber, particularly by copolymerization with 1,3-butadiene. The polymer formed in this reaction is called styrene-butadiene rubber, sometimes known as *SBR*, and is widely used as a component of automobile tires. Thus with ethylene, benzene, hydrogen chloride, and butadiene, one could produce the materials for components of our housing, the tires needed for transportation, and the containers to bring home groceries or prepared foods, in addition to making components of many household appliances and leisure goods.

The oxidation of ethylene by air results in the formation of ethylene oxide. The reaction occurs in the vapor phase at 300° in the presence of a silver catalyst. In a manner analogous to the production of formaldehyde, the hot product gases are contacted with cold water; distillation of the resulting aqueous solution yields ethylene oxide. Over half of current ethylene oxide production is used as an intermediate in the synthesis of 1,2-ethanediol (ethylene glycol). The basis of the ethylene glycol synthesis is the addition of water to ethylene oxide. The addition can be effected by two processes. The reaction of ethylene glycol with water can be driven by heat, although the reaction requires fairly stringent conditions (195°, 1.5 MPa, with reaction times of about an hour). An alternative method involves passing ethylene oxide into dilute aqueous sulfuric acid. The reaction proceeds at ambient pressure, temperatures of 50-70°, and is complete in about 30 minutes. Ethylene glycol has two major uses. It is a component of antifreeze used in automobile cooling systems. The other use is its reaction with dicarboxylic acids, particularly terephthalic acid, to form the so-called *polyester* polymers which are widely used as synthetic fabrics for clothing.

Ethylene reacts with acetic acid in the gas phase, in the presence of oxygen, to produce vinyl acetate.

$$CH_2=CH_2 + CH_3COOH + 0.5\ O_2 \rightarrow CH_3COOCH=CH_2 + H_2O$$

(We have previously discussed the formation of acetic acid by the carbonylation of methanol; an alternative route to acetic acid involves the two-step oxidation of ethylene, first to acetaldehyde and then oxidizing the acetaldehyde to acetic acid. Thus vinyl acetate could be made entirely from ethylene.) The oxidation of ethylene and acetic acid to vinyl acetate is performed at 150-200° and 0.5-1 MPa. Various salts of palladium, such as palladium(II) chloride, are used as catalysts and are dispersed on supports such as aluminum oxide or carbon. Vinyl acetate is polymerized to form poly(vinyl acetate), a polymer used in the manufacture of paints and adhesives.

Propylene

Like ethylene, propylene is produced by cracking either propane or naphtha. Propylene is also a very versatile intermediate in the synthesis of other useful compounds. Propylene is second only to ethylene as a feedstock for the

330

organic chemical industry, but the annual consumption of propylene is only half that of ethylene.

Propylene can be polymerized to produce polypropylene, in ways similar to the production of polyethylene. The production of polypropylene is the largest use of propylene. Polypropylene plastics are used to manufacture plastic bottles and other laboratory ware, many kinds of plastic toys, and plastic film used in packaging.

Propylene reacts in the gas phase with ammonia and oxygen in a process called *ammoxidation*. The product of this reaction is acrylonitrile.

$$2 \ CH_3CH=CH_2 + 2 \ NH_3 + 3 \ O_2 \rightarrow 2 \ CH_2=CHCN + 6 \ H_2O$$

Acrylonitrile is polymerized to form polyacrylonitrile, the component of acrylic fibers. Together with the polyesters mentioned above and nylon, acrylic fibers are among the most important of the synthetic fibers for clothing.

In another similarity to the chemistry of ethylene, propylene oxide is produced from propylene as an important intermediate for the eventual production of propylene glycol. However, the actual preparation of the oxide is considerably different from the production of ethylene oxide. The production of propylene oxide actually begins with the formation of t-butyl hydroperoxide by the mild oxidation of isobutane.

$$4 \ (CH_3)_3CH + 3 \ O_2 \rightarrow 2 \ (CH_3)_3COOH + 2 \ (CH_3)_3COH$$

(Notice that the first product is *not* a carboxylic acid; it is a hydroperoxide containing the functional group -O-O-H.) The t-butyl hydroperoxide then reacts with propylene to form the desired propylene oxide product:

$$(CH_3)_3COOH + CH_3CH=CH_3 \rightarrow (CH_3)COH + CH_3CHCH_2$$
$$O$$

In both the formation of t-butyl hydroperoxide and propylene oxide the by-product is t-butyl alcohol, valuable by-product for use as an anti-knock additive in gasoline, or as a starting material in the synthesis of methyl t-butyl ether, which is also an anti-knock agent. As with ethylene oxide, the propylene oxide can be reacted with water to form propylene glycol. Propylene glycol can also be used in the synthesis of polyesters. In addition, it is used as a starting material for polyurethanes, which are used in applications such as automobile seats.

The synthesis of 2-propanol (isopropanol) is an industrial adaptation of the hydration of double bonds in the presence of an acidic catalyst. Propylene dissolves in sulfuric acid to produce isopropyl sulfate

$$CH_3CH=CH_2 + H_2SO_4 \rightarrow CH_3CHCH_3$$
$$SO_4$$

In a subsequent step, the isopropyl sulfate is hydrolyzed with water to produce the desired isopropanol. On an industrial scale, the first step of the reaction is effected by contacting gaseous propylene with sulfuric acid in a countercurrent flow reactor at elevated pressures (usually 2-2.5 MPa) but near ambient temperatures (20-30°). Isopropyl sulfate is soluble in the sulfuric acid stream. The addition of water, in a subsequent step, hydrolyzes the sulfate to the alcohol. The addition of water to concentrated sulfuric acid is highly exothermic, and the reaction mixture actually gets hot enough to allow the isopropanol to distill out of the mixture. Currently the major use of isopropanol is as an industrial solvent. Isopropanol is familiar as the active ingredient in rubbing alcohol.

Like ethylene, propylene is also capable of undergoing Friedel-Crafts alkylation of benzene. The product is isopropylbenzene, commonly known as cumene. The reactive carbocation for attack on the benzene ring is generated by using a Bronsted acid catalyst, phosphoric acid. The reaction conditions are 175-225° with pressures of 2.5 - 4 MPa. Virtually all of the cumene is converted to phenol and acetone. Only a small fraction is used to produce α-methylstyrene, a manner analogous to the formation of styrene from ethylbenzene. The oxidation of cumene at mild conditions (100-110°) forms an intermediate cumene hydroperoxide

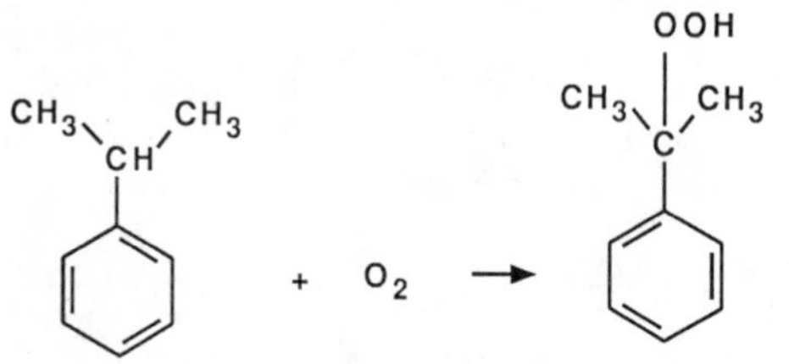

In a subsequent reaction, the hydroperoxide is decomposed to from phenol.

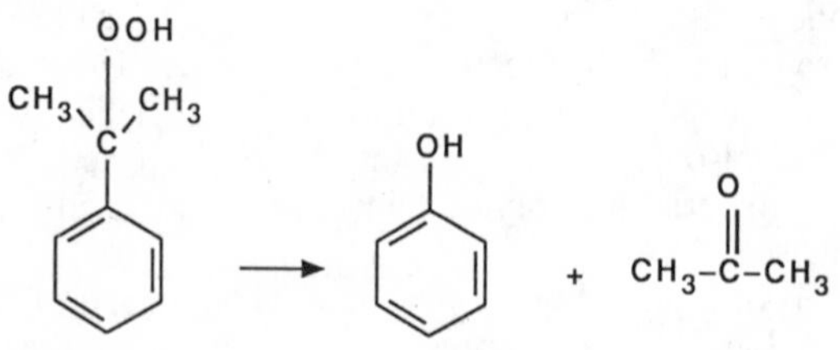

The other product arising from decomposition of cumene hydroperoxide is acetone. In this instance, both reaction products are valuable industrial commodities. Phenol is used to manufacture phenolic resins (by reaction with formaldehyde). The acetone is converted in a three step process into methyl methacrylate

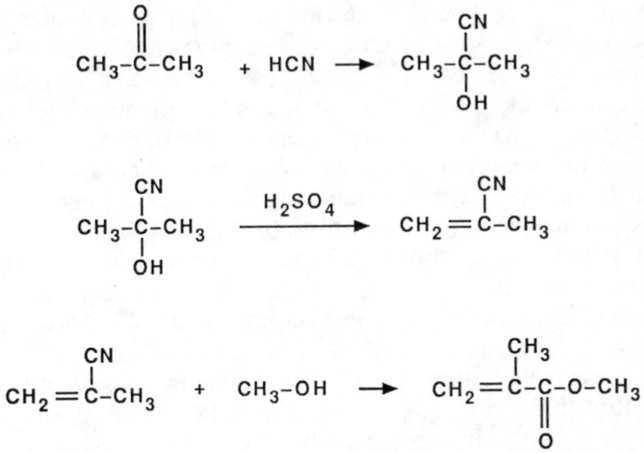

$$CH_3-\overset{\overset{\displaystyle O}{\|}}{C}-CH_3 \ + \ HCN \ \longrightarrow \ CH_3-\overset{\overset{\displaystyle CN}{|}}{\underset{\underset{\displaystyle OH}{|}}{C}}-CH_3$$

$$CH_3-\overset{\overset{\displaystyle CN}{|}}{\underset{\underset{\displaystyle OH}{|}}{C}}-CH_3 \ \xrightarrow{H_2SO_4} \ CH_2\!=\!\overset{\overset{\displaystyle CN}{|}}{C}-CH_3$$

$$CH_2\!=\!\overset{\overset{\displaystyle CN}{|}}{C}-CH_3 \ + \ CH_3-OH \ \longrightarrow \ CH_2\!=\!\overset{\overset{\displaystyle CH_3}{|}}{C}-\overset{\overset{\displaystyle }{}}{\underset{\underset{\displaystyle O}{\|}}{C}}-O-CH_3$$

The major use of methyl methacrylate is in the manufacture of poly(methyl methacrylate), a transparent plastic often sold under the trademarked name Plexiglas.

Benzene, toluene, and xylenes

In the discussion of catalytic reforming (Chapter 9) we have seen that a number of reactions potentially can take place on the catalyst surface. These reactions include dehydrocyclization, dehydroisomerization, and dehydrogenation. When the components of naphtha undergo these reactions, the eventual products are aromatic compounds containing six, seven, or eight carbon atoms - that is, benzene, toluene, and the isomers of xylene. We will see that each of these compounds has special uses; however, they are frequently referred to as a single product stream known as *BTX*.

The reforming reactions leading to the production of BTX are endothermic. In each case, there is an increase in the moles of gaseous product relative to reactants, and in each case hydrogen is a major by-product of the reaction. From equilibrium considerations (LeChatelier's Principle), we would predict that an endothermic reaction in which the number of moles of gas increases would be favored by high reaction temperatures and low pressures. Since hydrogen is a product, we would want the feed to the reactor

333

to be free of hydrogen, again from equilibrium considerations. In fact, the actual industrial practice is to run the process with high pressures (1 - 4 MPa) and very high partial pressures of hydrogen. This counter-intuitive choice of reaction conditions is done to prevent, or at least minimize, the formation of coke on the catalyst surface. It is preferable to accept some inefficiencies in production of the desired BTX, resulting from operating far from the conditions we might select from LeChatelier's Principle, to minimize the problems associated with the undesirable side reaction of coke formation.

We have already seen two of the major uses of benzene, the production of ethylbenzene as an intermediate for styrene, and the production of cumene as an intermediate for phenol. Another important industrial reaction of benzene is its hydrogenation to cyclohexane. The reaction is carried out using a supported metal catalyst, such as platinum on aluminum or silicon oxides. The reaction conditions require temperatures of 200 - 400° and pressures of 2.5 - 3 MPa. Cyclohexane is a useful solvent for some organic materials, but its most important use is in the synthesis of 1,6-hexanedioic acid (adipic acid), en route to nylon. The reaction of cyclohexane with air at 125 - 160° and pressures of 0.4 - 1.8 MPa results in the formation of a mixture of cyclohexanol and cyclohexanone. When these two compounds are then reacted with nitric acid at 50 - 90°, the strong oxidation results in ring opening to adipic acid. In a final reaction, the adipic acid is condensed with 1,6-diaminohexane (hexamethylenediamine) to produce a polyamide, familiarly known as nylon.

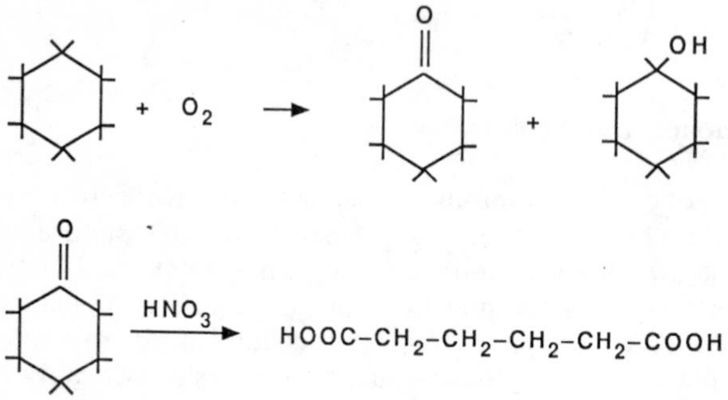

$$HOOC-CH_2-CH_2-CH_2-CH_2-COOH$$

$$HOOC-CH_2-CH_2-CH_2-CH_2-COOH +$$

$$H_2N-CH_2-CH_2-CH_2-CH_2-CH_2-CH_2-NH_2 \longrightarrow$$

The demand for benzene for production of styrene and cumene, as well as the other lesser uses of benzene, is so great that about 50% of toluene production is converted to benzene. The conversion occurs in the reaction of toluene with hydrogen at 550 - 650° and 3 - 8 MPa. The methyl group is removed from the ring as methane. The loss of an alkyl group effected by hydrogen is an example of a general reaction known as *hydrodealkylation*. Since the toluene is produced in catalytic reforming, which forms hydrogen as a by-product, the by-product hydrogen from reforming can be used as the reactant for the hydrodealkylation reaction. Toluene is a valuable additive for gasoline to increase the octane number. The octane number of pure toluene is is 120, and addition of toluene to gasoline can significantly increase its octane number. Benzene has been used as a solvent for many applications in industry and the laboratory, but in recent years its use has declined because benzene is a suspect carcinogen. Toluene is being used increasingly as a replacement for benzene, because the solvent behavior of the two compounds is similar, but toluene does not present the potential health hazard of benzene. A small amount of toluene is nitrated to produce the explosive 2,4,6-trinitrotoluene, commonly known as *TNT*.

The formation of xylenes during catalytic reforming produces all three isomers. It is possible to remove o-xylene by distillation, but the boiling points of the m- and p- isomers (139° and 138°, respectively) are too close for a distillation separation to be practical. For some applications, particularly use as a solvent or octane booster in gasoline, the specific isomer is not important, and a mixture of the isomers, called *mixed xylenes*, can be used satisfactorily. For use as an intermediate in chemical synthesis, p-xylene is by far the most important of the three isomers. Passing vapors of mixed xylenes over a zeolite catalyst at 400° facilitates the isomerization of o- and m-xylenes to the p-isomer. The oxidation of p-xylene by air at 200° and 2 MPa results in the formation of 1,4-benzenedicarboxylic acid (terephthalic acid). This acid is the major building block for the synthesis of poly(ethylene terephthalate), the polyester which is used for synthetic textiles and films.

Chemicals from coal and biomass

Today chemicals from petroleum satisfy an enormous range of our demand for consumer goods and transportation. Clothing, many common household articles, packaging, and many components of vehicles all can be manufactured from chemicals or polymers derived from petroleum. Unfortunately, the rate at which new reserves of petroleum are being discovered has lagged behind the rate at which petroleum is being used since the 1970's. If current annual rates of petroleum consumption are maintained, and no significant discoveries of new reserves, serious shortages of petroleum may occur in the early years of the twenty-first century. If that situation comes to pass, it will be necessary to find new sources of the organic compounds needed to sustain the industries and markets now based on petroleum. One choice is to return to coal as a source of chemicals. However, the days of the dominance of the coal tar industry are gone forever. A new era of chemicals from coal must be based on new processes. The other choice is biomass - wood, agricultural wastes, or other organic materials which have not undergone diagenesis and catagenesis.

Coal

An alternative approach to the manufacture of chemicals from coal, rather than using coal tar, relies on combining a sequence of known process technologies. For example, a process can be envisioned in which coal is converted to synthesis gas which, after the usual sweetening and shifting, is converted via the medium pressure F-T synthesis to naphtha. The naphtha can in turn be cracked via β-bond unzipping to ethylene. Once a supply of ethylene is assured, it can be used as the starting point for the manufacture of poly(ethylene), poly(vinyl chloride), polystyrene, ethanol (both a useful chemical and a fuel) and acetic acid. The acetic acid can be used to produce poly(vinyl acetate). Linear alkenes and alcohols that could also be manufactured via the F-T process can be the starting materials for such diverse materials as detergents and margarine.

All of the processes envisioned in the previous paragraph rely on known, demonstrated technology. However, if we consider the sequence from the point of view of carbon atoms, we begin with the macromolecular structure of coal, gasify it to compounds containing single carbon atoms, polymerize the single carbon atoms to compounds having ~10 carbon atoms (naphtha), degrade the naphtha to a compound of two carbon atoms (ethylene), and finally polymerize the ethylene to some desired product such as polyethylene. In essence, this sequence involves two "depolymerizations" and two "repolymerizations" to get to the desired product. A far superior scheme would take advantage of the molecular fragments already existing in the coal, and convert such fragments directly to useful materials. For example, recall that lignites have low values of the ring condensation, with R ~1.5, and thus must contain abundant single aromatic rings. If an oxidizing agent could be found to cut apart carefully the crosslinks in the structure, one could envision a reaction such as

The 1,4-benzenedicarboxylic acid (terephthalic acid) can be reacted in a separate step with 1,2-ethanediol (ethylene glycol) to produce poly(ethylene terephthalate), a useful polymer sold under such trade names as Dacron and Mylar. However, a process for the production of terephthalic acid in high yield from lignite has not yet been developed.

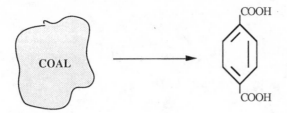

The current state-of-the-art in the production of chemicals from coal is represented by the *Eastman process*. The Eastman process is based on Texaco gasification and production of methanol from synthesis gas. The initial steps are the familiar reactions

$$C + H_2O \rightarrow CO + H_2$$
$$CO + H_2O \rightarrow CO_2 + H_2$$
$$CO + 2 H_2 \rightarrow CH_3OH$$

An interesting feature of the Eastman process is the large-scale use of cryogenic techniques: the use of cold methanol for sweetening and removal of carbon dioxide, and the cryogenic separation of carbon monoxide (which boils at -190°) and hydrogen (-253°). The latter operation is effected with liquid nitrogen.

Methanol is reacted with acetic acid to form methyl acetate

$$CH_3COOH + CH_3OH \rightarrow CH_3COOCH_3$$

which is then reacted further with carbon monoxide to form acetic anhydride

$$CH_3COOCH_3 + CO \rightarrow (CH_3CO)_2O$$

The production of acetic acid and acetic anhydride represent the routes to desirable commercial products. Acetic acid can be reacted with ethene

$$CH_2=CH_2 + CH_3COOH \rightarrow CH_2=CHOOCCH_3$$

to make vinyl acetate. Vinyl acetate can be polymerized to poly(vinyl acetate), used, for example, in the manufacture of water-based paints; the reaction with water can result in formation of poly(vinyl alcohol) which is a useful component of water-soluble adhesives.

The reaction of cellulose with acetic anhydride can proceed as any other esterification reaction

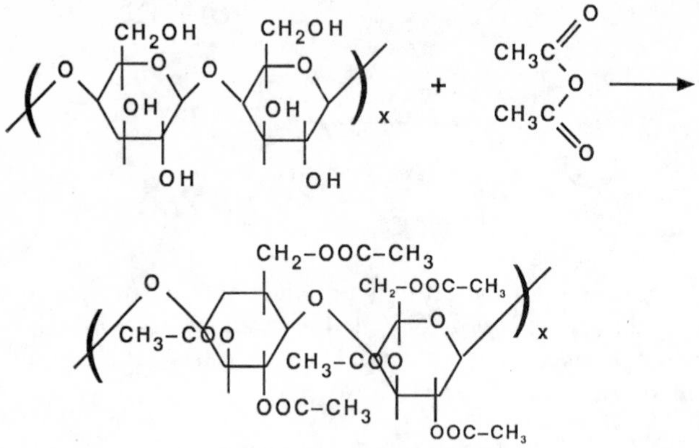

Each of the hexose monomers in the cellulose has three hydroxy groups, allowing the possibility of producing a mono-, di-, or tri-acetate. Cellulose acetate can be made into acetate rayon (a synthetic fiber), film, or can be extruded and molded to various plastic products.

Biomass

The production of adipic acid from petroleum-derived cyclohexane has been discussed previously. However, there is an alternative approach, using biomass as the starting material. In essence, this process eliminates the thousands of years of reaction time required for the conversion in nature of biological materials to petroleum. The reaction of pentoses results in the formation of the heterocyclic aldehyde furfural, C_4H_3OCHO:

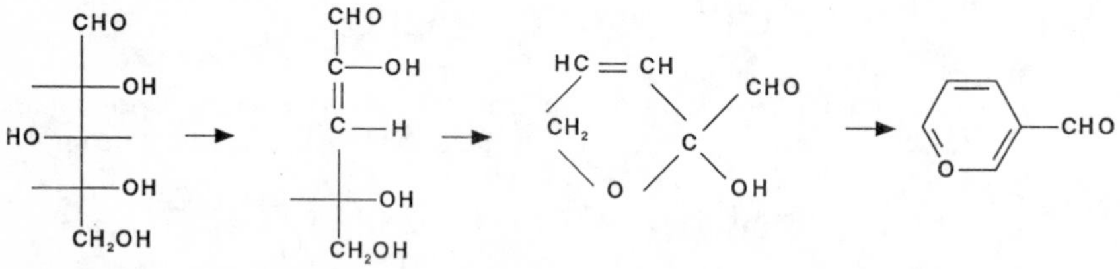

A series of reactions then converts the furfural, via tetrahydrofuran, to 1,4-dicyanobutane (adiponitrile). The synthesis of this di-nitrile is highly desirable because it is the starting point for the production of both components which are polymerized to produce nylon-6,6: adipic acid and 1,6-hexanediamine (hexamethylenediamine):

$$NCCH_2CH_2CH_2CH_2CN + 4H_2O \rightarrow HOOCCH_2CH_2CH_2CH_2COOH$$

$$NCCH_2CH_2CH_2CH_2CN + 4H_2 \rightarrow H_2NCH_2CH_2CH_2CH_2CH_2CH_2NH_2$$

One of the best sources of the pentoses, and hence furfural, is oat hulls. These materials are presently a waste product of the food industry.

Further reading

Chang, R. and Tikkanen, W. (1988). *The Top Fifty Industrial Chemicals*. New York: Random House.

Kent, J. A. (ed.) (1974). *Riegel's Handbook of Industrial Chemistry*, 7th edn. New York: VanNostrand Reinhold; Chapter 8.

McMurray, J. (1988). *Organic Chemistry*. Pacific Grove, CA: Brooks / Cole; Chapter 31.

Royal Dutch Shell (1983). *The Petroleum Handbook*, 6th edn. Amsterdam: Elsevier; Chapter 10.

Streiwieser, A. Jr. and Heathcock, C. H. (1985). *Introduction to Organic Chemistry*, 3rd edn. New York: Macmillan; Chapter 34.

Index

Formalin, 326
Fractional distillation, 177
Free radicals, 59
Free swelling index, 145
Friability, 256
Friedel-Crafts alkylation, 329, 332
Froth floatation, 262
Fuel NO_x, 313
Fuel oils, 182, 202
Fuel ratio, 119
Fuller's earth, 204
Furanose, 11
Furans, 32, 94
Furfural, 32, 203, 338, 339
Furnace blacks, 166
Fusinite, 117, 149, 298

Gallic acid, 15

1-Galloyl-β-D-glucoside, 15
Gas cap gas, 75
Gasohol, 314
Gas oil, 184, 197, 203
Gasoline, 180, 182, 184, 215, 216,
    235, 247, 283, 285, 286,
    309, 314, 317, 319, 322
Gas window, 65, 72
Gelification, 48
Geothermal gradients, 40
Geraniol, 18
Gieseler plastometer, 145
Gilsonite, 111
Glass transition temperature, 149
Global carbon cycle, 3
Glucose, 3, 10, 15
Glucosides, 12-14
Glucopyranose, 11
Glutaric acid, 110
Glycerol, 16, 33, 289
Glyceryl palmitate, 33
Glycons, 14
Glycosides, 11, 14, 25, 30, 33,
    34
Graphite, 38, 39, 45, 133, 293
Greases, 204, 286
Greenhouse effect, 7
Grindability, 256

Hardgrove grindability index, 256
Heat of combustion, 82, 93, 94, 126,
    127
Heavy diesel fuel, 182
Heavy gas oil, 182
Heavy oils, 66, 95, 251, 309
Helium, 76, 79, 142
Helium density (coals), 142, 144

Ring condensation, 126, 128, 134, 138, 144
Ring enlargement, 238, 321
Rotting, 28
Ruthenium, 238, 285, 286

Salicin, 14
Salicylic acid, 14
Salt, 154
Saponification, 57
Sapropelic coals, 69
SBR, 329
Schulz-Flory distribution, 286
Sclerotinite,117
Scouring rush, 153
Secondary tower, 182
Semianthracite, 120
Semifusinite, 117, 149
Sesquiterpenes, 18
Sesterterpenes, 19
Seyler chart, 121
Shot coke, 209
Side stream stripper, 181
Silica, 154
Silica-alumina, 231, 248
Sinapyl alcohol, 23, 43, 44, 132
Single stage distillation, 179
Sinks, 259
Slot-type coke oven, 300
Sodium, 155
Sodium metasilicate, 226
Solvent extraction, 203
Softening temperature, 145
Soot, 165, 196
Sour gas, 78
Specific fuel consumption, 314
Sponge coke, 209
Sporinite, 117, 298
Squalene, 20
Stabilizer, 182
Starch, 12
Steam reforming, 168, 271, 273
Steam stripping, 204
Stearic acid, 17, 101
Steroids, 20
Storax, 23
Straight run gasoline, 184, 190, 193, 246
Stripped gas, 160
Styrene, 131, 307
Styrene-butadiene rubber, 167, 329
Subbituminous coals, 120
Substitute natural gas, 281
Sulfates, 32, 78
Sulfides, 249
Sulfur, 79, 107, 158
Sulfuric acid, 206, 219, 331, 332

346

Triterpenes, 19
True density, 142
Tungsten, 250
Two stage distillation, 182

Ultimate analysis, 120
Undecane, 51, 198
Unzipping, 61, 205, 327
Urea complexation, 204
Urea-formaldehyde resins, 283, 326, 327

Vacuum gas oil, 182
Vacuum tower, 183
Vanadium, 235
Vanadium oxide, 326
van der Waals forces, 80
van Krevelen diagram, 35, 36, 38, 46, 55
    70, 114, 117
Vapor lock, 194
Vinyl acetate, 330, 337
Vinyl chloride, 171
Visbreaking, 214, 215
Viscosity, 98
Vitrnites, 50, 53, 71, 116, 117, 149, 298
Volatile matter, 118, 119

Washability curve, 260
Water gas, 273, 303
Water gas shift reaction, 168, 274, 279, 280
Waxes, 16, 25, 33, 55, 204, 205, 285, 286
Wet gas, 76

Xylan, 12
o-Xylene, 51
Xylenes, 93, 102, 326, 333, 335

Young-deep oils, 103
Young-shallow oils, 101

Zeolites, 226, 228, 231, 236, 316, 320
    322, 335
Ziegler-Natta catalysts, 328
Zinc oxide, 281, 320, 329
ZSM-5, 320, 322

B.C.

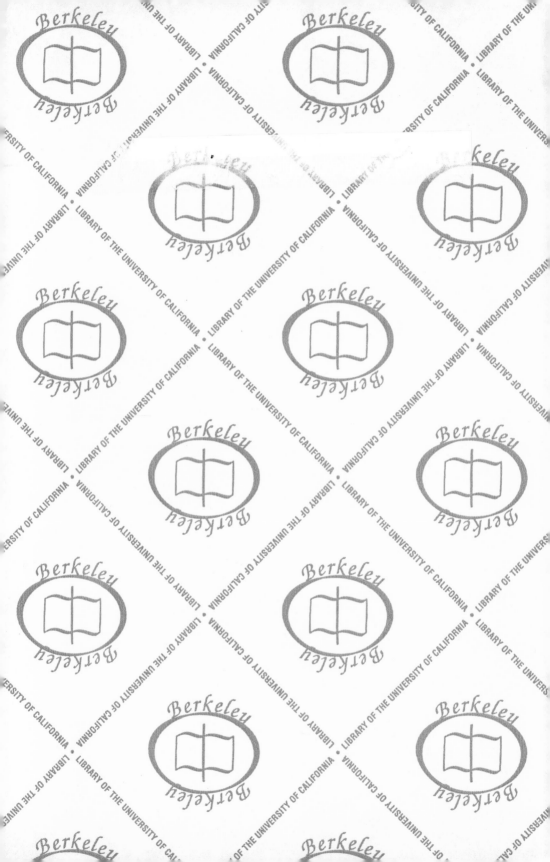